Stress Signaling in Plants: Genomics and Proteomics Perspective, Volume 1

Maryam Sarwat • Altaf Ahmad • M.Z. Abdin
Editors

Stress Signaling in Plants: Genomics and Proteomics Perspective, Volume 1

Springer

Editors
Maryam Sarwat
Pharmaceutical Biotechnology
Amity Institute of Pharmacy
Amity University, NOIDA, India

Altaf Ahmad
Department of Botany, Faculty of Science
Jamia Hamdard
New Delhi, India

M.Z. Abdin
Centre for Transgenic
Plant Development
Department of Biotechnology
Faculty of Science
Jamia Hamdard
New Delhi, India

ISBN 978-1-4614-6371-9 ISBN 978-1-4614-6372-6 (eBook)
DOI 10.1007/978-1-4614-6372-6
Springer New York Heidelberg Dordrecht London

Library of Congress Control Number: 2012956287

Printed on acid-free paper

Springer is part of Springer Science+Business Media (www.springer.com)

Preface

The most important prerequisites imperative for survival of human kind; food, fodder and fuel depend upon the performance of plants. Due to worldwide temperature inversions, the weather conditions have become completely hostile and unpredictable. Thus, there is a growing need of such plants which are better adapted to these adverse climatic conditions. A good understanding of the signalling mechanism within the plant system during these climatic conditions will certainly going to help in raising plants which are better suited for these adverse conditions.

In this book, we have put together both genomics and proteomics approach to further our understanding in this direction. The chapters in this book expand our understanding from bioinformatical approaches to develop the models, as well as proving the ideas up to field conditions. Hence, this book contains comprehensive knowledge of stress signalling useful for graduate students, researchers as well as scientists working in this area.

The ten chapters written by international dignitaries give much weightage to this book.

NOIDA, India — Maryam Sarwat
New Delhi, India — Altaf Ahmad
New Delhi, India — M.Z. Abdin

Contents

Contributors

M.Z. Abdin Centre for Transgenic Plant Development, Department of Biotechnology, Faculty of Science, Jamia Hamdard, New Delhi, India

Atlaf Ahmad Department of Botany, Faculty of Science, Jamia Hamdard, New Delhi, India

Pravej Alam Department of Biotechnology, Centre for Transgenic Plant Development, Jamia Hamdard, New Delhi 110062, India

Athar Ali Department of Biotechnology, Centre for Transgenic Plant Development, Jamia Hamdard, New Delhi 110062, India

Essa Ali Institute of Crop Science, College of Agriculture and Biotechnology, Zhejiang University, Hangzhou, People's Republic of China

Alejandra A. Covarrubias Department of Biología Molecular de Plantas, Instituto de Biotecnología, Universidad Nacional Autónoma de México, Cuernavaca, Morelos, Mexico

Renu Deswal Molecular Physiology and Proteomics Laboratory, Department of Botany, University of Delhi, Delhi 110007, India

Priyanka Giri Department of Molecular Biology & Genetic Engineering, College of Basic Sciences & Humanities, GB Pant University of Agriculture & Technology, Pantnagar, U S Nagar, India

Lixi Jiang Institute of Crop Science, College of Agriculture and Biotechnology, Zhejiang University, Hangzhou, People's Republic of China

Chongwei Jin MOE Key Laboratory of Environment Remediation and Ecosystem Health, College of Environmental and Resource Sciences, Zhejiang University, Hangzhou, People's Republic of China

Prakriti Kashyap Molecular Physiology and Proteomics Laboratory, Department of Botany, University of Delhi, Delhi 110007, India

Gurjeet Kaur Department of Botany, Faculty of Science, Jamia Hamdard, New Delhi 110062, India

Mather Ali Khan Department of Biotechnology, Centre for Transgenic Plant Development, Jamia Hamdard, New Delhi 110062, India

Anil Kumar Department of Molecular Biology & Genetic Engineering, College of Basic Sciences & Humanities, GB Pant University of Agriculture & Technology, Pantnagar, U S Nagar, India

Gitto Thomas Kuruthukulangarakoola Institute of Biochemical Plant Pathology, Helmholtz Zentrum München, Ingolstädter Landstrasse 1, 85764 Neuherberg, Germany

Xianyong Lin MOE Key Laboratory of Environment Remediation and Ecosystem Health, College of Environmental and Resource Sciences, Zhejiang University, Hangzhou, People's Republic of China

Christian Lindermayr Institute of Biochemical Plant Pathology, Helmholtz Zentrum München, Ingolstädter Landstrasse 1, 85764 Neuherberg, Germany

Wenjing Liu MOE Key Laboratory of Environment Remediation and Ecosystem Health, College of Environmental and Resource Sciences, Zhejiang University, Hangzhou, People's Republic of China

Gowher Nabi Genetics and Molecular Biology, Jazan University, College of Applied Medical Sciences, Jazan, Kingdom of Saudi Arabia

Shadab Nizam National Institute of Plant Genome Research, Aruna Asaf Ali Marg, New Delhi, India

Dinesh Pandey Department of Molecular Biology & Genetic Engineering, College of Basic Sciences & Humanities, GB Pant University of Agriculture & Technology, Pantnagar, U S Nagar, India

Preeti Rathore Pharmaceutical Biotechnology, Amity Institute of Pharmacy, NOIDA, India

José Luis Reyes Department of Biología Molecular de Plantas, Instituto de Biotecnología, Universidad Nacional Autónoma de México, Cuernavaca, Morelos, Mexico

Maryam Sarwat Pharmaceutical Biotechnology, Amity Institute of Pharmacy, Amity University, NIODA, India

Imran Haider Shamsi MOE Key Laboratory of Environment Remediation and Ecosystem Health, College of Environmental and Resource Sciences, Department of Agronomy, Key Laboratory of Crop Germplasm Resource, College of Agriculture and Biotechnology, Institute of Crop Science, Zhejiang University, 310058, Hangzhou, People's Republic of China

Sugandha Sharma Department of Molecular Biology & Genetic Engineering, College of Basic Sciences and Humanities, GB Pant University of Agriculture & Technology, Pantnagar, Uttarakhand, India

Prabhjeet Singh Department of Biotechnology, Guru Nanak Dev University, Amritsar, Punjab, India

Guadalupe Sosa-Valencia Department of Biología Molecular de Plantas, Instituto de Biotecnología, Universidad Nacional Autónoma de México, Cuernavaca, Morelos, Mexico

Chengliang Sun MOE Key Laboratory of Environment Remediation and Ecosystem Health, College of Environmental and Resource Sciences, Zhejiang University, Hangzhou, People's Republic of China

Gohar Taj Department of Molecular Biology & Genetic Engineering, College of Basic Sciences & Humanities, G B Pant University of Agriculture & Technology, Pantnagar, U S Nagar, India

Praveen K. Verma National Institute of Plant Genome Research, Aruna Asaf Ali Marg, New Delhi, India

Sandhya Verma National Institute of Plant Genome Research, Aruna Asaf Ali Marg, New Delhi, India

Amardeep Singh Virdi School of Life Sciences, Jawaharlal Nehru University, New Delhi, India

Asha Wadhwa Department of Botany, Faculty of Science, Jamia Hamdard, New Delhi 110062, India

Chapter 1
Ca^{2+}, Calmodulin and Plant-Specific Calmodulin-Binding Proteins: Implications in Abiotic Stress Adaptation

Prabhjeet Singh and Amardeep Singh Virdi

Abbreviations

a.a.	Amino acid
Ca^{2+}	Calcium
CaM	Calmodulin
CaMBD	Calmodulin-binding domain
CaMBOT	Calmodulin-binding gel overlay technique
CaMBP	Calmodulin-binding protein
CAMTA	Calmodulin-binding transcription activator
CBK	Calmodulin-binding kinase
CRCK	Cytoplasmic-localized Ca^{2+}-CaM regulated kinase
CRLK	Plasma membrane-localized Ca^{2+}-CaM regulated kinase
HS	Heat stress
HSF	Heat shock factor
Hsp	Heat shock protein
MAPK	Mitogen activated protein kinase
MEK	MAPK kinase
MEKK	MAPK kinase kinase
PCD	Programmed cell death
SA	Salicylic acid

P. Singh (✉)
Department of Biotechnology, Guru Nanak Dev University, Amritsar 1430 05, Punjab, India
e-mail: singhprabhjeet62@gmail.com

A.S. Virdi
School of Life Sciences, Jawaharlal Nehru University, New Delhi, India

M. Sarwat et al. (eds.), *Stress Signaling in Plants: Genomics and Proteomics Perspective, Volume 1*, DOI 10.1007/978-1-4614-6372-6_1,

Summary

Plants, being sessile, are frequently exposed to different types of abiotic stresses, which may affect their growth and development adversely. Plants adapt to stress conditions by activation of cascade(s) of molecular mechanisms, which result in alterations in gene expression and synthesis of protective proteins/compounds. From the perception of the stimulus to transduction of the signal, followed by an appropriate response, the plants employ a complex network of primary and secondary messenger molecules, of which Ca^{2+} is one of the most well studied. Decoding of spatial and transient changes in Ca^{2+} levels is accomplished by different Ca^{2+}-binding proteins (CaBPs), which act as Ca^{2+} sensors. Calmodulin (CaM) is one of the best characterized primary transducer of cytosolic Ca^{2+} changes in all eukaryotes. CaM is an acidic, heat stable and multifunctional protein consisting of two globular domains, each with two Ca^{2+}-binding EF-hand motifs. After binding to Ca^{2+}, the CaM undergoes a conformational change and binds to diverse range of proteins. The CaM-binding proteins (CaMBPs) have been identified and characterized from different plants and recent studies suggest the involvement of several of the CaMBPs in abiotic stress adaptation. Further, different isoforms of CaM have been reported to modulate the activity of CaMBPs differentially, thus, implying intricate mechanisms of regulation by CaM. Therefore, in the following section, the likely role of Ca^{2+}, CaM and plant-specific CaMBPs in abiotic stress adaptation will be discussed.

Introduction

Plants are frequently exposed to different types of abiotic stresses, thus, affecting their growth and development adversely. Adaptation to stress conditions by plants involves activation of cascade(s) of molecular mechanisms that result in alterations in gene expression and synthesis of protective proteins/compounds (Wang et al. 2003). For conveying the signal in response to stress, the plants employ a complex network of primary and secondary messenger molecules (Shinozaki et al. 2003; Bartels and Sunkar 2005). Ca^{2+} is one of the most well studied secondary messengers involved in signal transduction in eukaryotes (Clapham 2007). The resting cytosolic concentration of Ca^{2+} (100–200 nM) increases up to several folds in response to specific stimulus (Reddy 2001). The intracellular changes in Ca^{2+} are sensed by different Ca^{2+}-Binding Proteins (CaBPs) (Reddy 2001; Bouche et al. 2005), which are characterized by the presence of helix-loop-helix motif called EF-hand motif, that typically occur in pairs and facilitate high affinity binding of Ca^{2+} (Gifford et al. 2007). Four broad categories of CaBPs *viz.*, Ca^{2+}-dependent protein kinases, calmodulin (CaM), CaM-like proteins and calcineurin B-like proteins have been reported (Bouche et al. 2005 and references therein).

Of the different Ca^{2+} sensors, CaM has been characterized most extensively (Roberts and Harmon 1992; Snedden and Fromm 1998, 2001). The CaM, though primarily a cytosolic protein, is also detected in the mitochondrion, chloroplast, peroxisome, nucleus, and extracellular matrix (Roberts et al. 1983; Van der Luit et al. 1999), thus, signifying the versatility in its roles. CaM is an acidic protein of approximately 150 amino acid (a.a.) residues. CaM consists of two globular domains connected by a long flexible helix, with each of the globular domains containing two EF-hand motifs that bind to Ca^{2+} cooperatively (Babu et al. 1988). After binding to Ca^{2+}, CaM undergoes conformation change, thus, exposing two hydrophobic sites surrounded by negative charges in each of the globular domains, which interact with several target proteins, thereby, regulating their activities (Crivici and Ikura 1995).

Calmodulin is a highly conserved protein in eukaryotes and as compared to animals, which contain only few genes of CaM (e.g. three in humans), there are multiple genes in plants. A total of nine true CaM genes encoding for four different isoforms have been identified in *Arabidopsis thaliana* (McCormack et al. 2005). On the contrary, the wheat genome contains up to 13 genes of CaM and ten cDNAs encoding three different isoforms of CaM have been isolated (Yang et al. 1996). Rice genome has been reported to show the presence of five true CaM genes, which encode two sets of CaM, with three of the genes coding for identical isoforms (Boonburapong and Bauboocha 2007). The different CaM isoforms differ in their ability to regulate target proteins (Lee et al. 1999, 2000), which suggests that multiple CaM isoforms may be enabling the plants to respond differentially to different environmental and developmental signals.

Detection of CaM-Binding Proteins

Calmodulin targets a vast array of diverse proteins such as metabolic enzymes, transcription factors, kinases, cytoskeletol proteins, chaperones, etc., in plants and animals (Defalco et al. 2010). Since the primary sequence of CaM-binding domains (CaMBDs) among different CaM-binding proteins (CaMBPs) (except within protein families) is not conserved, therefore, the CaMBPs are identified by CaM-binding gel overlay technique (CaMBOT) (O' Day 2003), which employs labelled CaM (Fromm and Chua 1992; Lydan and O' Day 1994; O' Day 2003). This assay identifies the CaMBPs on the basis of protein-protein interactions and provides a crucial beginning point for identifying CaMBPs.

Molecular Basis of CaM-CaMBPs Interaction

Structural analysis, which has been carried out for some CaMBPs, reveals that two bulky hydrophobic a.a. residues (Phe, Trp, Ile, Leu or Val), situated at specific distance apart, in target proteins play an important role in interaction with CaM (Rhoads and Friedberg 1997). The different motifs, identified to be involved in CaM-binding, are IQ motif, and 1–10, 1–14 and 1–16 motifs, since the key anchor a.a. residues are 8, 12, and 14 residues apart, respectively (Alexander et al. 1988; Cheney and Mooseker 1992; Osawa et al. 1999). The CaM-binding motifs from different CaMBPs form characteristic basic amphipathic α-helices with several positive residues on one side and hydrophobic residues on the other side (Rhoads and Friedberg 1997). The propensity of a protein to bind to CaM can be analyzed using software (http://calcium.uhnres.utoronto.ca/ctdb/ctdb/browse.html) in which a score of probability ranging from 0 (unlikely) to 9 (very likely) is calculated per a. a. residue (Yap et al. 2000). A stretch of a.a. residues with a score of 7–9 signify the presence of a putative CaM recruitment signal. Although the accuracy of this programme is 80%, it provides a useful tool to determine the CaM-binding property of proteins by *in silico* analysis.

Role of CaMBPs in Abiotic Stress Response in Plants

The CaMBPs in plants have been implicated in various aspects, such as regulation of ion transport, metabolism, cytoskeleton, protein folding, transcription, protein phosphorylation and dephosphorylation, phospholipid metabolism, disease resistance, cell division, etc. (Reddy et al. 2011). However, the focus of this chapter will be on their role in abiotic stress adaptation since several CaMBPs have been identified which are regulated by different abiotic stress conditions (Table 1.1) (Singh and Virdi 2010).

Regulation by CaM of Enzymes Involved in Generation of Reactive Oxygen Species and Programmed Cell Death

Plants often respond to environmental stresses by producing the reactive oxygen intermediates (ROIs) and their levels in cell are tightly regulated to avoid cellular damage. Exposure to oxidative stress with H_2O_2, which results in enhanced Ca^{2+} levels (Lecourieux et al. 2002), also caused an increase in the expression of oxidative stress-responsive genes, including some specific CaMBPs like catalases and superoxide dismutases (SODs) (Gong and Li 1995). Catalases are protective enzymes, which are involved in degradation of H_2O_2 to water and oxygen. Regulation of catalases by CaM appears to be plant-specific phenomenon, since animal counterparts of catalases do not show CaM-binding properties and this feature might have evolved in plants due to their sessile nature (Bouche et al. 2005). SODs are another class of ROIs-scavenging enzymes, which show binding to CaM in Ca^{2+}-dependent manner (Gong and Li 1995). However, the regulation of SODs by Ca^{2+}/CaM in plants needs to be explored further.

Table 1.1 Abiotic stress modulated calmodulin-binding proteins in plants

S. no.	Protein	Gene	Source	Stress response	Reference
1	Calmodulin-binding protein kinase	*NtCBK*	*Nicotiana tabacum*	SS, GA	Hua et al. (2004)
2	Ca^{2+}-dependent calmodulin-binding cytoplasmic receptor-like kinase	*CRCK1*	*Arabidopsis thaliana*	SS, CS, H_2O_2, ABA	Yang et al. (2004)
3	Calmodulin-binding receptor-like kinase	*CaMRLK*	*A. thaliana*	CS	Charpenteau et al. (2004)
4	Ca^{2+}/calmodulin-regulated receptor-like kinase	*AtCRLK1*	*A. thaliana*	CS	Yang et al. (2010b)
5	Ca^{2+}-dependent calmodulin-binding recepotor-like kinase	*GsCBRLK*	*Glycine soja*	CS, SS, OS, ABA	Yang et al. (2010a)
6	Calmodulin-binding protein kinase 3	*AtCBK3*	*A. thaliana*	HS	Liu et al. (2008)
7	Mitogen-activated protein kinase phosphatase	*NtMKP1*	*N. tabacum*	SA, wounding	Yamakawa et al. (2004)
8	Calmodulin-binding Ser/Thr phosphatase	*AtPP7*	*A. thaliana*	HS	Liu et al. (2007)
9	Calmodulin-binding transcription activator	*CAMTA1* *CAMTA3*	*Brassica napus* *A. thaliana*	CS Plant immunity	Bouche et al. (2002) Du et al. (2009)
10	Calmodulin-binding transcription factor	*OsCBT*	*Oryza sativa*	Negative regulator of plant defence gene expression	Choi et al. (2005)
11	MYB2 transcription factor	*AtMYB2*	*A. thaliana*	SS	Yoo et al. (2005)
12	Calmodulin-binding BTB and TAZ domain protein	*AtBT1-5*	*A. thaliana*	H_2O_2, SA	Du and Poovaiah (2004)
13	Calmodulin-binding protein	*NtCBP4*	*N. tabacum*	Heavy metal tolerance	Arazi et al. (1999)
14	Heat shock protein 90	*Hsp90*	*Sorghum bicolor*	HS	Virdi et al. (2009)
15	Heat shock protein 70	*Hsp70*	*Zea mays*	HS	Sun et al. (2000)
16	FK506-binding protein	*FKBP77*	*Triticum aestivum*	HS, DS	Kurek et al. (1999)
17	Glyoxalase 1	*GLX1*	*B. juncea*	Heavy metal (Zn^{2+}, Cd^{2+}) tolerance, OS, SS, ABA	Deswal and Sopory (1991)

(continued)

Table 1.1 (continued)

S. no.	Protein	Gene	Source	Stress response	Reference
18	Soybean Ca^{2+}-ATPase 1	*SCA1*	*Glycine max*	SS, fungal elicitor	Chung et al. (2000)
19	Glutamate decarboxylase	*GAD*	Petunia *Z. mays*	ABA, MJ, SS, OS, CS, anoxia, mechanical damage	Baum et al. (1993) Zhuang et al. (2010)
20	Apyrase	Apyrase	*Pisum sativam* *A. thaliana*	Tolerence to xenobiotic compounds	Hsieh et al. (2000) Steinebrunner et al. (2000)
21	Catalase	Catalase	*A. thaliana*	H_2O_2 homeostasis	Yang and Poovaiah (2002)
22	Calmodulin-binding protein 25 kDa	*AtCAMBP25*	*A. thaliana*	Negative regulator of OS	Perruc et al. (2004)
23	Superoxide dismutase	SOD	*Z. mays*	H_2O_2 homeostasis	Gong and Li (1995)
24	BCL2-associated athanogene protein (BAG)	*AtBAG6*	*A. thaliana*	PCD, SA, H_2O_2, HS	Kang et al. (2006)

Adapted from Singh and Virdi (2010). Reprinted with permission from Journal of Plant Biology
ABA abscisic acid, *CS* cold stress, *DS* drought stress, *GA* gibberellic acid, *HS* heat stress, *MJ* methyl jasmonate, *OS* osmotic stress, *SA* salicylic acid, *SS* salt stress, *PCD* programmed cell death

The production of ROIs under abiotic stress conditions is followed by necrosis and plant cell death (PCD) (Mittler 2002). ROIs act as secondary messengers for execution of cell death during hypersensitivity responses and also play key role in ozone-mediated cell death and PCD (Rao and Davis 1999; Mittler 2002). The BAG (BCL2-associated athanogene) proteins were implicated in antiapoptotic activity (Takayama et al. 1995). Eight genes that encode proteins with the BAG domain have been identified in Arabidopsis genome, with four (AtBAG5, AtBAG6, AtBAG7, and AtBAG8) exhibiting the presence of a CaMBD (IQ motif) close to the conserved BAG domain (Kang et al. 2006). Heterologous expression of AtBAG6, which interacts canonically with different CaM isoforms in Ca^{2+}-independent manner, in yeast cells resulted in induced cell death and its expression was enhanced in response to SA, H_2O_2 and high temperature stress, all of which are known to be involved in plant PCD processes (Kang et al. 2006). Although the role of CaM-BAG complex in PCD has been well characterized, the precise function of this complex in downstream components *in planta* is yet to be elucidated.

Regulation of Stress-Modulated Kinases by CaM

Kinases constitute an indispensible component of the signal transduction pathways and this is achieved by alteration in autophosphorylation status and/or formation of multi-component complex (Charpenteau et al. 2004). CaM-binding protein kinases (CBKs) have been cloned and characterized from several plant species and elaborately reviewed by Zhang and Lu (2003). The presence of N-terminal CaMBD and C-terminal protein kinase catalytic domain of variable length and sequence is an important feature of plant CBKs. The CBKs from *A. thaliana* (AtCBK1) and tobacco (NtCBK2) bind to CaM in Ca^{2+}-dependent manner (Zhang and Lu 2003). The autophosphorylation and substrate phosphorylation activities of these proteins were Ca^{2+}-dependent and enhanced by CaM up to 4- to 5-fold. The autophosphorylation activity of CBKs in lily and tobacco was, on the contrary, inhibited in the presence of CaM (Takezawa et al. 1996; Liu et al. 1998). The autophosphorylation activity of maize CBK (ZmCCaMK) was unaffected by CaM (Pandey and Sopory 1998, 2001), though its substrate phosphorylation activity was Ca^{2+}-CaM dependent. The autophosphorylation and substrate phosphorylation activities of rice CBK (OsCBK), which showed higher affinity for CaM, were, however, CaM-independent (Zhang et al. 2002). It is, thus, evident that regulation of activity of different CBKs by CaM is differential. The expression of CBKs in plants is reported to be modulated by different stress conditions and, hence, have been implicated in abiotic stress adaptation response (Hua et al. 2004).

Large number of receptor-like serine/threonine kinases (RLKs) are reported in plants and at least 600 RLK homologs have been identified in *Arabidopsis* (Hardie 1999). RLKs are transmembrane proteins, which recognize an extracellular signal that results in autophophorylation on the cytoplasmic kinase domain, thus, leading to transduction of signal (Stone and Walker 1995). About 3/4th of all the RLK homologs known are localized to plasma membrane and rest are cytoplasmic. The presence of cytoplasmic kinase domain, a single membrane spanning domain and an extracellular ligand binding domain, most commonly the leucine rich repeat domain (LRR Domain), are the important features of plasma membrane-localized plant RLKs (Torii 2000; Barre et al. 2002). On the contrary, the cytoplasmic RLKs contain only a kinase domain (Yang et al. 2004). A cytoplasmic-localized Ca^{2+}-CaM regulated kinase (CRCK1), isolated from *Arabidopsis*, showed binding to CaM in Ca^{2+}-dependent manner and the CaM-binding site was localized around the kinase domain. Both autophosphorylation and substrate phosphorylation activities of CRCK1 were enhanced by CaM. The autophosphorylation activity of CRCK1 was Mg^{2+}-dependent and no activity was observed in the presence of Ca^{2+}. The CaM-induced enhancement of autophosphorylation and substrate phosphorylation activity of CRCK1 was attributed to direct interaction of the former with a.a. residues 160–183 of this protein. *CRCK1* expression in the seedlings was enhanced at both transcript and protein levels in response to H_2O_2, salt, cold, and ABA treatment (Yang et al. 2004), thus, suggesting its role in multiple stress pathways.

The members of RLK family, which are localized to plasma membrane, have also been proposed to play a role in stress response. Ca^{2+}-regulated RLK from *Glycine soja (GsCBLRK)* (Yang et al. 2010a) and *Arabidopsis* (*CRLK1*) (Yang et al. 2010b) were demonstrated to act as positive regulators of cold- and salt stress adaptation, respectively. Both GsCBRLK and CRLK1 exhibited binding to CaM, which was Ca^{2+}-dependent, and their kinase activity was regulated through Ca^{2+}-CaM interaction (Yang et al. 2010a, 2010b). The autophosphorylation activity of another plasma membrane-localized RLK of *A. thaliana* (*AtCaMRLK*), whose binding to CaM is Ca^{2+}-dependent, on the contrary, was Ca^{2+}-CaM-independent (Charpenteau et al. 2004). The expression of *CRLK1* was observed in roots and leaves. CRLK1 protein levels were enhanced under cold stress (4 °C) without a significant increase in the mRNA level, thus, suggesting that this gene may be regulated at post-transcriptional level. Though *crlk1* mutant knockout plants showed no observable difference as compared to wild type under normal growth conditions, but imposition of cold stress resulted in decrease in root and shoot growth, early signs of senescence and more severe damage due to chilling in the *crlk1* mutants. The *CRLK1*-induced chilling stress tolerance appeared to be through modulation of cold regulated genes viz; *CBF1, RD29A, COR15a,* and *KIN1*, since cold-induced expression of these genes was delayed in *crlk1* mutant plants (Yang et al. 2010b). It, therefore, appears that *CRLK1* is likely to be an important component of cold stress signal transduction pathway in plants.

The role of CaM-binding CBKs in stress tolerance in plants was further emphasized by the cloning of a receptor-like protein kinase, *GsCBRLK,* from a salt tolerant plant, *Glycine soza,* and its over-expression in *A. thaliana*. The *GsCBRLK* binds to CaM in the presence of Ca^{2+}. The *GsCBRLK* has been proposed to act as a master regulator of salt stress response (Yang et al. 2010a). The expression of *GsCBRLK* was elevated by different abiotic stress conditions *viz.*, salt-, cold-, and osmotic stress. Constitutive over-expression of *GsCBRLK* in the transgenic *Arabidopsis* plants conferred tolerance, as was evident from enhanced germination, higher root and shoot growth, and increased levels of chlorophyll under salt stress, and in response to ABA treatment.

The differential regulation of phosphorylation activity of the plant CBKs by CaM may be the result of evolutionary divergence resulting from adaptation to different environmental conditions. The stress-inducibility of different CaM-binding kinases in response to diverse stresses implies that these proteins are playing an important role in stress signal transduction pathways and the differential regulation of different kinases by CaM may be enabling the plants to respond in a stimulus-specific manner. Comparative analysis of upstream sequences is, however, required to understand the molecular basis of differential stress-inducibility of the different kinase genes.

Regulation of Stress-Modulated Phosphatases by CaM

Mitogen activated protein kinases (MAPKs) constitute another class of protein kinases that play an important role in signal transduction in eukaryotes. Each MAPK signaling cascade consists of a functionally interlinked pre-kinase module, an MAPK kinase (MEK) and an MAPK kinase kinase (MEKK). MEK carries out activation of MAPKs by phosphorylation of threonine and tyrosine residues with in a conserved TxY motif (Katou et al. 2007). The phosphorylated MAPKs are dephosphorylated by MAPK phosphatases (MKP) thus resulting in inactivation of MAPKs. Modulation of MKP activity, thus, is an important regulatory point in signal transduction in plants.

CaM regulates the activity of plant MKPs, which have been implicated in different abiotic- and biotic stress responses (Ulm et al. 2002). The CaM-binding property appears to be a unique and conserved feature of plant MKPs (Katou et al. 2007). MKPs have been cloned and characterized from tobacco (*NtMKP1*) (Yamakawa et al. 2004), Rice (*OsMKP1*) (Katou et al. 2007), and *Arabidopsis* (*AtMKP1*) (Lee et al. 2008). NtMKP1 and OsMKP1 are orthologs of AtMKP1. Though high similarity is observed in a.a. sequence of NtMKP1 and AtMKP1 but the protein structures, particularly the CaMBDs, are different. The NtMKP1 and OsMKP1 contain a single putative CaMBD. On the contrary, AtMKP1 consists of two different CaMBDs, both of which bind to CaM in Ca^{2+}-dependent manner, though the binding affinity of CaMBD2 is higher than CaMBD1 (Lee et al. 2008). The CaMBD2 is absent in NtMKP1 but CaMBD1 of the two correspond with each other. Studies carried out showed that phosphatase activity of AtMKP1 was positively regulated by CaM in a Ca^{2+}-dependent manner (Lee et al. 2008). AtMKP1, NtMKP1 and OsMKP1, through their phosphatase activities, were implicated in regulation of wound and defence response in plants. This was supported by the observation that over-expression of *NtMKP1* in transgenic tobacco plants attenuated the kinase activity of several defence-related MAPKs and wound-induced protein kinases (Yamakawa et al. 2004). These studies demonstrated that NtMKP1 may be acting as a negative regulator of MAPKs. The effect of different CaM isoforms on phosphatase activity of MKPs needs to be investigated in order to determine whether the regulation of wound response in plants is mediated through differential expression of different CaM isoforms (Yamakawa et al. 2004). The Ca^{2+}-CaM-regulated MKPs, therefore, may provide a critical link between two important signaling pathways in plants i.e., Ca^{2+}-signaling and MAPK signaling cascades, which may enable the plants to withstand stressful conditions.

Calmodulin-Binding Transcription Factors

Recent studies suggest that the gene expression at transcriptional level is also regulated by CaM through modulation of activity of transcriptional factors

(Finkler et al. 2007). Various transcription factors, which are involved in cold stress tolerance (Doherty et al. 2009), modulation of plant immune response (Du et al. 2009), auxin signaling (Galon et al. 2010), etc., have been reported to show CaM-binding property. The expression of cold-regulated genes under chilling stress conditions is mediated through an increase in [Ca^{2+}] (Minorsky 1989; Knight et al. 1991). Three regulatory genes *viz., CBF1* (C-repeat binding factor), dehydration responsive element binding factor *(DREB1b), CBF2 (DREB1c),* and *CBF3 (DREB1a)* are rapidly expressed (with in 15 min) in response to low temperature. The product of these genes further induce the expression of ~100 genes by binding to their RT/DRE regulating elements in their promoters (Gilmour et al. 1998; Vogel et al. 2005). Recent studies have demonstrated that the CaM-binding transcriptional activator (CAMTA) proteins constitute the molecular link between [Ca^{2+}] spike and cold stress-regulated genes in Arabidopsis (Doherty et al. 2009).

Six CAMTA members have been identified in Arabidopsis (Bouche et al. 2002). These proteins carry an IQ domain for CaM-Binding, along with a CG1-1 domain that binds to core sequence VCGLGB (da Costa e Silva 1994; Bouche et al. 2002), which is similar to the conserved motif CM2 sequence [CCGCGT]. The CM2 motif overlaps with ICEr2 (inducer of CBF expression region 1 and 2) in CBF2 and is responsible for cold induction of CBF2. CAMTA proteins, 1, 2, 3 and 5 exhibited binding to CM2 sequence but analysis of T-DNA mutant of *A. thaliana* showed that only *CAMTA3* was responsible for regulating the expression of *CBF1*, *CBF2* and *ZAT12*. CBF3 expression, on the contrary, was not affected in any of the *CAMTA* mutants (Doherty et al. 2009). Studies by Doherty et al. (2009) also demonstrated that as contrary to *camta3* single mutants, *camta1 camta3* double mutants showed significant reduction in cold-induced expression of only *CBF1* but *CBF2* and *ZAT12* (zinc-finger protein) levels were not affected, thus, implying interaction of CAMTA1 and CAMTA2 in regulation of cold-induced gene expression. Analysis of *camta1* and *camta 3* single mutants, and *camta1 camta3* double mutants further revealed the role of *CAMTA1* and *CAMTA2* in cold acclimation of plants but not in the cold tolerance process *per se*. It is likely that both CAMTA1 and CAMTA3 may be required for stabilizing the proteins synthesized during acquisition of chilling tolerance, as was reported for a heat shock-associated protein HSA32 in Arabidopsis (Charng et al. 2006).

The structural homolog of AtCAMTA3 in rice, OsCBT (*Oryzae sativa* Cam-binding transcriptional factor) was also reported to bind to CaM in a Ca^{2+}-dependent/independent manner through two distinct types of CaMBDs (Choi et al. 2005). OsCBT was demonstrated to act as a negative regulator in plant defence related gene expression (Koo et al. 2009). CaM is a negative regulator of OsCBT since co-transformation of *OsCBT* and rice CaM genes *(OsCaM)* in *Arabidopsis* resulted in inhibition of transcriptional activation activity of OsCBT. Ca^{2+}-CaM, which plays an important role in plant defence signaling (Ali et al. 2003), may be responsible for triggering the plant defence response by inhibiting the OsCBT. The role of *OsCBT1* also need to be investigated in cold stress tolerance since this gene has been proposed as a functional orthologue of *CAMTA3* (Koo et al. 2009), which was implicated in regulation of cold-responsive genes (Doherty et al. 2009). The diverse functions of CAMTA3 and OsCBT1, as observed between *A. thaliana* and rice, signify the versatility in regulation of transcriptional factors by CaM.

A nuclear-localized protein of 25 kDa in Arabidopsis (AtCAMBP25) was also proposed to act as a negative regulator of salt- and osmotic stress tolerance since over-expression of this gene increased the sensitivity of transgenic plants to these stress conditions. On the contrary, suppression of this gene resulted in higher levels of tolerance (Perruc et al. 2004). However, information on the role of AtCaMBP25 as a transcriptional factor is lacking.

MYB proteins constitute another important class of transcriptional regulators in plants. The DNA-binding activity of MYB proteins is also modulated by Ca^{2+}-CaM in an isoform-dependent manner (Yoo et al. 2005). The DNA-binding activity of a MYB protein from *Arabidopsis* (AtMYB2) was affected differentially by different *Glycine max* CaM isoforms. Whereas, GmCaM4 enhanced the DNA-binding activity of AtMYB32 in a Ca^{2+}-dependent manner, another isoform, GmCaM1, had an inhibitory effect (Yoo et al. 2005). Transgenic Arabidopsis plants, which overexpressed *GmCaM4*, showed higher expression of *AtMYB2* and were more tolerant to salt stress. Higher level of salt stress tolerance, observed in the *GmCaM4* over-expressing plants, was attributed to an increase in expression of dehydration-responsive gene (*RD22*), alcohol dehydrogenase 1 (*ADH1*) and Delta (1)-pyrroline-5-carboxylate synthetase 1 (*P5CS1*), along with elevated levels of proline. On the contrary, over-expression of isoform GmCaM1 had no significant effect on the expression of stress-inducible genes. These observations suggested that salt stress tolerance through AtMYB2 activity is regulated through CaM in an isoform-specific manner. Since different isoforms are regulated differentially by different stimuli (Botella and Arteca 1994; Heo et al. 1999), it may enable the cell to fine tune the response under different environmental conditions.

CaM also regulates the activity of transcriptional factors through mediator proteins. A group of proteins in *Arabidopsis*, designated as AtBT1-5 (*A. thaliana* BTB and TAZ proteins), which bind to CaM in a Ca^{2+}- dependent manner, interact with two proteins, *Arabidopsis thaliana* Bromodomain and Extra Terminal domain proteins (AtBET10 and AtBET9), which belong to the family of fsh/ring3 class transcriptional regulators. *In vivo* activation of transcriptional function of AtBET10 ensues after interaction of this protein with AtBT through BTB domain (Du and Poovaiah 2004). The studies carried out till now suggest that some of the responses mediated by messenger molecules like Ca^{2+}, SA and H_2O_2 are through regulation of expression and modulation of conformation of AtBTs, which in turn facilitate the downstream responses of the cell by activating transcriptional activators such as AtBET10 (Du and Poovaiah 2004).

Role of CaM in Regulation of Transport of Ca^{2+}, Heavy Metal Ions and Xenobiotic Compounds

Plasma membrane-localized channel proteins are involved in transport of heavy metals in plants (Arazi et al. 1999). These proteins are characterized by the presence of transmembrane domains and a putative cyclic nucleotide monophosphate domain that overlaps with a CaM-binding domain located at C-terminus (Köhler et al. 1999). A gene for an 81 kDa plasma membrane-localized CaMBP in *Nicotiana tabacum* (NtCBP4), which showed homology to cyclic nucleotide gated channel protein, CNGC1, of *Arabidopsis,* was cloned (Arazi et al. 1999). Over expression of *NtCBP4* in tobacco resulted in enhanced tolerance to Ni^{2+} but hypersensitivity to Pb^{2+} in the transgenic plants. Tolerance to Ni^{2+} in transgenic plants was due to reduced uptake of this ion and hypersensitivity to Pb^{+} was attributed to increased accumulation of Pb^{2+} in shoots of transgenic plants (Arazi et al. 1999). Deletion of the CaMBD and cyclic nucleotide-binding domains resulted in improved tolerance to Pb^{2+} in transgenic plants, which was primarily the result of decrease in uptake of this metal (Sunkar et al. 2000). Ca^{2+}-permeable channels have been identified as a pathway of Pb^{2+} entry into animal and plant cells (Tomsig and Suszkiw 1991; Huang and Cunningham 1996). It is, therefore, likely that transport of Pb^{2+} into the plant cells may be regulated by the CaM through regulation of plasma membrane-localized proteins. This study demonstrates that it may be possible to confer tolerance to heavy metal ions in crop plants by engineering the CaM-binding property of the channel proteins.

Presence of apyrases, which hydrolyse nucleosides di- and tri-phosphates, is an ubiquitous feature of all eukaryotes (Hsieh et al. 2000). Hydrolysis of nucleoside tri- and di-phosphates by apyrases in animals is implicated in neurotransmission (Todorov et al. 1997) and also in preventing thrombosis by inhibition of ADP-induced platelet aggregation (Marcus et al. 1997). The role of apyrases in plants is, however, not very well defined. The activity of animal apyrases has not been reported to be affected by CaM. On the contrary, CaM modulates the activity of plant apyrases (Hsieh et al. 2000), therefore, suggesting the role of CaM in the regulation of these enzymes. An endogenous apyrase from *Pisum sativam* (PsNTP9) was demonstrated to bind to CaM in a Ca^{2+}-independent manner (Hsieh et al. 2000) and its activity was reported to be stimulated by Ca^{2+}-CaM (Chen and Roux 1986). Transgenic expression of *PsNTP9* in Arabidopsis resulted in enhanced resistance to toxic concentrations of different xenobiotic compounds like cyclohexane, plant growth regulators (Thomas et al. 2000) and different herbicides (Windsor et al. 2003). These studies, thus, supported the role of pea apyrase in multidrug resistance mechanism. Identification and characterization of novel CaM-regulated apyrases from different sources, therefore, may provide versatile tools for exploring strategies for introducing herbicide tolerance in plants.

CaM was also demonstrated to be involved in tolerance to methylglyoxal, a toxic metabolite, which is accomplished through regulation of glyoxalase 1 (Espartero et al. 1995). Glyoxalase I catalyses the conversion of toxic methylglyoxal to a nontoxic metabolite and was reported to be induced by NaCl, mannitol or ABA (Espartero et al. 1995). Glyoxaylase I, isolated from *Brassica juncea* (*BjGly I*), exhibited binding to CaM and its activity was also stimulated by Ca^{2+}/CaM (Deswal and Sopory 1991). The *BjGly I* over-expressing transgenic plants showed higher levels of tolerance to salt stress (Veena and Sopory 1999).

The stimulus-induced increase in intracellular Ca^{2+} levels must be restored to basal levels so as to maintain homeostasis. This is achieved by efflux of [Ca^{2+}] from the cytosol, which is mediated by two Ca^{2+} transporters viz., Ca^{2+} pumps and Ca^{2+}/H^+ antiporters (Bush 1995). The Ca^{2+}-ATPases catalyze the efflux of Ca^{2+} by using the energy of ATP hydrolysis and the proton motive force generated due to H^+ gradient built up by the H^+-ATPase (Palmgren 2001). In plants, two CaMBDs were reported in the plasma membrane-localized Ca^{2+}-ATPases, SCA1 (soybean Ca^{2+}-ATPase) (Chung et al. 2000) and AtACA8 (*A. thaliana* Ca^{2+}-ATPase) (Luoni et al. 2006). The expression of *SCA1* was induced by salt and fungal elicitor but not by osmotic stress (Chung et al. 2000). Though the expression of *AtACA8* under abiotic stress conditions is not reported but the differential binding of different CaMs to this protein, which is affected by Ca^{2+} (Luoni et al. 2006), may allow the fine tuning of the cellular response to different external stimuli.

Regulation of GABA Shunt by Calmodulin

A non-protein amino acid GABA (γ-aminobutyric acid) is an important inhibitory neurotransmitter in animals (Turano and Fang 1998). Glutamate decorboxylase (GAD) is a key enzyme responsible for the synthesis of GABA in plants. The role of GABA is still a matter of conjecture but enhancement in its levels by different abiotic stress conditions in different plants (Shelp et al. 1999) implies its role in stress response. The bacterial and animal GADs have not been reported to bind CaM but all the plant GADs studied till now, except OsGAD2 (Akama and Takaiwa 2007), show the presence of CaMBDs (Oh et al. 2005). Deletion of CaMBD domain from the petunia GAD and its overexpression in transgenic tobacco plants resulted in severe morphological abnormalities (Baum et al. 1993, 1996), therefore, implicating the role of Ca^{2+}-CaM in regulation of glutamate and GABA metabolism through control of GAD activity. The lack of CaMBD in rice OsGAD2 suggests that plants have evolved both Ca^{2+}-CaM-dependent and independent strategies for regulation of GAD activities.

Regulation of Heat Shock Response by Calmodulin

Studies have been carried out to investigate the possible role of CaM in heat stress-mediated signal transduction pathways. Transient increase in Ca^{2+} concentration has been well characterized in response to HS in plants (Gong et al. 1997). Ca^{2+}-channel blockers and CaM-antagonists are widely used to understand the possible role of Ca^{2+} and CaM, respectively, in various signaling events that occur in cell. Maize seedlings, which were subjected to heat stress after treatment with Ca^{2+}-channel blockers and CaM-antagonists alone or in combinations, showed decreased survivability after exposure to sublethal and lethal temperatures (Gong et al. 1997). On the contrary, the exogenous application of Ca^{2+} significantly enhanced the plant survival both under sublethal and lethal temperatures, thus, providing evidence for the direct involvement of Ca^{2+}-CaM in heat stress signal transduction pathways (Gong et al. 1997).

The plants are likely to face frequent incidents of elevated temperature stress due to global warming (Angilletta 2009). Accumulation of heat shock proteins (HSPs) is one of the adaptive responses observed in plants subjected to high temperature stress. Due to their chaperonic functions, the Hsps protect cellular proteins from heat-induced damage, besides assisting in refolding of the damaged proteins. Plant Hsps are divided into five different families viz., small Hsps (12–40 kDa), Hsp60 (chaperonin), Hsp70, Hsp90 and Hsp100 (Wang et al. 2004). Studies carried out to investigate the role of Ca^{2+} in synthesis of HSPs under control and heat stress conditions demonstrated that the presence of Ca^{2+} significantly enhanced the synthesis of HSPs in cultured sugar beet cells under heat stress (Trofimova et al. 1999). Sun et al. (2000) provided evidence that CaM interacts with maize Hsp70 in the presence of Ca^{2+} and inhibits its intrinsic ATPase activity. Our studies have also shown that as compared to salt- or cold stress, imposition of heat stress has more profound effect on the expression of CaMBPs in young seedlings of sorghum, which is a crop of hot and dry regions (Singh and Virdi 2010). The CaMBPs, which were observed in the sorghum seedlings in response to HS (Fig. 1.1), ranged from low to high molecular weight and showed difference in their kinetics of induction. The expression of a wide array of CaMBPs and their differential regulation by thermal stress in the sorghum seedlings signify the importance of these proteins in heat shock adaptation. Two of the heat-modulated CaMBPs in sorghum, HSP85 and HSP73, were identified as members of HSP90 and HSP70 families, respectively (Virdi et al. 2009). Though binding of HSP70 to CaM was also reported earlier in maize (Sun et al. 2000), this was the first study to provide evidence for the Ca^{2+}-dependent binding of CaM to a HSP90 member in plants. These studies also demonstrated that application of Ca^{2+}-channel blockers and CaM antagonists decreased the steady state levels of HSP85, whereas the steady state levels of HSP73 were not affected significantly under similar conditions (Virdi et al. 2011), thus, providing evidence that accumulation of only HSP90 member (HSP85) is mediated through Ca^{2+}-CaM, at least in sorghum. These observations were at variance with the studies of Liu et al. (2003), who reported that the expression of *HSP70* at transcript level in wheat was modulated through Ca^{2+}-CaM. The variability in response of HSP70 between sorghum and wheat may either be due to genotypic differences and/or due to regulation of wheat HSP70 at post-transcriptional level (Leborgne-Castel et al. 1999).

Nitric oxide (NO) has been proposed as a signal transducer of heat stress, which functions upstream to CaM-regulated signal transduction (Xuan et al. 2010). It was reported that NO mediates the Ca^{2+}-channel function in plasma membrane (Delledonne et al. 1998). A model explaining the role of NO in heat stress signal transduction pathway was proposed (Liu et al. 2008). According to this model, heat stress imposition results in NO production, which regulates the expression of AtCaM3 (*A. thaliana* CaM isoform 3). The AtCaM3 interacts with a CaM-binding protein kinase (AtCBK3) that phosphorylates heat shock factor (HSF), which then binds to heat shock elements (HSE) of different HSP genes, thereby, inducing

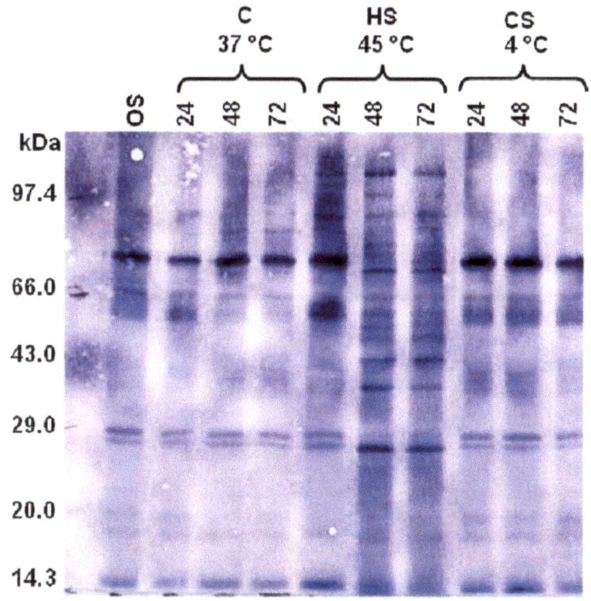

Fig. 1.1 Changes in calmodulin-binding proteins in the 24 h-old sorghum seedlings in response to different durations of cold- (4 °C) and heat stress (45 °C). 50 μg of crude proteins, isolated from 24 h-old sorghum seedlings after exposure to heat- (45 °C) and cold (4 °C) stress, were resolved on 12.5 % SDS-PAGE gel, transferred onto Hybond-C membrane and incubated with biotinylated calmodulin (1.5 μg) in the presence of $CaCl_2$, followed by probing with streptavidin-alkaline phosphate conjugate. 0S: 24 h-old seedlings grown at 37 °C at the beginning of treatment; C: control; HS: heat shock; CS: cold stress. From Singh and Virdi. Calmodulin-binding proteins ni plants: implications in abiotic stress adaptation. J Plant Biol 2010;37(1): 1–17. Journal of Plant Biology

the production of these proteins. Of the 24 HSFs identified in *A. thaliana* (Nover et al. 2009), AtCBK3 phosphorylates AtHSF1 in the presence of Ca^{2+}-CaM. The role of *CBK3* in heat shock adaptation was also supported by transgenic studies. The *cbk3* mutants of *Arabidopsis*, lacking in expression of *AtCBK3,* were impaired in thermotolerance. The rescue of *cbk3* mutants by transgenic *AtCBK3* overexpression improved the basel thermotolerance. The rescued mutants accumulated higher levels of HSPs as compared to wild type plants. Higher levels of HSPs in the rescued *cbk3* mutants were attributed to enhanced binding of HSF to HSE in the presence of AtCBK3 (Liu et al. 2008). Although these observations do implicate the role of *AtCBK3* in HSP expression but the fact that expression of these proteins was still observed in *cbk3* mutants, albeit at 50% level of the wild type, implied the presence of redundant/alternate regulatory pathway(s) that control the expression of HSPs.

Virdi et al. (2011) proposed a novel model which explained the role of CaM in regulation of heat shock response in plants (Fig. 1.2). According to this model, imposition of thermal stress results in elevation of intracellular Ca^{2+}. The CaM binds to Ca^{2+} and undergoes change in conformation, followed by binding to

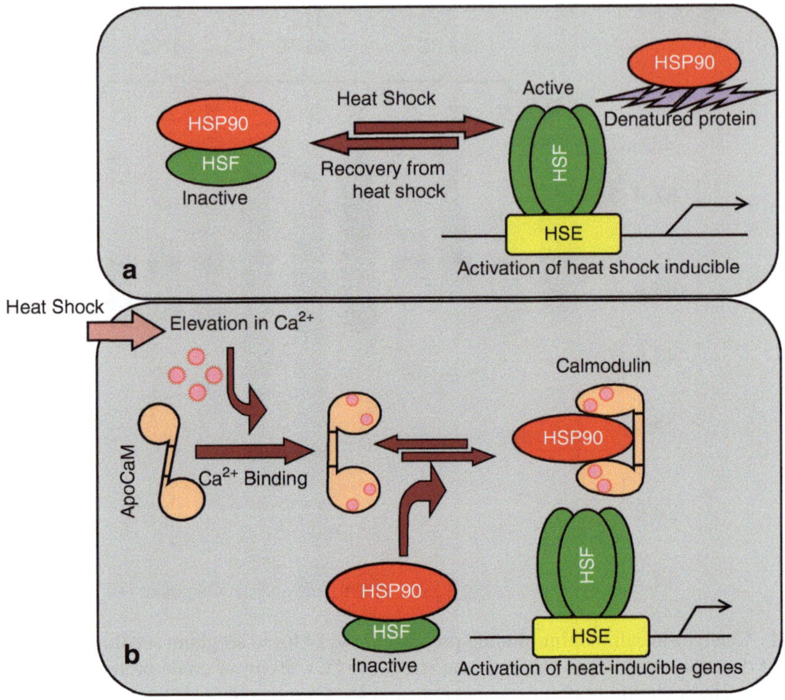

Fig. 1.2 Modification of the hypothetical model (**a**), proposed by Yamada and Nishimura (2008), to understand the molecular mechanisms underlying the regulation of HSP90 family members (HSP85 here) in the absence and presence of heat shock. It is proposed that besides denatured proteins, calmodulin (CaM) may also be regulating the heat shock response (**b**). Heat shock-induced transient increase in intracellular Ca^{2+} activates apocalmodulin (ApoCaM; inactive form) to its Ca^{2+}-bound active form. The Ca^{2+}-CaM binds to HSP90 and dissociates the monomeric heat shock factor (HSF), which trimerizes and interacts with heat shock elements (HSE), thus, resulting in upregulation of the heat shock-inducible genes. From Virdi et al. Evidence for the possible involvement of calmodulin in regulation of steady state levels of hsp90 family members (hsp87 and hsp85) in response to heat shock in sorghum. Plant Signal Behav 2011 Mar;6(3):393–399. Reprinted with permission from Plant Signaling and Behavior

HSP90. Since HSP90 was identified as inhibitor of HSF (Zou et al. 1998), therefore, the interaction of CaM with HSP90 results in dissociation of HSP90 and HSF. The HSF trimerizes and interacts with HSEs thus resulting in upregulation of HS-induced genes (Yamada et al. 2007; Yamada and Nishimura 2008).

CaM-binding protein phosphatases may also be playing a role in heat shock response since it was demonstrated that *A. thaliana* protein phosphatase (AtPP7) binds to CaM in a Ca^{2+}-dependent manner (Kutuzov et al. 1998; Liu et al. 2007). The role of AtPP7 in thermotolerance is validated by the observation that the Arabidopsis T-DNA insertion lines of *AtPP7* were impaired in thermotolerance. Further, as compared to wild type, over-expression of *AtPP7* resulted in enhanced induction of both *HSP70* and *HSP101* at both transcript and protein levels. As contrary to AtCBK3, which is cytoplasmic (Liu et al. 2008), the AtPP7 is detected in nucleus (Liu et al. 2007). It is, therefore, likely that it may be involved in activation of HSF through its phosphatase activity since dephosphorylation in some sites of HSF is reported to enhance transcriptional activities (Høj and Jakobsen 1994).

Apart from HSPs, FK-506-binding proteins, FKBP73 and FKBP77 in wheat, which possess peptidyl prolyl *cis-trans* isomerase activity, were also reported to contain CaMBDs (Kurek et al. 1999). FKBP77 is the heat shock-induced isoform (Kurek et al. 1999) and binds to HSP90 (Reddy et al. 1998; Kurek et al. 1999; Kamphausen et al. 2002). Although overexpression of *FKBP77* in transgenic wheat plants caused morphological abnormalities, which were associated with higher level of *HSP90* mRNA, the role of CaM in regulation of FKBP77 is still a matter of conjecture.

During the past decade, though substantial progress has been made in understanding the molecular mechanisms of abiotic stress tolerance in plants, it is only recently that the role of CaMBPs in abiotic stress adaptation of plants has begun to emerge. CaM has been demonstrated to regulate the activities of a diverse set of proteins, which further regulate the plant response to different abiotic stress conditions (Table 1.1) (Singh and Virdi 2010). Whereas, some CaMBPs have been shown to act as positive regulators of stress response, others act as negative regulators, thus, suggesting that under adverse conditions the CaM is enabling the plant to maintain homeostasis through regulation of CaMBPs.

Concluding Remarks and Future Perspectives

In view of emerging role of CaMBPs in stress response in plants, it is important that stress-modulated CaMBPs should be identified and functionally characterized in crop plants. Application of protein microarray technology for studying the CaM-interactome is likely to reveal novel CaMBPs. The dynamics of interaction of CaMBPs with CaM should be analysed *in vivo* by implying strategies like FRET and BIFCA. These studies will also elucidate the role of different CaM isoforms in the regulation of different CaMBPs. Concerted efforts are also needed for genetically engineering the different CaMBPs, particularly with respect to Ca^{2+}-CaM domain. The results of these studies may provide a tool for generating novel phenotypes, some of which could be highly desirable for agricultural improvement. Further, the regulation of stress-modulated CaMBPs in plants needs to be studied in context of the different hormones, ABA in particular, by using mutants so that questions regarding the cross-talk among different pathways, that are regulated by stress, can also be addressed.

Acknowledgements We gratefully acknowledge the financial assistance provided by the Department of Biotechnology, Government of India, New Delhi. ASV is thankful to Council of Scientific and Industrial Research, Government of India, New Delhi for the award of Senior Research Fellowship.

References

Akama K, Takaiwa F (2007) C-terminal extension of rice glutamate decarboxylase (OsGAD2) functions as an autoinhibitory domain and overexpression of a truncated mutant results in the accumulation of extremely high levels of GABA in plant cells. J Exp Bot 58:2699–2707

Alexander KA, Wakim BT, Doyle GS, Walsh KA, Storm DR (1988) Identification and characterization of the calmodulin-binding domain of neuromodulin, a neurospecific calmodulin-binding protein. J Biol Chem 263:7544–7549

Ali GS, Reddy VS, Lindgren PB, Jakobek JL, Reddy ASN (2003) Differential expression of genes encoding calmodulin-binding proteins in response to bacterial pathogens and inducers of defense responses. Plant Mol Biol 51:803–815

Angilletta MJ Jr (2009) Looking for answers to questions about heat stress: researchers are getting warmer. Funct Ecol 23:231–232

Arazi T, Sunkar R, Kaplan B, Fromm H (1999) A tobacco plasma membrane calmodulin-binding transporter confers Ni^{2+} tolerance and Pb^{2+} hypersensitivity in transgenic plants. Plant J 20:171–182

Babu YS, Bugg CE, Cook WJ (1988) Structure of calmodulin refined at 2.2 Å resolution. J Mol Biol 204:191–204

Barre A, Herve C, Lescure B, Rouge P (2002) Lectin receptor kinases in plants. Crit Rev Plant Sci 21:379–399

Bartels D, Sunkar R (2005) Drought and salt tolerance in plants. Crit Rev Plant Sci 24:23–58

Baum G, Chen Y, Arazi T, Takatsuji H, Fromm H (1993) A plant glutamate decarboxylase containing a calmodulin-binding domain. Cloning, sequence, and functional analysis. J Biol Chem 268:19610–19617

Baum G, Lev-Yadun S, Fridmann Y, Arazi T, Katsnelson H, Zik M, Fromm H (1996) Calmodulin-binding to glutamate decarboxylase is required for regulation of glutamate and GABA metabolism and normal development in plants. EMBO J 15:2988–2996

Boonburapong B, Bauboocha T (2007) Genome-wide identification and analysis of the rice calmodulin and related potential calcium sensor proteins. BMC Plant Biol 7:4

Botella JR, Arteca RN (1994) Differential expression of two calmodulin genes in response to physical and chemical stimuli. Plant Mol Biol 24:757–766

Bouche N, Scharlat A, Snedden W, Bouchez D, Fromm H (2002) A novel family of calmodulin-binding transcription activators in multicellular organisms. J Biol Chem 277:21851–21861

Bouche N, Yellin A, Snedden WA, Fromm H (2005) Plant-specific calmodulin-binding proteins. Annu Rev Plant Biol 56:435–466

Bush DS (1995) Calcium regulation in plant cells and its role in signaling. Annu Rev Plant Physiol Plant Mol Biol 46:95–122

Charng YY, Liu HC, Liu NY, Hsu FC, Ko SS (2006) Arabidopsis Hsa32, a novel heat-shock protein, is essential for acquired thermotolerance during a long recovery period after acclimation treatment. Plant Physiol 140:1297–1305

Charpenteau M, Jaworski K, Ramirez BC, Tretyn A, Ranjeva R, Ranty BT (2004) A receptor-like kinase from *Arabidopsis thaliana* is a calmodulin-binding protein. Biochem J 379:841–848

Chen YR, Roux SJ (1986) Characterization of nucleoside triphosphatase activity in isolated pea nuclei and its photoreversible regulation by light. Plant Physiol 81:609–613

Cheney RE, Mooseker MS (1992) Unconventional myosins. Curr Opin Cell Biol 4:27–35

Choi MS, Kim MC, Yoo JH, Moon BC, Koo SC, Park BO, Lee JH, Koo YD, Han HJ, Lee SY, Chung WS, Lim CO, Cho MJ (2005) Isolation of a calmodulin-binding transcription factor from rice (*Oryza sativa* L.). J Biol Chem 280:40820–40831

Chung WS, Lee SH, Kim JC, Do HW, Kim MC, Park CY, Park HC, Lim CO, Kim WB, Harper JF, Cho MJ (2000) Identification of a calmodulin-regulated soybean Ca^{2+}-ATPase (SCA1) that is located in the plasma membrane. Plant Cell 12:1393–1408

Clapham DE (2007) Calcium signaling. Cell 131:1047–1058

Crivici A, Ikura M (1995) Molecular and structural basis of target recognition by calmodulin. Annu Rev Biophys Biomol Struct 24:85–116
da Costa e Silva O (1994) CG-1, a parsley light-induced DNA-binding protein. Plant Mol Biol 25:921–924
DeFalco TA, Bender KW, Snedden WA (2010) Breaking the code: Ca^{2+} sensors in plant signaling. Biochem J 425:27–40
Delledonne M, Xia Y, Dixon RA, Lamb C (1998) Nitric oxide functions as a signal in plant disease resistance. Nature 394:585–588
Deswal R, Sopory SK (1991) Purification and partial characterization of glyoxalase I from a higher plant *Brassica juncea*. FEBS Lett 282:277–280
Doherty CJ, Van Buskirk HA, Myers SJ, Thomashow MF (2009) Roles for Arabidopsis CAMTA transcription factors in cold-regulated gene expression and freezing tolerance. Plant Cell 21:972–984
Du L, Poovaiah BW (2004) A novel family of Ca^{2+}/calmodulin-binding proteins involved in transcriptional regulation: Interaction with fsh/Ring3 class transcription activators. Plant Mol Biol 54:549–569
Du L, Ali GS, Simons KA, Hou J, Yang T, Reddy ASN, Poovaiah BW (2009) Ca^{2+}/calmodulin regulates salicylic-acid-mediated plant immunity. Nature 457:1154–1158
Espartero J, Sanchez AI, Pardo JM (1995) Molecular characterization of glyoxalase-I from a higher plant; up-regulation by stress. Plant Mol Biol 29:1223–1233
Finkler A, Ashery PR, Fromm H (2007) CAMTAs: calmodulin-binding transcription activators from plants to human. FEBS Lett 581:3893–3898
Fromm H, Chua N-H (1992) Cloning of plant cDNAs encoding calmodulin-binding proteins using ^{35}S -labeled recombinant calmodulin as a probe. Plant Mol Biol Rep 10:199–206
Galon Y, Finkler A, Fromm H (2010) Calcium-regulated transcription in plants. Mol Plant 3:653–669
Gifford JL, Walsh MP, Vogel HJ (2007) Structures and metal-ion-binding properties of the Ca^{2+}-binding helix-loop-helix EF-hand motifs. Biochem J 405:199–221
Gilmour SJ, Zarka DG, Stockinger EJ, Salazar MP, Houghton JM, Thomashow MF (1998) Low temperature regulation of the Arabidopsis CBF family of AP2 transcriptional activators as an early step in cold-induced COR gene expression. Plant J 16:433–442
Gong M, Li Z-G (1995) Calmodulin-binding proteins from *Zea mays* germs. Phytochemistry 40:1335–1339
Gong M, Li YJ, Dai X, Tian M, Li ZG (1997) Involvement of calcium and calmodulin in the acquisition of heat-shock induced thermotolerance in maize seedlings. J Plant Physiol 150:615–621
Hardie DG (1999) Plant protein serine/threonine kinases: classification and functions. Annu Rev Plant Physiol Plant Mol Biol 50:97–131
Heo WD, Lee SH, Kim MC, Kim JC, Chung WS, Chun HJ, Lee KJ, Park CY, Park HC, Choi JY, Cho MJ (1999) Involvement of specific calmodulin isoforms in salicylic acid-independent activation of plant disease resistance responses. Proc Natl Acad Sci USA 96:766–771
Høj A, Jakobsen BK (1994) A short element required for turning off heat shock transcription factor: evidence that phosphorylation enhances deactivation. EMBO J 13:2617–2624
Hsieh HL, Song CJ, Roux SJ (2000) Regulation of a recombinant pea nuclear apyrase by calmodulin and casein kinase II. Biochim Biophys Acta 1494:248–255
Hua W, Li RJ, Wang L, Lu YT (2004) A tobacco calmodulin-binding protein kinase (NtCBK2) induced by high-salt/GA treatment and its expression during floral development and embryogenesis. Plant Sci 166:1253–1259
Huang JW, Cunningham SD (1996) Lead phytoextraction: species variation in lead uptake and translocation. New Phytol 134:75–84
Kamphausen T, Fanghänel J, Neumann D, Schulz B, Rahfeld JU (2002) Characterization of *Arabidopsis thaliana* AtFKBP42 that is membrane-bound and interacts with Hsp90. Plant J 32:263–276

Kang CH, Jung WY, Kang YH, Kim JY, Kim DG, Jeong JC, Baek DW, Jin JB, Lee JY, Kim MO, Chung WS, Mengiste T, Koiwa H, Kwak SS, Bahk JD, Lee SY, Nam JS, Yun DJ, Cho MJ (2006) AtBAG6, a novel calmodulin-binding protein, induces programmed cell death in yeast and plants. Cell Death Diff 13:84–95

Katou S, Kuroda K, Seo S, Yanagawa Y, Tsuge T, Yamazaki M, Miyao A, Hirochika H, Ohashi Y (2007) A calmodulin-binding mitogen-activated protein kinase phosphatase is induced by wounding and regulates the activities of stress-related mitogen-activated protein kinases in rice. Plant Cell Physiol 48:332–344

Knight MR, Campbell AK, Smith SM, Trewavas AJ (1991) Transgenic plant aequorin reports the effects of touch and cold-shock and elicitors on cytoplasmic calcium. Nature 352:524–526

Köhler C, Merkle T, Neuhaus G (1999) Characterisation of a novel gene family of putative cyclic nucleotide- and calmodulin-regulated ion channels in *Arabidopsis thaliana*. Plant J 18:97–104

Koo SC, Choi MS, Chun HJ, Shin DB, Park BS, Kim YH, Park HM, Seo HS, Song JT, Kang KY, Yun DJ, Chung WS, Cho MJ, Kim MC (2009) The calmodulin-binding transcription factor OsCBT suppresses defense responses to pathogens in rice. Mol Cells 27:563–570

Kurek I, Aviezer K, Erel N, Herman E, Breiman A (1999) The wheat peptidyl prolyl cis-trans-isomerase FKBP77 is heat induced and developmentally regulated. Plant Physiol 119:693–704

Kutuzov MA, Evans DE, Andreeva AV (1998) Expression and characterization of PP7, a novel plant protein Ser/Thr phosphatase distantly related to RdgC/PPEF and PP5. FEBS Lett 440:147–152

Leborgne-Castel N, Jelitto-Van Dooren EP, Crofts AJ, Denecke J (1999) Overexpression of BiP in tobacco alleviates endoplasmic reticulum stress. Plant Cell 11:459–470

Lecourieux D, Mazars C, Pauly N, Ranjeva R, Pugin A (2002) Analysis and effects of cytosolic free calcium increases in response to elicitors in *Nicotiana plumbaginifolia* cells. Plant Cell 14:2627–2641

Lee SH, Kim MC, Heo WD, Kim JC, Chung WS, Park CY, Park HC, Cheong YH, Kim CY, Lee S-H, Lee KJ, Bahk JD, Lee SY, Cho MJ (1999) Competitive binding of calmodulin isoforms to calmodulin-binding proteins: implication for the function of calmodulin isoforms in plants. Biochim Biophys Acta 1433:56–67

Lee SH, Johnson JD, Walsh MP, Van Lierop JE, Sutherland C, Xu A, Snedden WA, Kosk-Kosicka D, Fromm H, Narayanan N, Cho MJ (2000) Differential regulation of Ca^{2+}/calmodulin-dependent enzymes by plant calmodulin isoforms and free Ca^{2+} concentration. Biochem J 350:299–306

Lee K, Song EH, Kim HS, Yoo JH, Han HJ, Jung MS, Lee SM, Kim KE, Kim MC, Cho MJ, Chung WS (2008) Regulation of MAPK phosphatase 1 (AtMKP1) by calmodulin in Arabidopsis. J Biol Chem 283:23581–23588

Liu Z, Xia M, Poovaiah BW (1998) Chimeric calcium/calmodulin-dependent protein kinase in tobacco: differential regulation by calmodulin isoforms. Plant Mol Biol 38:889–897

Liu HT, Li B, Shang ZL, Li XZ, Mu RL, Sun DY, Zhou RG (2003) Calmodulin is involved in heat shock signal transduction in wheat. Plant Physiol 132:1186–1195

Liu HT, Li GL, Chang H, Sun DY, Zhou RG, Li B (2007) Calmodulin-binding protein phosphatase PP7 is involved in thermotolerance in Arabidopsis. Plant Cell Environ 30:156–164

Liu HT, Gao F, Li GL, Han JL, Liu DL, Sun DY, Zhou RG (2008) The calmodulin-binding protein kinase 3 is part of heat-shock signal transduction in *Arabidopsis thaliana*. Plant J 55:760–773

Luoni L, Bonza MC, De Michelis MI (2006) Calmodulin/Ca^{2+}-ATPase interaction at the *Arabidopsis thaliana* plasma membrane is dependent on calmodulin isoform showing isoform-specific Ca^{2+} dependencies. Physiol Plant 126:175–186

Lydan MA, O' Day DH (1994) Production of ^{35}S-labeled proteins in *E. coli* and their use as molecular probes. Methods Mol Biol 31:389–396

Marcus AJ, Broekman MJ, Drosopoulos JH, Islam N, Alyonycheva TN, Safier LB, Hajjar KA, Posnett DN, Schoenborn MA, Schooley KA, Gayle RB, Maliszewski CR (1997) The endothelial cell ecto-ADPase responsible for inhibition of platelet function is CD39. J Clinic Invest 99:1351–1360

McCormack E, Tsai YC, Braam J (2005) Handling calcium signaling: Arabidopsis CaMs and CMLs. Trends Plant Sci 10:383–389
Minorsky PV (1989) Temperature sensing by plants: A review and hypothesis. Plant Cell Environ 12:119–135
Mittler R (2002) Oxidative stress, antioxidants and stress tolerance. Trends Plant Sci 7:405–410
Nover L, Bharti K, Döring P, Mishra SK, Ganguli A, Scharf KD (2009) Arabidopsis and the heat stress transcription factor world: how many heat stress transcription factors do we need? Cell Stress Chaperones 6:177–189
O' Day DH (2003) CaMBOT: Profiling and characterizing calmodulin-binding proteins. Cell Signal 15:347–354
Oh SH, Choi WG, Lee IT, Yun SJ (2005) Cloning and characterization of a rice cDNA encoding glutamate decorboxylase. J Biochem Mol Biol 38:595–601
Osawa M, Tokumitsu H, Swindells MB, Kurihara H, Orita M, Shibanuma T, Furuya T, Ikura M (1999) A novel target recognition revealed by calmodulin in complex with Ca^{2+}-calmodulin-dependent kinase kinase. Nat Struc Mol Biol 6:819–824
Palmgren MG (2001) Plant plasma membrane H^{+}-ATPases: Powerhouses for nutrient uptake. Annu Rev Plant Physiol Plant Mol Biol 52:817–845
Pandey S, Sopory SK (1998) Biochemical evidence for a calmodulin-stimulated calcium-dependent protein kinase in maize. Eur J Biochem 255:718–726
Pandey S, Sopory SK (2001) *Zea mays* CCaMK: autophosphorylation-dependent substrate phosphorylation and down-regulation by red light. J Exp Bot 52:691–700
Perruc E, Charpenteau M, Ramirez BC, Jauneau A, Galaud JP, Ranjeva R, Ranty B (2004) A novel calmodulin-binding protein functions as a negative regulator of osmotic stress tolerance in *Arabidopsis thaliana* seedlings. Plant J 38:410–420
Rao MV, Davis KR (1999) Ozone-induced cell death occurs via two distinct mechanisms in Arabidopsis: the role of salicylic acid. Plant J 17:603–614
Reddy ASN (2001) Calcium: silver bullet in signaling. Plant Sci 160:381–404
Reddy RK, Kurek I, Silverstein AM, Chinkers M, Breiman A, Krishna P (1998) High-molecular-weight FK506-binding proteins are components of heat-shock protein 90 heterocomplexes in wheat germ lysate. Plant Physiol 118:1395–1401
Reddy ASN, Ali GS, Celesnik H, Day IS (2011) Coping with stresses: Roles of calcium- and calcium/calmodulin-regulated gene expression. Plant Cell 23:2010–2032
Rhoads AR, Friedberg F (1997) Sequence motifs for calmodulin recognition. FASEB J 11:331–340
Roberts DM, Harmon AC (1992) Calcium-modulated proteins: targets of intracellular calcium signals in higher plants. Annu Rev Plant Physiol Plant Mol Biol 43:375–414
Roberts DM, Zielinski RE, Schleicher M, Watterson DM (1983) Analysis of suborganellar fractions from spinach and pea chloroplasts for calmodulin-binding proteins. J Cell Biol 97:1644–1647
Shelp BJ, Bown AW, McLean MD (1999) Metabolism and functions of gamma-aminobutyric acid. Trends Plant Sci 4:446–452
Shinozaki K, Yamaguchi-Shinozaki K, Seki M (2003) Regulatory network of gene expression in the drought and cold stress responses. Curr Opin Plant Biol 6:410–417
Singh P, Virdi AS (2010) Calmodulin binding proteins: implications in abiotic stress adaptation. J Plant Biol 37(1):1–17
Snedden WA, Fromm H (1998) Calmodulin, calmodulin-related proteins and plant responses to the environment. Trends Plant Sci 3:299–304
Snedden WA, Fromm H (2001) Calmodulin as a versatile calcium signal transducer in plants. New Phytol 151:35–66
Steinebrunner I, Jeter C, Song C, Roux SJ (2000) Molecular and biochemical comparison of two different apyrases from *Arabidopsis thaliana*. Plant Physiol Biochem 38:913–922
Stone JM, Walker JC (1995) Plant protein kinase families and signal transduction. Plant Physiol 108:451–457

Sun XT, Li B, Zhou GM, Tang WQ, Bai J, Sun DY, Zhou RG (2000) Binding of the maize cytosolic Hsp70 to calmodulin, and identification of calmodulin-binding site in Hsp70. Plant Cell Physiol 41:804–810

Sunkar R, Kaplan B, Bouché N, Arazi T, Dolev D, Talke IN, Maathuis FJ, Sanders D, Bouchez D, Fromm H (2000) Expression of a truncated tobacco NtCBP4 channel in transgenic plants and disruption of the homologous Arabidopsis CNGC1 gene confer Pb^{2+} tolerance. Plant J 24:533–542

Takayama S, Sato T, Krajewski S, Kochel K, Irie S, Milian JA, Reed JC (1995) Cloning and functional analysis of BAG-1: a novel Bcl-2-binding protein with anti-cell death activity. Cell 80:279–284

Takezawa D, Ramachandiran S, Paranjape V, Poovaiah BW (1996) Dual regulation of a chimeric plant serine/threonine kinase by calcium and calcium/calmodulin. J Biol Chem 271:8126–8132

Thomas C, Rajagopal A, Windsor B, Dudler R, Lloyd A, Roux SJ (2000) A role for ectophosphatase in xenobiotic resistance. Plant Cell 12:519–534

Todorov LD, Mihaylova-Todorova S, Westfall TD, Sneddon P, Kennedy C, Bjur RA, Westfall DP (1997) Neuronal release of soluble nucleotidases and their role in neurotransmitter inactivation. Nature 387:76–79

Tomsig J, Suszkiw JB (1991) Permeation of Pb^{2+} through calcium channels: Fura-2 measurements of voltage- and dihydropyridine-sensitive Pb^{2+} entry in isolated bovine chromaffin cells. Biochim Biophys Acta 1069:197–200

Torii KU (2000) Receptor kinase activation and signal transduction in plants: an emerging picture. Curr Opin Plant Biol 3:361–367

Trofimova MS, Andreev IM, Kuznetsov VV (1999) Calcium is involved in regulation of the synthesis of Hsps in suspension-cultured sugar beet cells under hyperthermia. Physiol Plant 105:67–73

Turano FJ, Fang TK (1998) Characterization of two glutamate decarboxylase cDNA clones from Arabidopsis. Plant Physiol 117:1411–1421

Ulm R, Ichimura K, Mizoguchi T, Peck SC, Zhu T, Wang X, Shinozaki K, Paszkowski J (2002) Distinct regulation of salinity and genotoxic stress responses by Arabidopsis MAP kinase phosphatase 1. EMBO J 21:6483–6493

Van der Luit AH, Olivari C, Haley A, Knight MR, Trewavas AJ (1999) Distinct calcium signaling pathways regulate calmodulin gene expression in tobacco. Plant Physiol 121:705–714

Veena RVS, Sopory SK (1999) Glyoxalase I from *Brassica juncea*: molecular cloning, regulation and its over expression confer tolerance in transgenic tobacco under stress. Plant J 17:385–396

Virdi AS, Thakur A, Dutt S, Kumar S, Singh P (2009) A sorghum 85 kDa heat stress-modulated protein shows calmodulin-binding properties and cross-reactivity to anti-Neurospora crassa Hsp 80 antibodies. FEBS Lett 583:767–770

Virdi AS, Pareek A, Singh P (2011) Evidence for the possible involvement of calmodulin in regulation of steady state levels of Hsp90 family members (Hsp87 and Hsp85) in response to heat shock in sorghum. Plant Signal Behav 6:393–399

Vogel JT, Zarka DG, Van Buskirk HA, Fowler SG, Thomashow MF (2005) Roles of the CBF2 and ZAT12 transcription factors in configuring the low temperature transcriptome of Arabidopsis. Plant J 41:195–211

Wang W, Vinocur B, Altman A (2003) Plant responses to drought, salinity and extreme temperatures: Towards genetic engineering for stress tolerance. Planta 218:1–14

Wang W, Vinocur B, Shoseyov O, Altman A (2004) Role of plant heat-shock proteins and molecular chaperones in the abiotic stress response. Trends Plant Sci 9:244–252

Windsor B, Roux SJ, Lloyd A (2003) Multiherbicide tolerance conferred by AtPgp1 and apyrase overexpression in *Arabidopsis thaliana*. Nature Biotech 21:428–433

Xuan Y, Zhou S, Wang L, Cheng Y, Zhao L (2010) Nitric oxide functions as a signal and acts upstream of AtCaM3 in thermotolerance in Arabidopsis seedlings. Plant Physiol 153:1895–1906

Yamada K, Nishimura M (2008) Cytosolic heat shock protein 90 regulates heat shock transcription factor in *Arabidopsis thaliana*. Plant Signal Behav 3:660–662

Yamada K, Fukao Y, Hayashi M, Fukazawa M, Suzuki I, Nishimura M (2007) Cytosolic Hsp90 regulates the heat shock response that is responsible for heat acclimation in *Arabidopsis thaliana*. J Biol Chem 282:37794–37804

Yamakawa H, Katou S, Seo S, Mitsuhara I, Kamada H, Ohashi Y (2004) Plant MAPK phosphatase interacts with calmodulins. J Biol Chem 279:928–936

Yang T, Poovaiah BW (2002) Hydrogen peroxide homeostasis: activation of plant catalase by calcium/calmodulin. Proc Natl Acad Sci USA 99:4097–4102

Yang T, Segal G, Abbo S, Feldman M, Fromm H (1996) Characterization of the calmodulin gene family in wheat: structure, chromosomal location, and evolutionary aspects. Mol Gen Genet 252:684–694

Yang T, Chaudhuri S, Yang L, Chen Y, Poovaiah BW (2004) Calcium/calmodulin up-regulates a cytoplasmic receptor-like kinase in plants. J Biol Chem 279:42552–42559

Yang L, Ji W, Zhu Y, Gao P, Li Y, Cai H, Bai X, Guo D (2010a) GsCBRLK, a calcium/calmodulin-binding receptor-like kinase, is a positive regulator of plant tolerance to salt and ABA stress. J Exp Bot 61:2519–2533

Yang T, Chaudhuri S, Yang L, Du L, Poovaiah BW (2010b) A calcium/calmodulin-regulated member of the receptor-like kinase family confers cold tolerance in plants. J Biol Chem 285:7119–7126

Yap K, Kim J, Truong K, Sherman M, Yuan T, Ikura M (2000) Calmodulin target database. J Struct Funct Genomics 1:8–14

Yoo JH, Park CY, Kim JC, Do HW, Cheong MS, Park HC, Kim MC, Moon BC, Choi MS, Kang YH, Lee JH, Kim HS, Lee SM, Yoon HW, Lim CO, Yun DJ, Lee SY, Chung WS, Cho MJ (2005) Direct interaction of a divergent CaM isoform and the transcription factor, MYB2, enhances salt tolerance in Arabidopsis. J Biol Chem 280:3697–3706

Zhang L, Lu YT (2003) Calmodulin-binding protein kinases in plants. Trends Plant Sci 8:123–127

Zhang L, Liu BF, Liang S, Jones RL, Lu YT (2002) Molecular and biochemical characterization of a calcium/calmodulin-binding protein kinase from rice. Biochem J 368:145–157

Zhuang Y, Ren G, He C, Li X, Meng Q, Zhu C, Wang R, Zhang J (2010) Cloning and characterization of a maize cDNA encoding glutamate decarboxylase. Plant Mol Biol Rep 28:620–626

Zou J, Guo Y, Guettouche T, Smith DF, Voellmy R (1998) Repression of heat shock transcription factor HSF1 activation by HSP90 (HSP90 complex) that forms a stress-sensitive complex with HSF1. Cell 94:471–480

Chapter 2
Biotic and Abiotic Stress Signaling in Plants

Sandhya Verma, Shadab Nizam, and Praveen K. Verma

Introduction

Plants relentlessly encounter a wide range of environmental stresses which limits the agricultural productivity. The environmental stresses conferred to plants can be categorized as 1) Abiotic stress, and 2) Biotic stress. Abiotic stresses include salinity, drought, flood, extremes in temperature, heavy metals, radiation etc. It is a foremost factor that causes the loss of major crop plants worldwide. This situation is going to be more rigorous due to increasing desertification of world's terrestrial area, increasing salinization of soil and water, shortage of water resources and environmental pollution. On the other hand, biotic stress includes attack by various pathogens such as fungi, bacteria, oomycetes, nematodes and herbivores. Diseases caused by these pathogens accounts for major yield loss worldwide. Being sessile plants have no choice to escape from these environmental cues. Expertise in tolerating these stresses is crucial for completing the lifecycle successfully. Therefore, to combat these threats plants have developed various mechanisms for getting adapted to such conditions for survival. They sense the external stress environment, get stimulated and then generate appropriate cellular responses. These cellular responses work by relaying the stimuli from sensors, located on the cell surface or cytoplasm to the transcriptional machinery which is situated in the nucleus, with the help of various signal transduction pathways. This leads to differential transcriptional changes making the plant tolerant against the stress. The signaling pathways play an indispensable role and acts as a connecting link between sensing the stress environment and generating an appropriate physiological and biochemical response (Zhu 2002). Recent studies using genomics and proteomics approach

S. Verma • S. Nizam • P.K. Verma (✉)
National Institute of Plant Genome Research, Aruna Asaf Ali Marg, New Delhi-67, India
e-mail: praveen_verma@nipgr.res.in

M. Sarwat et al. (eds.), *Stress Signaling in Plants: Genomics and Proteomics Perspective, Volume 1*, DOI 10.1007/978-1-4614-6372-6_2,

to decode and understand these signaling networks has increased our knowledge regarding signaling pathways. In this review, we have discussed the recent findings in plant signaling pathways during various stresses and the specificity and points of cross-talk between these signaling pathways.

Abiotic Stresses

Drought

Drought is a condition when there is water deficit in the atmosphere and soil. Plants undergo a series of physiological and biochemical events during drought stress. These responses result in closing of stomata, fall in transpiration rate, suppression of photosynthesis and cell growth, decrease in osmotic potential of plant tissues and activation of respiration. Additionally, there are also adaptations at cellular and molecular levels. Plants accumulate different kinds of osmolytes (viz. mannitol, proline, sorbitol, trehalose, fructans), phytohormone abscisic acid (ABA) and also synthesize new stress tolerating proteins. Studies using microarray analysis identified a number of stress responsive genes in *Arabidopsis* and rice. A total of 299 drought-inducible genes were identified after expression profiling of 1,300 genes using a full-length cDNA microarray in *Arabidopsis* (Seki et al. 2002). More than 50 % of these drought-inducible genes were found induced in response to high salinity and/or ABA treatments. Microarray analysis in rice using 1700 independent cDNAs revealed 73 stress inducible genes (Rabbani et al. 2003). Approximately 40 % of drought-inducible genes were also found induced during cold stress. Moreover, almost all the genes expressed during high salinity and ABA treatment were also induced during drought stress. These results clearly show highly substantial cross-talk between the signaling pathways generated during various stresses, such as, drought, cold, high salinity and ABA treatment. The drought-responsive genes can be divided into two groups on the basis of their biological functions. The first group consists of the genes encoding functional proteins involved in membrane and protein stabilization and cellular homeostasis, such as heat shock proteins (HSPs), late embryogenesis abundant (LEA) proteins and lipid transfer proteins (LTPs). The second group includes Calcineurin-B-Interacting Protein Kinases (CIPK16) and Protein Phosphatases class 2C (PP2C) which function as signal molecules in response to stress (Boominathan et al. 2004; Ok et al. 2005). Therefore, the putative mechanism of drought tolerance can be understood on the basis of the expression patterns of these genes encoding functional and regulatory proteins. Recently, high-throughput Roche 454 sequencing was combined with SuperSAGE and revealed the drought-responsive transcriptome of chickpea roots (Molina et al. 2008). This was quite helpful to overcome the problem of lacking EST and genomic data in a non-model crop plant chickpea. From drought-stressed and

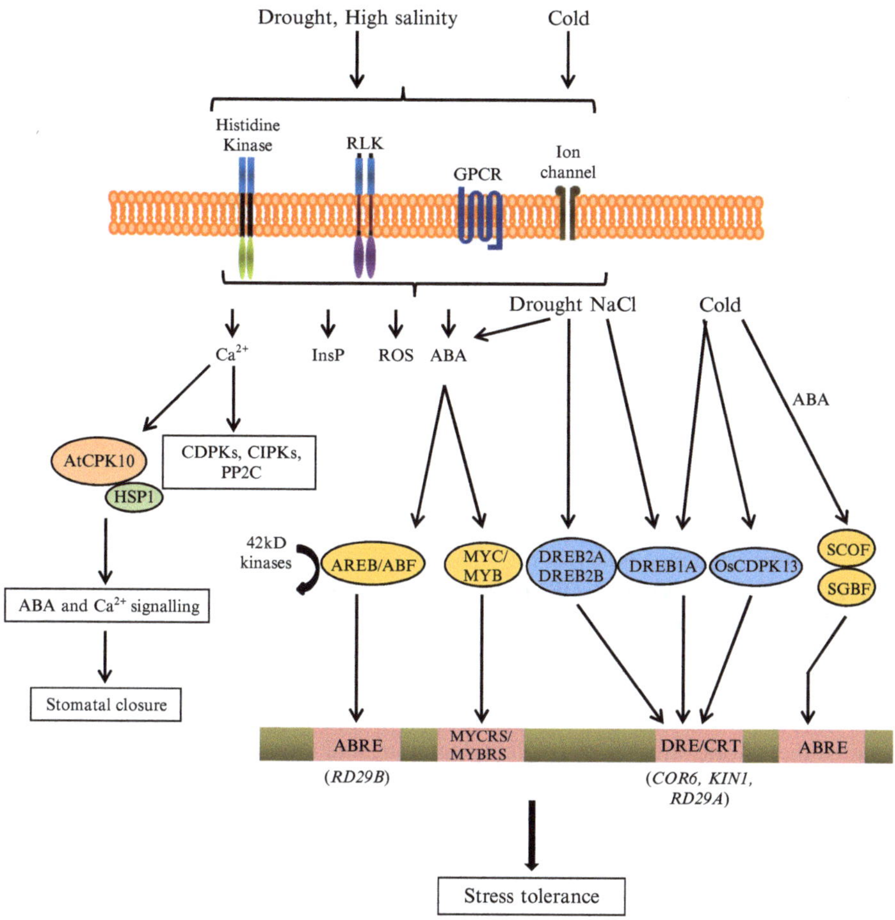

Fig. 2.1 Schematic representation of signaling during abiotic stresses. Upon stress signal perception, various signaling cascades get activated leading to differential gene expression for generating appropriate stress tolerance

non-stressed control roots, a total of 17,493 unique transcripts were obtained which included genes involved in stress-perception, signaling and transcription initiation.

The signaling pathways induced during drought are either ABA-dependent or ABA-independent (Fig. 2.1). ABA is a phytohormone that is produced during water scarcity. Under the effect of ABA plants become responsive and tolerant to drought and high salinity. Various genes induced during drought and cold stress also gets induced by application of ABA exogenously, indicating ABA-dependent signal transduction. On the other hand, several genes are there which respond to drought and cold stress but does not get induced by application of ABA exogenously, indicating the existence of ABA-independent signal transduction in *Arabidopsis* (Zhu 2002; Yamaguchi-Shinozaki and Shinozaki 2006).

ABA-Independent Signal Transduction

Generally, transcription factors are involved in regulating the signaling pathways. Several transcription factors have been identified in *Arabidopsis* that gets activated during drought stress in ABA-independent manner. Recent studies depicted major role of DREB1 (dehydration-responsive element binding factor)/CBF (C-repeat binding factor) and DREB2 transcription factors during drought. Molina et al. (2008) identified 124 UniTags from transcription factor transcripts in chickpea against drought which were classified into 26 transcription factor families. Majority of them belonged to bZIP family of transcription factors along with HDZ, HMG and WRKY families. Moreover, transcripts of DREB2 transcription factors were also highly induced.

There are six *DREB1/CBF* genes and eight *DREB2* genes present in *Arabidopsis* genome (Sakuma et al. 2002). Among these genes, *DREB2A* and *DREB2B* get induced during dehydration and high salinity (Nakashima et al. 2000). However, they do not get induced during cold stress. This indicates the important role being played by DREB2/CBF protein in developing the stress tolerance against drought. Rice genome sequence analysis revealed presence of ten *DREB1/CBF* genes and four *DREB2* genes. When *OsDREB1A* was overexpressed in *Arabidopsis*, it showed a similar function in terms of gene expression and stress tolerance (Dubouzet et al. 2003). Overexpression of *OsDREB1* or *Arabidopsis DREB1* in rice made the transgenic plants more tolerant to drought and cold (Ito et al. 2006). This suggests that the similar transcription factors function in same way to improve the tolerance against the abiotic stress both in dicotyledonous and monocotyledonous plants.

ABA-Dependent Signal Transduction

There are various drought and high salinity-inducible genes in *Arabidopsis* and rice that get induced in response to ABA (Seki et al. 2002; Rabbani et al. 2003). AREB (ABA-Responsive Element Binding factor) or ABFs (which are basically bZIP transcription factors) regulates transcription by binding to ABA-responsive complex present in the promoter region of most of the ABA-inducible genes. Overexpression of the phosphorylated active form of AREB1 leads to the expression of ABA-inducible genes (like *RD29B*) even in the absence of the ABA treatment (Furihata et al. 2006). This results in improved drought tolerance in transgenic plants suggesting that multisite phosphorylation of AREB1 in an ABA-dependent manner is required for its own activation. Its phosphorylation- and dephosphorylation-directed signal transduction is essential for ABA signaling.

Therefore, various studies carried out using transcriptomic approach concludes the 3generation of appropriate response against drought stress which begins with the perception of drought signals by various membrane receptors such as GPCRs, RLKs, histidine kinases and ion channels (Fig. 2.1). This brings changes in the cytoplasmic Ca^{2+} levels and generates secondary signaling molecules (ABA, ROS, inositol phosphates, etc.). These secondary messengers initiate a protein phosphorylation cascade by various kinases like CDPKs, CIPKs, protein kinases and protein phosphatases. They further activate transcription factors such as DREB/CBF, AREB/ABF, bZIP, MYC/MYB, WRKY etc. through phosphorylation and dephosphorylation events. In turn, these transcription factors lead to the activation of numerous stress-responsive genes which includes genes encoding HSPs, LEA proteins, LTPs, antioxidants, osmolytes, ion transporters, etc. Finally, tolerance against drought stress is generated and the damages are repaired.

To investigate the initial response of rice against drought, a proteomic approach was used by Ali and Komatsu (2006). It revealed four novel drought-responsive mechanisms in the rice leaf blade; up-regulation of an S-like RNase homologue, an actin depolymerizing factor and RuBisCO activase, and down-regulation of an isoflavone reductase-like protein. Dehydration-responsive extracellular matrix (ECM) proteins were analysed in chickpea and rice (Bhushan et al. 2007; Pandey et al. 2010). ECM functions as a depot for numerous components of the cell signaling process and acts as a frontline defense. More than 100 differentially expressed ECM proteins were identified which consisted of predicted as well as novel dehydration-responsive proteins. There proteins were involved in various cellular functions such as cell wall modification, metabolism, signal transduction, cell defense and rescue.

Protein analysis is the most direct approach to describe the function of its associated gene. Therefore, proteome analysis linked to genome-sequence information proves to be a very powerful tool in functional genomics. In bread wheat, differential expression of 114 proteins in response to drought stress was revealed by proteomic analysis (Peng et al. 2009). Recent report in alfalfa leaf indicated that drought causes impairment of photosynthetic activity which was evident by down-regulation of Rubisco and proteins involved in Rubisco assembly (Aranjuelo et al. 2011). Recently, first shotgun proteomics study was performed to gain insights into protein responses to drought in wheat (Ford et al. 2011). This identified the largest number of differentially expressed wheat proteins (1,299) in a single study. Significant increase in the expression level of superoxide dismutases and catalases suggested an increase in oxidative stress metabolism and reactive oxygen species (ROS) scavenging capacity. There was decrease in the proteins involved in photosynthesis and Calvin cycle which suggest ROS avoidance by plants in response to drought. Therefore, there is a massive transcriptional and translational reprogramming which results in protecting and repairing the cell wall, modification of metabolism and signal transduction to generate appropriate response in order to make the plants tolerant against stress.

Cold

Another very common environmental stress that unfavorably affects the plant growth and crop yield is cold i.e. low temperature. Although most of the plants are not quite tolerant to freezing temperatures, however, they can increase their tolerance against freezing temperatures by getting exposed to the chilling, non-freezing temperatures. This process is known as cold acclimation. It restructures the cell membranes physically and biochemically by changing the lipid composition and inducing other non-enzymatic proteins. Recently, substantial transcriptomic studies have been carried out to elucidate the sensing and regulatory mechanisms of plants against cold that enables them to develop cold acclimation response.

Cold stress triggers multiple transcriptional cascades because many early cold-responsive genes encode transcription factors which subsequently activate the genes induced late in the cold response. There is a differential regulation of genes involved in biosynthesis or signaling of plant hormones such as ABA, gibberellic acid and auxin. The DNA replication, spliceosome and mismatch repair pathways were also found to be associated with cold stress (Rong et al. 2011). Transcriptome analysis of wheat suggests that there is a massive transcriptome reprogramming during the course of cold acclimation and gaining the freezing tolerance. It requires coordination of many different physiological and biochemical changes and is regulated through the differential expression of many genes (Winfield et al. 2010). Microarray analysis of the mitochondrial transcriptome in germinating wheat embryos and seedlings revealed differential regulation of an array of genes which results in increased transpiration rate both through cytochrome and alternative pathways. Few nuclear-encoded mitochondria-targeted genes suppressed while transcripts of *MnSOD* and *AOX* induced (Naydenov et al. 2010). This suggests involvement of both nuclear and mitochondrial genes in response to cold stress. In rice, microarray analyses revealed that few microRNA are derived from transposable element sequence. They get clustered within an intron and are co-transcribed with the host gene only under cold conditions. In the upstream regions of the cold-responsive microRNAs, hormone-responsive elements are present which indicates the importance of hormones in this microRNA mediated defense system (Lv et al. 2010). Therefore, in response to cold conditions the stress signals are transmitted to transcription factors which are encoded by early cold-responsive genes (Fig. 2.1). This activates various signaling cascades mediated by hormones and Ca^{2+}. The MAPK signal transduction also comes into action to relay the signals further. It leads to massive nuclear, mitochondrial and chloroplast transcriptome reprogramming which finally results in cold acclimation by bringing different physiological and biochemical changes. In signal transduction, CDPKs also play an important role. Studies shows that *OsCDPK13* (*Oryza sativa CDPK 13*) gets induced at gene and protein level in response to cold and gibberellin (GA) but not in response to drought, high salinity and ABA treatment (Abbasi et al. 2004). In low temperature signal transduction pathways, calcium plays a role as an important messenger. In response to low temperature there is rapid increase in cytoplasmic calcium levels. This is mainly due to an influx of calcium from extracellular reservoirs. It has been shown that calcium is necessary for the complete expression of some of the cold induced genes like the *CRT/DRE* controlled *COR6* and *KIN1* genes in *Arabidopsis*.

Proteomic analysis in rice against cold stress showed high expression of molecular chaperones, proteases, energy pathway-linked proteins, enzymes involved in protein biosynthesis, cell wall biosynthesis, detoxification and signal transduction components. Chaperones and proteases control the protein quality and enhance cell wall components which play important roles in developing the tolerance to cold. Two-dimensional difference gel electrophoresis (2-D DIGE) analysis of sub-cellular fractions of *Arabidopsis* proteome identified a number of early cold-responsive proteins such as HSPs, enzymes involved in starch degradation and sucrose metabolism (Li et al. 2011). During cold stress, some proteins show enhanced degradation especially the photosynthetic proteins such as large subunit of Rubisco (Yan et al. 2006). Similarly, there is down-regulation of the enzymes involved in Krebs cycle and Calvin cycle while ascorbate recycling gets increased (Rinalducci et al. 2011). Cold brings rapid metabolic changes which develop instant protection before temperature drops below freezing point. In sunflower, a total of 243 differentially expressed, non-redundant proteins were identified during cold acclimation (Balbuena et al. 2011). Therefore, in response to cold stress a large set of cellular proteins get differentially expressed and establish cold acclimation in plant.

High Salinity

Salinity affects a large area of the land every year globally and the affected area is increasing day by day. This makes salinity a major environmental stress that highly limits the crop production. Due to high salinity plants becomes unable to take up water. It rapidly leads to reductions in growth rate, together with a set of metabolic changes. These changes are identical to those caused by water stress. Salt specific impact can be seen when excessive salt enters inside the plant. The amount of salt rises above the toxic levels in the older transpiring leaves. This results in premature senescence and also reduces the photosynthetic leaf area of the plant, as a result of which it cannot sustain growth.

Barley is a well known salt-tolerant crop species. Transcript profiling during salinity stress was performed using a microarray containing $\sim$22,750 probe sets (Walia et al. 2006). One of the prominent features of salinity stress response was the induction of jasmonic acid (JA) pathway genes and genes responsive to JA and ABA. Also, a large number of abiotic stress (drought, heat and cold) related genes were induced which support the existence of cross talk among certain components of abiotic stress responses. Suppression subtractive hybridization (SSH) and microarray approaches were also combined to identify the early salt stress response genes in tomato root (Ouyang et al. 2007). A total of 201 non-redundant genes were found differentially expressed upon 30 min of rigorous salt stress. These genes included the stress sensors, inducible transcriptional activators, upstream signal pathway components and HSPs. Several key components, such as PP2C and MAPKKK were also up-regulated. Recently, a novel CBF transcription factor MtCBF4 has been found highly induced during drought in *Medicago truncatula* (Li et al. 2011). In *Arabidopsis*, MEKK1 (a MAPKKK) mRNA gets highly accumulated in response to stresses including salinity stress (Mizoguchi et al. 1996). There is protein-protein interactions between MEKK1 and MKK2/MEK1 (MAPKKs), between MKK2/MEK1 and MPK4 (a MAPK), and between MPK4 and MEKK1 (Ichimura et al. 1998). In response to environmental stress the signals are conveyed to at least two MAPK cascades. One is the MPK4 cascade (MEKK1-MEK1/MKK2-MPK4) and the other comprises MPK6 and p44MAPK (Ichimura et al. 2000). During salt or cold stress, MAPK pathway involves MEKK1 as an upstream activator of MKK2, however, MAPKs-MPK4 and MPK6 acts downstream (Teige et al. 2004). MAP kinase phosphatases (MKP1) play a negative role through MAPKs-MPK6 and MPK4 (Ulm et al. 2002). Therefore, the JA pathway gets induced as soon as the plant senses the salt stress. The cytosolic Ca^{2+} levels also get elevated which activates various signal transduction pathways including SOS (Salt Overly Sensitive) pathway. The transcription factors viz. DREB/CBF, AREB/ABF, bZIP, MYC/MYB also gets activated resulting in further downstream signaling cascade (Fig. 2.1). They help in maintaining the ion homeostasis and improve the tolerance of plants against salt stress.

Proteomic analysis of potato under salt stress revealed 47 differentially expressed proteins in shoot (Aghaei et al. 2008). Photosynthetic genes and protein synthesis-related proteins showed drastic down-regulation. In contrast, there was up-regulation in levels of osmotine-like proteins, TSI-1 protein, HSPs, protein inhibitors, calreticulin, and five novel proteins suggesting that induction of defense associated proteins provide relative salt tolerance to potato plants. In rice roots, changes in plasma-membrane-associated proteins were investigated (Cheng et al. 2009). Change of more than 1.5-fold in the expression of 18 proteins was also reported. The nine up-regulated proteins were found to be involved in membrane stabilization, ion homeostasis and signal transduction. Alterations in the protein phosphorylation patterns have also been detected (Chitteti and Peng 2007). Recently, a salt stress-responsive protein with a putative function in stress signaling has been located in the apoplast of the rice root (Zhang et al. 2009). It indicates the importance of ion uptake, transportation and regulation of signal transduction in the root. The proteome of the root differs for the tolerant and susceptible genotypes of a plant (Manaa et al. 2011). The metabolism-related proteins also play an important role in each organ (e.g. leaves, hypocotyls and roots) in order to achieve adaptation to salinity stress (Sobhanian et al. 2010). Even during the course of seed germination, salt stress induces an array of proteins which generate adequate tolerance (Xu et al. 2011). Therefore, the plants improve tolerance against salinity stress by developing a more efficient osmotic and ionic homeostasis, superior capacity to remove toxic byproducts, and finally a better potential for growth recovery. All these processes involve proteins regulating energy metabolism, ROS scavenging, cytoskeleton stability, protein processing and folding, photosynthesis, photorespiration and signal transduction.

Other Abiotic Stresses

Together with above mentioned stresses, other stresses encountered by plants include high temperature, flood and presence of heavy metals. Heat stress has detrimental effects on plant physiology and development by affecting cellular components and metabolism. A number of transcriptomic and proteomic studies have been carried out in order to gain better insights. GeneChip Wheat Genome Array of heat susceptible and tolerant variety showed differential expression of large number genes which includes HSPs, transcription factors and other stress related genes (Qin et al. 2008). Involvement of these genes suggests that multiple processes and mechanisms are responsible for developing heat tolerance in a plant. Genomic organization and transcript profiling of heat shock factors (HSFs) were studied in rice in response to high temperature (Mittal et al. 2009). Several *OsHsf* genes were highly induced suggesting that they are critical for the transcriptional induction of heat shock genes. Another approach using cDNA-AFLP and microarray analysis in tomato plants showed different constitutive gene expression profile in heat-tolerant genotype as compared to the heat-sensitive genotype (Bita et al. 2011). These genes consist of genes related to heat shock, metabolism, antioxidant and development pathways indicating substantial genetic differences in adaptation in response to high temperatures. Proteomic analysis of wheat grain revealed up-regulation of various enzymes involved in starch synthesis pathway, carbohydrate metabolism, small HSPs and translation initiation factors (Majoul et al. 2004). Similarly, proteomic approach in rice detected several low molecular weight small HSPs and antioxidant enzymes with high induction (Lee et al. 2007). Heat stress also induces the dephosphorylation of RuBisCo and the phosphorylation of ATP-β in rice resulting in decreased activities of RuBisCo and ATP synthase (Chen et al. 2011). This shows involvement of phosphoproteins in the transduction of heat-stress signaling.

Flooding is another recurrent natural disaster causing plant hypoxia which limits the plant yield worldwide. Transcriptome analysis of maize roots showed up-regulation of a large number of genes in response to waterlogging which were related to signal transduction, protein degradation, carbon and amino acid metabolism, ion transport, and transcriptional and translational regulation (Zou et al. 2010). These genes were grouped into two response processes i.e. defense at the early stage and adaption at the late stage. Crosstalk between carbon and amino acid metabolism provides evidence for the two main roles of amino acid metabolism at the late stage. These are the regulations of cytoplasmic pH and energy supply. During prolonged waterlogging, tolerance-related genes are activated for survival via signal transduction. Another study using microarray analysis of roots of soybean seedlings revealed more than 6,000 flooding-responsive genes which were related to photosynthesis, glycolysis, amino acid synthesis, transcriptional regulation, ubiquitin-mediated protein degradation and cell death (Nanjo et al. 2011). On the other hand, there was down-regulation of genes involved in cell wall synthesis, secondary metabolism, metabolite transport, cell organization and chromatin structure synthesis.

These findings suggest that transcriptional and post-transcriptional regulations resulting in enhanced acclimative responses against flooding protect the soybean seedlings from flooding injury. Proteomic analysis of soybean roots revealed up-regulation of storage proteins and energy producing proteins (Komatsu et al. 2010b). Comparative proteomics analysis of soybean cell wall showed differential expression of many proteins. It concludes that the lignification in the soybean roots and hypocotyls gets suppressed by down-regulation of reactive oxygen species and JA biosynthesis during flooding (Komatsu et al. 2010a). Several proteomics studies have been carried out on specific cellular organelles in plants in response to various abiotic stresses, such as mitochondria, nucleus, chloroplasts, cell wall and the plasma membrane. Hossain et al. (2012) has provided an excellent detailed review in this regard.

Plants require heavy metals, such as iron (Fe), copper (Cu), manganese (Mn), molybdenum (Mo), cobalt (Co) and zinc (Zn), as essential micronutrients in small quantities. They are involved in essential biological functions like redox reactions and acts as enzyme cofactors. However, higher concentrations of these metals can be toxic. Some other heavy metals have no function as nutrients and are very toxic. For example cadmium (Cd), mercury (Hg), lead (Pb), aluminum (Al) and arsenic (As) lead to ROS production and inactivation of enzymes. Microarray analysis in rice against cadmium stress revealed differential expression of several genes encoding cytochrome P450 family proteins, HSPs, glutathione S-transferase, protein kinases, ion transporters and transcription factors such as DREB and NAC. This suggests the role of transporters in Cd detoxification by exporting Cd from the cytoplasm (Ogawa et al. 2009). Similarly, microarray analysis of *B. carinata* seedlings after lithium chloride exposure revealed differential expression of genes involved in defense, primary metabolism, transcription, transportation and secondary metabolism (Li et al. 2009). Various proteomics studies carried out in different plants against cadmium stress revealed high induction of proteins involved in photosynthetic pathways, transcription, translation and encoding molecular chaperones (Zhao et al. 2011).

Biotic Stresses

Plants are constantly exposed to a variety of potential microbial pathogens such as fungi, bacteria, oomycetes, nematodes and herbivores. In order to defend themselves plants have developed a variety of defense responses many of which are induced by pathogen attack. Penetration of the cell wall exposes the microbes to the plant plasma membrane, where they encounter extracellular surface receptors that recognize pathogen-associated molecular patterns (PAMPs). Recognition a microbe at the cell surface initiates PAMP-triggered immunity (PTI), which usually halts infection before the pathogen gains a hold in the plant. However, pathogenic microbes have evolved the means to suppress PTI by secreting specialized proteins, called as effectors, into the plant cell cytosol that alter resistance signaling or manifestation of resistance responses. Once pathogens acquired the capacity to suppress primary defenses, plants developed a more specialized mechanism to detect microbes, called as effector triggered immunity (ETI). In the case of ETI, products of major resistance (R) genes, usually intracellular receptors, recognize corresponding effector molecules delivered by the pathogen into the host cell. The interaction between effector and receptors triggers a complex response network aimed at determining resistance to infection (McDowell and Dangl 2000). R gene-mediated resistance is usually accompanied by an oxidative burst, which is, rapid production of reactive oxygen species (ROS). ROS production is required for another component of the response known as hypersensitive response (HR), a type of programmed cell death thought to limit the access of the pathogen to water and nutrients. Indeed, studies of the complex network properties of plant immunity have illustrated that it is comprised of distinct signaling sectors that interrelate with each other in an intricate fashion (Sato et al. 2010).

Bacteria

Metabolomic and transcriptomic analysis of rice in response to bacterial blight pathogen *Xanthomonas oryzae* pv. *oryzae* revealed global metabolic and transcriptomic changes in leaf tissues (Sana et al. 2010). Ethylene response element binding protein (EREBP) transcription factor gets significantly expressed together with ROS scavenging system and lower expression of alcohol dehydrogenase gene. These factors lead to hypersensitive cell death in the resistant cultivar upon bacterial infection. Stimulation of glutathione-mediated detoxification and flavonoid biosynthetic pathways in combination with up-regulation of defense genes during infection inhibits pathogen from further spreading in the host tissues (Kottapalli et al. 2007). Transcripts encoding disease resistance proteins via JA/ET signaling as well as osmotic regulation via proline synthesis genes were found differentially expressed when microarray analysis was performed in cotton associated with *Bacillus subtilis* induced tolerance (Medeiros et al. 2011).

The major protein of bacterial flagella is flagellin which is a well characterized PAMP. In *Arabidopsis*, flagellin is shown to be recognized by the LRR receptor kinase FLS2 (Zipfel et al. 2004). FLS2 is plasma membrane localized and is supposed to be involved in early bacterial-plant interaction. This further activates a downstream mitogen activated protein kinase pathway, composed of *At*MEKK1, *At*MKK4/*At*MKK5 and *At*MPK3/*At*MPK6 (Asai et al. 2002). The resulting PTI is quite sufficient to stop the infection before the establishment of microbe (Chisholm et al. 2006). In contrast to this, *At*MEKK1-*At*MKK1/2-*At*MPK4 pathway negatively regulates defence responses by inhibiting SA and H_2O_2 production (Pitzschke et al. 2009). To overcome PTI, bacteria introduce effector proteins (e.g. avirulence (Avr) factors) inside the host cells via type III secretion system (T3SS), where they interact with their corresponding R proteins (Dangl and McDowell 2006). Some conserved bacterial effectors of *Pseudomonas syringae* pv. tomato (*Pst*) strain

DC3000, such as AvrE and HopPtoM, play a dual role by inhibiting SA-mediated basal immunity and promoting disease necrosis (DebRoy et al. 2004).

Recent studies demonstrate that whenever the cellular fatty acid composition experiences even slight agitations, it helps to increase the levels of phytohormone JA (Savchenko et al. 2010). JA is known to be involved in activation of plant defense responses to a range of phytopathogens, including fungi, oomycetes and insects. Plants have developed the ability to perceive and respond against biotic challengers as part of their global defense-responsive network. This signal functions equally on the JA and SA pathways. It enhances JA levels by enhancing the expression of JA biosynthetic genes. This result in increased resistance to all biotic threats examined except to *Pst* (a phytopathogen sensitive to SA-activated defense responses). On the same hand, it suppresses SA levels indirectly through the well-known antagonism between these two pathways.

GeneChip experiments of *Arabidopsis-Pst* DC3000 infection revealed differential expression of 52 unique proteins. These proteins included defense-related antioxidants and metabolic enzymes. Modifications in protein spot density occurred before significant transcriptional reprogramming. Many proteins represented by more than one spot suggest posttranscriptional modifications of the proteome (Jones et al. 2006). The protein profiles of *Sinorhizobium meliloti* 1021 inoculated rice indicated up- or down-regulation of proteins involved in nine different functional categories. These proteins were mainly related to photosynthesis and defense. This beneficial plant–endophyte interaction favours plant by activating defense mechanisms to minimize negative effects of environmental factors, improving anabolism (e.g. Photosynthesis) to increase plant biomass; and regulating the auxin level to promote growth (Chi et al. 2010). Moreover, an interactome of 100 proteins was constructed by yeast two-hybrid (Y2H) assays to elucidate the signaling networks in rice during biotic and abiotic stress responses (Seo et al. 2011).

Fungi

On the basis of their lifestyles, plant pathogenic fungi have been divided into two classes: the biotrophs and the necrotrophs. Biotrophs feed on living host tissue, whereas necrotrophs first kill the host tissue and then feed on the dead tissues. However, there are many plant pathogenic fungi which behave both as biotrophs and necrotrophs, depending on the conditions in which they find themselves or the stages of their life cycles. Such pathogens are called hemi-biotrophs. Earlier, many fungi were commonly considered as necrotrophs whereas they had a biotrophic stage early in the infection process and hence were basically hemi-biotrophs. In general, SA signaling is involved for resistance against biotrophic and hemibiotrophic pathogens whereas the JA and ET signaling is important for immunity towards necrotrophs (Pieterse et al. 2009).

Biotrophic Fungi

For resistance against biotrophs, gene-for-gene mechanism is important. According to gene-for-gene hypothesis, given by Flor, for every gene in the plant that confers resistance, there is a corresponding gene in the pathogen that confers avirulence. It leads to activation of SA-dependent signaling and SAR. In *Arabidopsis*, overexpression of *ADR1* (NBS-LRR resistance gene) provides resistance against *Erysiphe cichoracearum* (Grant et al. 2003). Another example is of barley and *Blumeria graminis* (Schulze-Lefert and Vogel 2000) where gene-for-gene resistance response is evident. Various studies show that SA signaling has important role in resistance whereas JA- and ET-signaling may not be involved. Therefore, during biotrophic pathogen attack, SA-dependent defense responses are effective together with gene-for-gene resistance. There is no induction in JA-dependent responses, but if they are artificially induced, then they are quite effective (Glazebrook. 2005). Transcriptome data from microarray experiments suggest that during defense responses the photosynthesis-related genes are highly down-regulated which is required to support the induction of a defence response (Bilgin et al. 2010). The nitrogen invested in photosynthetic proteins, primarily Rubisco, is lowered or even withdrawn to provide nitrogen for the induction of defensive compounds.

During *Peronospora viciae* infection of a susceptible cultivar of pea, several proteins were abundantly expressed. These included stress-responsive proteins, several cytosolic and chloroplastic proteins (Amey et al. 2008). Quantitative proteomic analysis revealed the changes in the bean leaf (*Phaseolus vulgaris*) proteome during the course of a rust infection. It showed that some basal defense proteins are potential regulators of the strong defense which is weakened by the fungus and that the R-gene controls proteins similar to the basal system proteins. It proposes a new model picturing R-genes as a part of the basal system and involved in repairing disabled defenses to restore strong resistance (Lee et al. 2009).

Necrotrophic Fungi

Transcript profiling of various plant-pathogen systems suggest differential regulation of a large number of transcripts in response to pathogen attack. These transcripts included those which are associated with JA biosynthesis and signaling, ROS metabolism, and cell wall structure and function. Isolation of early responsive genes of chickpea infected with blight fungus *Ascochyta rabiei* was carried out using PCR based suppression subtractive hybridization (SSH) strategy and ~250 unique genes were identified. These genes belonged to eleven different categories viz. stress, signaling, gene regulation, cellular metabolism and genes of unknown functions (Jaiswal et al. 2012). Chitin, which is a major component of fungal cell wall, serves as a PAMP. Therefore, chitosan (the deacetylated form of chitin) plays important role in inducing defense responses against pathogens in many plant species. GeneChip microarrays and quantitative RT-PCR of *Botrytis cinerea* infected *Arabidopsis* leaves revealed that chitosan has inductive role on several genes involved in defense responses and camalexin biosynthesis (Povero et al. 2011).

AP2/ERF-type transcription factor family which regulate the expression of JA- and ET-response genes was found to be highly induced during infection in *Arabidopsis* (Pieterse et al. 2009). Recent studies showed that *ERF59/ORA59* acts as a vital mediator that neutralizes the SA-mediated suppression of JA/ET-response genes (Leon-Reyes et al. 2010). The function of ERF59/ORA59 represents the interplay between SA, JA and ET-signaling (Birkenbihl and Somssich 2011).

The zinc-finger-type plant specific transcription factor WRKY70 functions as a convergent point of SA and JA signaling and acts as negative regulator of resistance against necrotrophy (Li et al. 2004). WRKY33 is a significant positive regulator of defense against necrotrophic fungi (Zheng et al. 2006). Later, the importance of WRKY33 was also found in camalexin biosynthesis (Mao et al. 2011). For this the *WRKY33* expression and WRKY33 phosphorylation were MPK3 and MPK6 dependent. Recent studies show that MPK3 is a key player in maintaining the basal defence against *B. cinerea* while MPK6 has a major role in PAMP-triggered resistance against the necrotrophic pathogen (Galletti et al. 2011). Therefore, the MAPK signaling cascade generated in response to fungal PAMPs is *At*MEKK1-*At*MKK4/5-*At*MPK3/6 which ultimately leads to tryptophan biosynthesis resulting in camalexin production.

Proteomic studies of *B. napus–A. brassicae* interaction showed that in tolerant line levels of 48 proteins were considerably affected and most of them were up-regulated. While 23 proteins were affected with majority of them showing down-regulation in susceptible line. Up-regulated proteins from tolerant line consisted of enzymes involved in the generation of ROS, ROS mediated signaling, auxin signal transduction and metabolic pathways. Therefore, ROS mediated auxin signaling plays an important role in this pathosystem (Sharma et al. 2007). During *Arabidopsis–A. brassicicola* interaction also at least 11 protein spots exhibited reproducible differences in abundance. The pathogenesis-related protein PR4, a glycosyl hydrolase, and the antifungal protein osmotin were strongly up-regulated (Mukherjee et al. 2010). Therefore, proteomics effectively contribute to the knowledge of life cycle, infection mechanisms, and virulence of the plant pathogenic fungi (Fernandez and Jorrin-Novo 2012).

Cross-Talk and Specificity Between Signaling Pathways

Till now we have gone through the signaling pathways in plants against various abiotic and biotic stresses. This makes quite apparent that there is existence of signaling network which is interconnected at many levels (Fig. 2.2). During cross-talk, with the help of common components there is transfer of important information between signaling pathways. The purpose of this information transfer is to modify the output of the different signaling pathways. But at the same time, there is clear demarcation between two or more potential consequences of a signaling pathway.

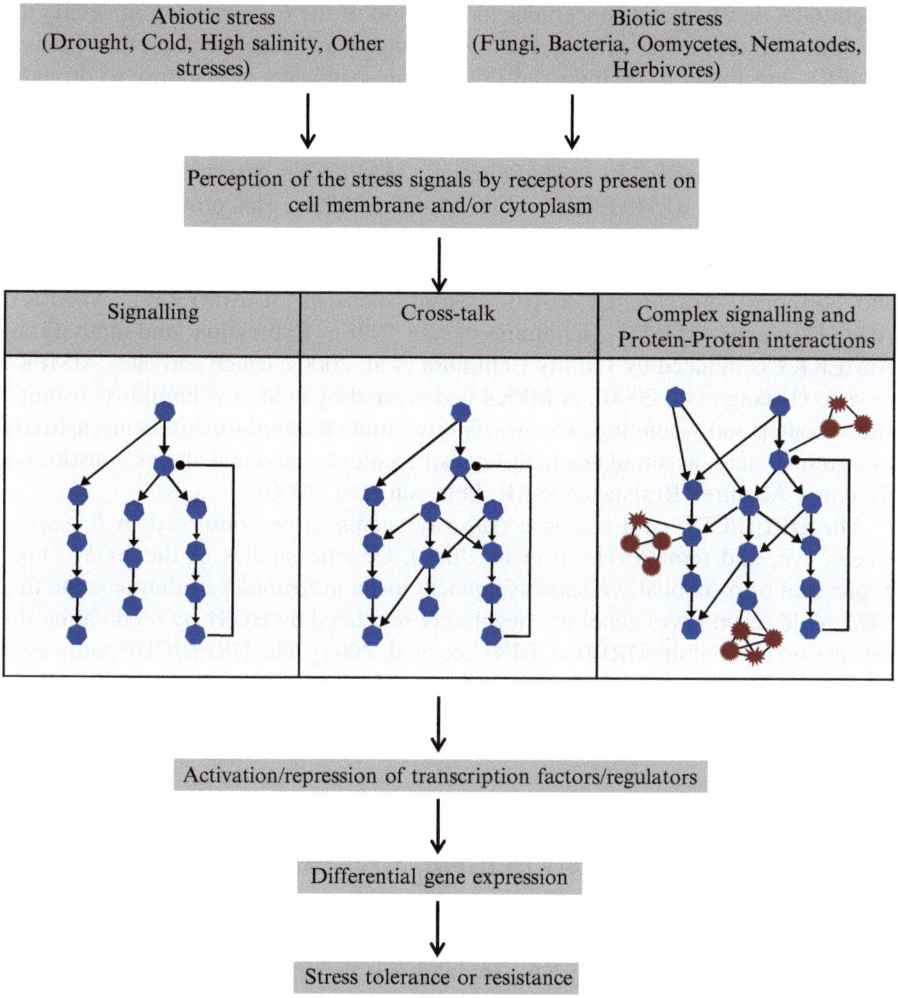

Fig. 2.2 Schematic representation of cross-talk among various signaling pathways. Large-omics experiments has revealed more complex pathway maps with a large degree of cross-talk and interaction with proteins other than signaling components

Therefore, a particular stimulus ends up at a particular end response and not at some other responses. This is termed as 'specificity' (Knight and Knight 2001). In any signaling pathway, there are chances of existence of cross-talk and specificity.

Cross-Talk and Specificity in Abiotic Stress Signaling

During abiotic stress signaling, Ca^{2+} serves as a second messenger in plant cells. Ca^{2+} can be stimulated by various abiotic, biotic, developmental and hormonal signals and, therefore, is a major point of signaling cross-talk. Factors such as magnitude, duration and subcellular localization of the Ca^{2+} oscillation decide the specificity and/or cross-talk in Ca^{2+} signaling. Ca^{2+}-dependent protein kinases (CDPKs) are found to be involved in signaling pathways in response to drought, wounding and cold stress.

MAPK signaling cascade offers the strongest evidence for cross-talk during abiotic stress signaling. In *Arabidopsis*, 60 MAPKKKs perceive the signals and transduce them to 10 MAPKKs and then to 20 MAPKs. This provides great scope for cross-talk between different stress signals. MAPK signaling cascades are stimulated by more than one type of stresses, such as, biotic, abiotic, developmental and hormonal. *Arabidopsis* MAPK cascade consists of AtMEKK1, AtMEK1/AtMKK2, and AtMPK4 (Ichimura et al. 1998). Expression and activity of AtMEKK1 is induced by salinity (Ichimura et al. 2000), which activates AtMPK4 *in vitro* (Huang et al. 2000). AtMPK4 is activated by cold, low humidity, osmotic stress, touch, and wounding. The *Arabidopsis* mutant *atmpk4* exhibits insensitivity to jasmonic acid, accumulates high levels of salicylic acid and shows constitutive Systemic Acquired Resistance (SAR, Peterson et al. 2000).

The DREB/CBF cold-response pathway seems to be conserved in *B. napus*, wheat, rye, and tomato (Hsieh et al. 2002). Genetic studies of the *HOS1* (high expression of osmotically responsive genes) locus in *Arabidopsis* demonstrate that *COR* (cold responsive) genes are negatively regulated by HOS1 by modulating the expression level of the DREBs/CBFs (Lee et al. 2001). The DREB/CBF pathway is thought to be regulated in an ABA-independent manner, but *DREB4/CBF4* gene gets induced by drought and ABA, but not by cold stress indicating a potential cross-talk.

Cross-Talk and Specificity in Biotic Stress Signaling

During pathogen attack plants execute specific primary defense responses by generating the alarm signals SA, JA, and ET in appropriate balance (De Vos et al. 2005). Endogenous accumulation of these signals initiates an array of signaling pathways which generate different defense responses against partially distinct classes of attackers (Koornneef and Pieterse 2008). Generally, biotrophic pathogens are more sensitive to SA-mediated induced defenses, whereas necrotrophic pathogens and herbivorous insects are sensitive to JA/ET-mediated defenses (Glazebrook 2005), although few exceptions are there (Thaler et al. 2004). To cope up with simultaneous or subsequent invasion by multiple challengers, a powerful regulatory potential is provided by cross-talk between induced defense-signaling pathways.

SA- and JA-dependent signaling mostly exhibit a mutually antagonistic interaction, but there are reports demonstrating their synergistic interactions also (Mur et al. 2006). Along with SA/JA cross-talk, there are interactions between SA and ET, JA and ABA, and JA and ET that function in the adaptive response of plants against various pathogens. There are several regulatory components which act as key molecular players of SA/JA cross-talk, such as, NONEXPRESSOR OF PR GENES1 (NPR1), WRKY transcription factors, MAP kinases and glutaredoxin GRX480 etc. The regulatory protein NPR1 is essential for SA signal transduction which activates the genes encoding pathogenesis-related (PR) proteins (Van Loon et al. 2006). In *Arabidopsis*, when *NPR1* was mutated, the suppression of JA-inducible gene expression by SA was obstructed. This indicates a crucial role of NPR1 in the cross-talk between SA and JA signaling (Spoel et al. 2007). Rayapuram and Baldwin (2007) proposed that in wild-type plants of tobacco, during herbivore attack NPR1 negatively regulates SA production and thus results in suppression of SA/JA cross-talk to allow induction of JA-mediated defenses against herbivores.

WRKY transcription factors are one of the significant regulatory components. They regulate SA-dependent defense responses and some of them are involved in SA/JA cross talk (Wang et al. 2006). WRKY70 positively regulates SA-mediated defenses and inhibits the JA response (Li et al. 2004). Interestingly, WRKY11 and WRKY17 acts as negative regulators of WRKY70 and, thus, of SA-mediated defense responses, thereby positively regulating the JA response (Journot-Catalino et al. 2006). In another study, Mao et al. (2007) found that WRKY62 functions as a repressor of JA response.

MAP kinases relay information from sensors to the cellular responses and are involved in plant defense signaling (Nakagami et al. 2005). In *Arabidopsis*, Peterson et al. (2000) demonstrated MAP KINASE4 (MPK4) as a negative regulator of SA signaling which positively regulated JA signaling. In contrast to it, downstream effectors of MPK4 i.e. ENHANCED DISEASE SUSCEPTIBILITY1 (EDS1) and PHYTOALEXIN-DEFICIENT4 (PAD4) shows reverse action to that of MPK4 by acting as activators of SA signaling and repressors of JA signaling (Brodersen et al. 2006).

Cross-Talk Between Abiotic and Biotic Stress

Transcription factors and kinases are key players involved in cross-talk between signaling pathways. ABA, SA, JA, ethylene (ET) and ROS regulated signaling pathways play significant roles in the crosstalk between biotic and abiotic stress signaling. ABA is mainly involved in responses against abiotic stresses such as drought, low temperature, and osmotic stress, whereas SA, JA, and ET have important roles in biotic stress signaling. In certain cases, ABA is found to negatively regulate disease resistance (Mauch-Mani and Mauch 2005). In *Arabidopsis*, when ABA was exogenously applied the expression of JA- or ET-responsive defense gene get repressed (Anderson et al. 2004). This indicates that ABA functions antagonistically to JA and ET. A basic helix-loop-helix (bHLH) transcription factor, AtMYC2, functions as a transcriptional activator in the ABA-mediated drought stress signaling pathway (Abe et al. 2003). It acts as a negative regulator of JA/ET-mediated defense gene expression (Lorenzo et al. 2004).

Therefore, for cross-talk between biotic and abiotic stress responses, AtMYC2 might act as a key regulator via hormone signaling (Fujita et al. 2006). Another feasible candidate molecule that can regulate both biotic and abiotic signaling aspects is dehydration-responsive NAC transcription factor, RD26. The factors inducing the expression of *RD26* include JA, hydrogen peroxide (H_2O_2) and pathogen infections, as well as by drought, high salinity and ABA treatment (Fujita et al. 2004). RD26-regulated genes function in the detoxification of ROS, defense, or senescence and, hence, prove to be a convergence point for the pathogen defense, senescence, and ABA-mediated signaling pathways. In tobacco, CDPKs also regulate biotic and abiotic stress responses with the help of SA, ET, JA, and ABA-mediated signaling pathways (Ludwig et al. 2005).

In *Arabidopsis*, MEKK1-MAPK kinase 2 (MKK2)-MPK4/MPK6 cascade has been shown to activate in response to cold and salt stress signaling. In contrast to it, MEKK1-MKK4/MKK5-MPK3/MPK6 cascades were shown to be involved in the regulation of the pathogen defense response pathway via the expression of WRKY22 and WRKY29 (Asai et al. 2002). MPK3 and MPK6 are stimulated by abiotic stresses also and are involved in hormone signaling pathways (Droillard et al. 2002). In *Arabidopsis*, MPK3 and MPK6 are activated by oxidative stress (Yuasa et al. 2001). The components which affect their activities in ROS signaling are OXI1 (serine/threonine kinase), ANP1 (MAPKKK) and *NUCLEOSIDE-DIPHOSPHATE KINASE 2* (*AtNDPK2*) (Rentel et al. 2004; Moon et al. 2003). Therefore, MAPK cascades seem to intermediate ROS signaling and make the plant tolerant to environmental stresses by improving ROS scavenging capacity (Fujita et al. 2006). ROS is an important signaling molecule (Apel and Hirt 2012) and is a key player in both ABA signaling and disease resistance responses (Laloi et al. 2004). This suggests the role of ROS in mediating crosstalk between biotic and abiotic stress-responsive gene-expression networks.

Conclusion and Future Prospects

Plants have developed numerous mechanisms to respond to various environmental stresses. Whenever plant encounters any kind of stress, first of all there is perception of the signals by the potential sensors (Fig. 2.2). This further generates secondary messengers (e.g., ABA, ROS, Ca^{2+}, inositol phosphates etc.). The secondary messengers can alter the levels of intracellular Ca^{2+}. Any change in cytoplasmic Ca^{2+} is sensed by Ca^{2+} sensors i.e. calcium binding proteins and after interacting with corresponding interacting partners they initiate a protein phosphorylation cascade. This signaling cascade ultimately regulates the transcription factors that manage specific suite of stress-regulated genes and the functional proteins involved in protecting the cell against stress. The expression of stress-regulated genes leads to the production of various kinds of regulatory molecules like the phytohormones ABA, ET, and SA and transcription factors etc. A second round of signaling can get started by these regulatory molecules which may further enhance the initial signaling pathway. These signaling networks are interconnected at many levels for transferring information to generate robust plant response. Moreover, during signaling other components are also involved. A successful signal transduction pathway demands proper coordination of all signaling molecules. This is brought about by specific molecules whose functions are to transport, bring together or modify the signaling molecules (Fig. 2.2). They themselves are not involved in relaying the signal. These proteins consist of scaffolding proteins, protein modifiers (e.g., enzymes needed for protein glycosylation, methylation, ubiquitination and lipidation) and adaptors.

Transcriptome analyses based on microarrays and proteomic studies have given rise to rapid progress in the field of plant signal transduction and gene regulation. GeneChip and cDNA microarrays together with massive whole genome sequencing proved to be useful in identifying novel signaling determinants on a whole-genome scale in response to various stresses. This has increased our understanding towards the complex mechanisms involved in various aspects of plant responses to environmental cues. Proteomics approach is useful in studying post-translational modifications of the proteins. It can also be used to clone unique genes by means of differential analysis which will expedite our understanding of stress signaling mechanisms in plants. Therefore, stress-tolerant plants can be genetically engineered by the combined efforts of plant molecular biologists, physiologists, and breeders. However, further research is required for implying these strategies in crop improvement. We need to fully understand the whole stress-response system of the plants to minimize differences in the tolerance capacity of transgenic plants between laboratories and crop fields. But with the pace of progress in understanding the stress mechanisms in plants, we may hope that in the future these collective efforts and results of collaborative studies will add to supportable food production in the world and will help to prevent global-scale crop losses caused by various stresses.

Acknowledgements This work was partially supported by a research grant from the Department of Biotechnology, Government of India (BT/PR10605/PBD/16/791/2008). SV and SN acknowledge Council of Scientific and Industrial Research and University Grants Commission, respectively for the fellowship.

References

Abbasi F, Onodera H, Toki S, Tanaka H, Komatsu S (2004) OsCDPK 13, a calcium-dependent protein kinase gene from rice, is induced by cold and gibberellin in rice leaf sheath. Plant Mol Biol 55(4):541–552

Abe H, Urao T, Ito T, Seki M, Shinozaki K, Yamaguchi-Shinozaki K (2003) *Arabidopsis* AtMYC2 (bHLH) and AtMYB2 (MYB) function as transcriptional activators in abscisic acid signaling. Plant Cell 15(1):63–78

Aghaei K, Ehsanpour AA, Komatsu S (2008) Proteome analysis of potato under salt stress. J Proteome Res 7(11):4858–4868

Ali GM, Komatsu S (2006) Proteomic analysis of rice leaf sheath during drought stress. J Proteome Res 5(2):396–403

Amey RC, Schleicher T, Slinn J, Lewis M, Macdonald H, Neill SJ, Spencer-Phillips PTN (2008) Proteomic analysis of a compatible interaction between *Pisum sativum* (pea) and the downy mildew pathogen *Peronospora viciae*. Eur J Plant Pathol 122:41–55

Anderson JP, Badruzsaufari E, Schenk PM, Manners JM, Desmond OJ, Ehlert C, Maclean DJ, Ebert PR, Kazan K (2004) Antagonistic interaction between abscisic acid and jasmonate-ethylene signaling pathways modulates defense gene expression and disease resistance in Arabidopsis. Plant Cell 16(12):3460–3479

Apel K, Hirt H (2012) Reactive oxygen species: metabolism, oxidative stress, and signal transduction. Annu Rev Plant Biol 55:373–399

Aranjuelo I, Molero G, Erice G, Avice JC, Nogues S (2011) Plant physiology and proteomics reveals the leaf response to drought in alfalfa (*Medicago sativa* L.). J Exp Bot 62:111–123

Asai T, Tena G, Plotnikova J, Willmann MR, Chiu WL, Gomez-Gomez L, Boller T, Ausubel FM, Sheen J (2002) MAP kinase signaling cascade in *Arabidopsis* innate immunity. Nature 415(6875):977–983

Balbuena TS, Salas JJ, Martinez-Force E, Garces R, Thelen JJ (2011) Proteome analysis of cold acclimation in sunflower. J Proteome Res 10(5):2330–2346

Bhushan D, Pandey A, Choudhary MK, Datta A, Chakraborty S, Chakraborty N (2007) Comparative proteomics analysis of differentially expressed proteins in chickpea extracellular matrix during dehydration stress. Mol Cell Proteomics 6(11):1868–1884

Bilgin DD, Zavala JA, Zhu J, Clough SJ, Ort DR, DeLucia EH (2010) Biotic stress globally downregulates photosynthesis genes. Plant Cell and Environ 33:1597–1613

Birkenbihl RP, Somssich IE (2011) Transcriptional plant responses critical for resistance towards necrotrophic pathogens. Front Plant Sci 2:76

Bita CE, Zenoni S, Vriezen WH, Mariani C, Pezzotti M, Gerats T (2011) Temperature stress differentially modulates transcription in meiotic anthers of heat-tolerant and heat-sensitive tomato plants. BMC Genomics 12:384

Boominathan P, Shukla R, Kumar A, Manna D, Negi D, Verma PK, Chattopadhyay D (2004) Long term transcript accumulation during the development of dehydration adaptation in *Cicer arietinum*. Plant Physiol 135:1608–1620

Brodersen P, Petersen M, Bjorn Nielsen H, Zhu S, Newman MA, Shokat KM, Rietz S, Parker J, Mundy J (2006) Arabidopsis MAP kinase 4 regulates salicylic acid- and jasmonic acid/ethylene dependent responses via EDS1 and PAD4. Plant J 47(4):532–546

Chen X, Zhang W, Zhang B, Zhou J, Wang Y, Yang Q, Ke Y, He H (2011) Phosphoproteins regulated by heat stress in rice leaves. Proteome Sci 9:37

Cheng Y, Qi Y, Zhu Q, Chen X, Wang N, Zhao X, Chen H, Cui X, Xu L, Zhang W (2009) New changes in the plasma-membrane-associated proteome of rice roots under salt stress. Proteomics 9(11):3100–3114

Chi F, Yang P, Han F, Jing Y, Shen S (2010) Proteomic analysis of rice seedlings infected by *Sinorhizobium meliloti* 1021. Proteomics 10(9):1861–1874

Chisholm ST, Coaker G, Day B, Staskawicz BJ (2006) Host–microbe interactions: shaping the evolution of the plant immune response. Cell 124(4):803–814

Chitteti BR, Peng Z (2007) Proteome and phosphoproteome differential expression under salinity stress in rice (*Oryza sativa*) roots. J Proteome Res 6(5):1718–17127

Dangl JL, McDowell JM (2006) Two modes of pathogen recognition by plants. Proc Natl Acad Sci USA 103:8575–8576

De Vos M, Van Oosten VR, Van Poecke RM, Van Pelt JA, Pozo MJ, Mueller MJ, Buchala AJ, Metraux JP, Van Loon LC, Dicke M, Pieterse CM (2005) Signal signature and transcriptome changes of *Arabidopsis* during pathogen and insect attack. Mol Plant Microbe Interact 18(9):923–937

DebRoy S, Thilmony R, Kwack YB, Nomura K, He SY (2004) A family of conserved bacterial effectors inhibits salicylic acid-mediated basal immunity and promotes disease necrosis in plants. Proc Natl Acad Sci USA 101(26):9927–9932

Droillard M, Boudsocq M, Barbier-Brygoo H, Lauriere C (2002) Different protein kinase families are activated by osmotic stresses in *Arabidopsis thaliana* cell suspensions. Involvement of the MAP kinases AtMPK3 and AtMPK6. FEBS Lett 527(1–3):43–50

Dubouzet JG, Sakuma Y, Ito Y, Kasuga M, Dubouzet EG, Miura S, Seki M, Shinozaki K, Yamaguchi-Shinozaki K (2003) *OsDREB* genes in rice, *Oryza sativa* L, encode transcription activators that function in drought-, high-salt- and cold-responsive gene expression. Plant J 33:751–763

Fernandez RG, Jorrin-Novo JV (2012) Contribution of proteomics to the study of plant pathogenic fungi. J Proteome Res 11(1):3–16

Ford KL, Cassin A, Bacic A (2011) Quantitative proteomic analysis of wheat cultivars with differing drought stress tolerance. Front Plant Sci 2:44

Fujita M, Fujita Y, Maruyama K, Seki M, Hiratsu K, Ohme-Takagi M, Tran LS, Yamaguchi-Shinozaki K, Shinozaki K (2004) A dehydration induced NAC protein, RD26, is involved in a novel ABA-dependent stress-signaling pathway. Plant J 39(6):863–876

Fujita M, Fujita Y, Noutoshi Y, Takahashi F, Narusaka Y, Yamaguchi-Shinozaki K, Shinozaki K (2006) Crosstalk between abiotic and biotic stress responses: a current view from the points of convergence in the stress signaling networks. Curr Opin Plant Biol 9(4):436–442

Furihata T, Maruyama K, Fujita Y, Umezawa T, Yoshida R, Shinozaki K, Yamaguchi-Shinozaki K (2006) Abscisic acid-dependent multisite phosphorylation regulates the activity of a transcription activator AREB1. Proc Natl Acad Sci USA 103(6):1988–1993

Galletti R, Ferrari S, De Lorenzo G (2011) Arabidopsis MPK3 and MPK6 play different roles in basal and oligogalacturonide- or flagellin-induced resistance against Botrytis cinerea. Plant Physiol 157:804–814

Glazebrook J (2005) Contrasting mechanisms of defense against biotrophic and necrotrophic pathogens. Annu Rev Phytopathol 43:205–227

Grant JJ, Chini A, Basu D, Loake GJ (2003) Targeted activation tagging of the Arabidopsis NBS–LRR gene, ADR1, conveys resistance to virulent pathogens. Mol Plant Microbe Interact 16(8):669–680

Hossain Z, Nouri MZ, Komatsu S (2012) Plant cell organelle proteomics in response to abiotic stress. J Proteome Res 11(1):37–48

Hsieh TH, Lee JT, Yang PT, Chiu LH, Charng YY, Wang YC, Chan MT (2002) Heterology expression of the Arabidopsis C-repeat/dehydration response element binding factor 1 gene confers elevated tolerance to chilling and oxidative stresses in transgenic tomato. Plant Physiol 129(3):1086–1094

Huang Y, Li H, Gupta R, Morris PC, Luan S, Kieber JJ (2000) ATMPK4, an Arabidopsis homolog of mitogen-activated protein kinase, is activated in vitro by AtMEK1 through Threonine phosphorylation. Plant Physiol 122(4):1301–1310

Ichimura K, Mizoguchi T, Irie K, Morris P, Giraudat J, Matsumoto K, Shinozaki K (1998) Isolation of ATMEKK1 (a MAP Kinase kinase kinase)-interacting proteins and analysis of a MAP kinase cascade in Arabidopsis. Biochem Biophys Res Commun 253(2):532–543

Ichimura K, Mizoguchi T, Yoshida R, Yuasa T, Shinozaki K (2000) Various abiotic stresses rapidly activate *Arabidopsis* MAP kinases ATMPK4 and ATMPK6. Plant J 24(5):655–665

Ito Y, Katsura K, Maruyama K, Taji T, Kobayashi M, Seki M, Shinozaki K, Yamaguchi-Shinozaki K (2006) Functional analysis of rice DREB1/CBF-type transcription factors involved in coldresponsive gene expression in transgenic rice. Plant Cell Physiol 47(1):141–153

Jaiswal P, Cheruku JR, Kumar K, Yadav S, Singh A, Kumari P, Dube SC, Upadhyaya KC, Verma PK (2012) Differential transcript accumulation in chickpea during early phases of compatible interaction with a necrotrophic fungus *Ascochyta rabiei*. Mol Biol Rep 39(4):4635–46

Jones AM, Thomas V, Bennett MH, Mansfield J, Grant M (2006) Modifications to the *Arabidopsis* defense proteome occur prior to significant transcriptional change in response to inoculation with *Pseudomonas syringae*. Plant Physiol 142(4):1603–1620

Journot-Catalino N, Somssich IE, Roby D, Kroj T (2006) The transcription factors WRKY11 and WRKY17 act as negative regulators of basal resistance in *Arabidopsis thaliana*. Plant Cell 18(11):3289–3302

Knight H, Knight MR (2001) Abiotic stress signaling pathways: specificity and cross-talk. Trends in Plant Sci 6(6):262–267

Komatsu S, Kobayashi Y, Nishizawa K, Nanjo Y, Furukawa K (2010a) Comparative proteomics analysis of differentially expressed proteins in soybean cell wall during flooding stress. Amino Acids 39(5):1435–49

Komatsu S, Sugimoto T, Hoshino T, Nanjo Y, Furukawa K (2010b) Identification of flooding stress responsible cascades in root and hypocotyl of soybean using proteome analysis. Amino Acids 38(3):729–38

Koornneef A, Pieterse CMJ (2008) Cross talk in defense signaling. Plant Physiol 146(3):839–844

Kottapalli KR, Rakwal R, Satoh K, Shibato J, Kottapalli P, Iwahashi H, Kikuchi S (2007) Transcriptional profiling of indica rice cultivar IET8585 (Ajaya) infected with bacterial leaf blight pathogen *Xanthomonas oryzae* pv *oryzae*. Plant Physiol Biochem 45(10–11):834–50

Laloi C, Apel K, Danon A (2004) Reactive oxygen signaling: the latest news. Curr Opin Plant Biol 7(3):323–328

Lee H, Xiong L, Gong Z, Ishitani M, Stevenson B, Zhu JK (2001) The Arabidopsis HOS1 gene negatively regulates cold signal transduction and encodes a RING finger protein that displays cold-regulated nucleo-cytoplasmic partitioning. Genes Dev 15:912–924

Lee DG, Ahsan N, Lee SH, Kang KY, Bahk JD, Lee IJ, Lee BH (2007) A proteomic approach in analyzing heat-responsive proteins in rice leaves. Proteomics 7:3369–3383

Lee J, Feng J, Campbell KB, Scheffler BE, Garrett WM, Thibivilliers S, Stacey G, Naiman DQ, Tucker ML, Pastor-Corrales MA, Cooper B (2009) Quantitative proteomic analysis of bean plants infected by a virulent and avirulent obligate rust fungus. Mol Cell Proteomics 8(1):19–31

Leon-Reyes A, Du Y, Koornneef A, Proietti S, Korbes AP, Memelink J, Pieterse CM, Ritsema T (2010) Ethylene signaling renders the jasmonate response of *Arabidopsis* insensitive to future suppression by salicylic acid. Mol Plant Microbe Interact 23:187–197

Li J, Brader G, Palva ET (2004) The WRKY70 transcription factor: A node of convergence for jasmonate-mediated and salicylate-mediated signals in plant defense. Plant Cell 16:319–331

Li X, Gao P, Gjetvaj B, Westcott N, Gruber MY (2009) Analysis of the metabolome and transcriptome of *Brassica carinata* seedlings after lithium chloride exposure. Plant Science 177:68–80

Li D, Zhang Y, Hu X, Shen X, Ma L, Su Z, Wang T, Dong J (2011) Transcriptional profiling of *Medicago truncatula* under salt stress identified a novel CBF transcription factor MtCBF4 that plays an important role in abiotic stress responses. BMC Plant Biology 11:109

Li T, Xu SL, Oses-Prieto JA, Putil S, Xu P, Wang RJ, Lic KH, Maltby DA, An LH, Burlingame AL, Deng ZP, Wang ZY (2011) Proteomics analysis reveals post-translational mechanisms for cold-induced metabolic changes in *Arabidopsis*. Mol Plant 4(2):361–374

Lorenzo O, Chico JM, Sanchez-Serrano JJ, Solano R (2004) *JASMONATE-INSENSITIVE1* encodes a MYC transcription factor essential to discriminate between different jasmonate-regulated defense responses in *Arabidopsis*. Plant Cell 16(7):1938–1950

Ludwig AA, Saitoh H, Felix G, Freymark G, Miersch O, Wasternack C, Boller T, Jones JD, Romeis T (2005) Ethylene-mediated cross-talk between calcium-dependent protein kinase and MAPK signaling controls stress responses in plants. Proc Natl Acad Sci USA 102(30):10736–10741

Lv DK, Bai X, Li Y, Ding XD, Ge Y, Cai H, Ji W, Wu N, Zhu YM (2010) Profiling of cold-stress-responsive miRNAs in rice by microarrays. Gene 459:39–47

Majoul T, Bancel E, Triboï E, Hamida JB, Branlard G (2004) Proteomic analysis of the effect of heat stress on hexaploid wheat grain: characterization of heat-responsive proteins from non-prolamins fraction. Proteomics 4(2):505–513

Manaa A, Ahmed HB, Valot B, Bouchet JP, Smiti SA, Causse M, Faurobert M (2011) Salt and genotype impact on plant physiology and root proteome variations in tomato. J Exp Bot 62(8):2797–2813

Mao P, Duan M, Wei C, Li Y (2007) WRKY62 transcription factor acts downstream of cytosolic NPR1 and negatively regulates jasmonate responsive gene expression. Plant Cell Physiol 48(6):833–842

Mao G, Meng X, Liu Y, Zheng Z, Chen Z, Zhang S (2011) Phosphorylation of a WRKY transcription factor by two pathogen responsive MAPKs drives phytoalexin biosynthesis in *Arabidopsis*. Plant Cell 23:1639–1653

Mauch-Mani B, Mauch F (2005) The role of abscisic acid in plant–pathogen interactions. Curr Opin Plant Biol 8(4):409–414

McDowell JM, Dangl JL (2000) Signal transduction in the plant immune response. Trends Biochem Sci 25:79–82

Medeiros FHV, Souza RM, Medeiros FCL, Zhang H, Wheeler T, Payton P, Ferro HM, Paré PW (2011) Transcriptional profiling in cotton associated with Bacillus subtilis (UFLA285) induced biotic-stress tolerance. Plant Soil 347:327–337

Mittal D, Chakrabarti S, Sarkar A, Singh A, Grover A (2009) Heat shock factor gene family in rice: genomic organization and transcript expression profiling in response to high temperature, low temperature and oxidative stresses. Plant Physiol Biochem 47(9):785–795

Mizoguchi T, Irie K, Hirayama T, Hayashida N, Yamaguchi-Shinozaki K, Matsumoto K, Shinozaki K (1996) A gene encoding a mitogen-activated protein kinase kinase kinase is induced simultaneously with genes for a mitogen-activated protein kinase and an S6 ribosomal protein kinase by touch, cold and water stress in *Arabidopsis thaliana*. Proc Natl Acad Sci USA 93(2):765–769

Molina C, Rotter B, Horres R, Udupa SM, Besser B, Bellarmino L, Baum M, Matsumura H, Terauchi R, Kahl G, Winter P (2008) SuperSAGE: the drought stress-responsive transcriptome of chickpea roots. BMC Genomics 9:553

Moon H, Lee B, Choi G, Shin D, Prasad DT, Lee O, Kwak SS, Kim DH, Nam J, Bahk J, Hong JC, Lee SY, Cho MJ, Lim CO, Yun DJ (2003) NDP kinase 2 interacts with two oxidative stress-activated MAPKs to regulate cellular redox state and enhances multiple stress tolerance in transgenic plants. Proc Natl Acad Sci USA 100(1):358–363

Mukherjee AK, Carp MJ, Zuchman R, Ziv T, Horwitz BA, Gepstein S (2010) Proteomics of the response of *Arabidopsis thaliana* to infection with *Alternaria brassicicola*. J Proteomics 73:709–720

Mur LA, Kenton P, Atzorn R, Miersch O, Wasternack C (2006) The outcomes of concentration-specific interactions between salicylate and jasmonate signaling include synergy, antagonism, and oxidative stress leading to cell death. Plant Physiol 140(1):249–262

Nakagami H, Pitzschke A, Hirt H (2005) Emerging MAP kinase pathways in plant stress signaling. Trends Plant Sci 10(7):339–346

Nakashima K, Shinwari ZK, Sakuma Y, Seki M, Miura S, Shinozaki K, Yamaguchi-Shinozaki K (2000) Organization and expression of two Arabidopsis DREB2 genes encoding DRE-binding proteins involved in dehydration- and high-salinity-responsive gene expression. Plant Mol Biol 42:657–665

Nanjo Y, Maruyama K, Yasue H, Yamaguchi-Shinozaki K, Shinozaki K, Komatsu S (2011) Transcriptional responses to flooding stress in roots including hypocotyl of soybean seedlings. Plant Mol Biol 77:129–144

Naydenov NG, Khanam S, Siniauskaya M, Nakamura C (2010) Profiling of mitochondrial transcriptome in germinating wheat embryos and seedlings subjected to cold, salinity and osmotic stresses. Genes Genet Syst 85(1):31–42

Ogawa I, Nakanishi H, Mori S, Nishizawa NK (2009) Time course analysis of gene regulation under cadmium stress in rice. Plant Soil 325:97–108

Ok SH, Jeong HJ, Bae JM, Shin JS, Luan S, Kim KN (2005) Novel CIPK1-associated proteins in Arabidopsis contain an evolutionarily conserved C-terminal region that mediates nuclear localization. Plant Physiol 139:138–150

Ouyang B, Yang T, Li H, Zhang L, Zhang Y, Zhang J, Fei Z, Ye Z (2007) Identification of early salt stress response genes in tomato root by suppression subtractive hybridization and microarray analysis. J Exp Bot 58(3):507–520

Pandey A, Rajamani U, Verma J, Subba P, Chakraborty N, Datta A, Chakraborty S, Chakraborty N (2010) Identification of extracellular matrix proteins of rice (*Oryza sativa* L.) involved in dehydration-responsive network: a proteomic approach. J Proteome Res 9(7):3443–3464

Peng Z, Wang M, Li F, Lv H, Li C, Xia G (2009) A proteomic study of the response to salinity and drought stress in an introgression strain of bread wheat. Mol Cell Proteomics 8(12):2676–2686

Peterson M, Brodersen P, Naested H, Andreasson E, Lindhart U, Johansen B, Nielsen HB, Lacy M, Austin MJ, Parker JE, Sharma SB, Klessig DF, Martienssen R, Mattsson O, Jensen AB, Mundy J (2000) *Arabidopsis* MAP kinase 4 negatively regulates systemic acquired resistance. Cell 103(7):1111–1120

Pieterse CM, Leon-Reyes A, Van der Ent S, Van Wees SC (2009) Networking by small-molecules hormones in plant immunity. Nat Chem Biol 5:308–316

Pitzschke A, Schikora A, Hirt H (2009) MAPK cascade signaling networks in plant defence. Curr Opin Plant Biol 12(4):421–426

Povero G, Loreti E, Pucciariello C, Santaniello A, Di Tommaso D, Di Tommaso G, Kapetis D, Zolezzi F, Piaggesi A, Perata P (2011) Transcript profiling of chitosan-treated Arabidopsis seedlings. J Plant Res 124(5):619–629

Qin AD, Wu H, Peng H, Yao Y, Ni Z, Li Z, Zhou C, Sun Q (2008) Heat stress-responsive transcriptome analysis in heat susceptible and tolerant wheat (*Triticum aestivum* L.) by using wheat genome array. BMC Genomics 9:432

Rabbani MA, Maruyama K, Abe H, Khan MA, Katsura K, Ito Y, Yoshiwara K, Seki M, Shinozaki K, Yamaguchi-Shinozaki K (2003) Monitoring expression profiles of rice genes under cold, drought and high-salinity stresses, and abscisic acid application using cDNA microarray and RNA gel-blot analyses. Plant Physiol 133:1755–1767

Rayapuram C, Baldwin IT (2007) Increased SA in NPR1-silenced plants antagonizes JA and JA-dependent direct and indirect defenses in herbivore-attacked Nicotiana attenuata in nature. Plant J 52(4):700–715

Rentel MC, Lecourieux D, Ouaked F, Usher SL, Petersen L, Okamoto H, Knight H, Peck SC, Grierson CS, Hirt H, Knight MR (2004) OXI1 kinase is necessary for oxidative burst-mediated signaling in Arabidopsis. Nature 427:858–861

Rinalducci S, Egidi MG, Karimzadeh G, Jazii FR, Zolla L (2011) Proteomic analysis of a spring wheat cultivar in response to prolonged cold stress. Electrophoresis 32(14):1807–1818

Rong E, Zhao Z, Zhao Z, Ma J, Zang W, Wang L, Xie D, Yang W (2011) Wheat cold and light stress analysis based on the Arabidopsis homology protein–protein interaction (PPI) Network. J Med Plant Res 5(23):5493–5498

Sakuma Y, Liu Q, Dubouzet JG, Abe H, Shinozaki K, Yamaguchi-Shinozaki K (2002) DNA-binding specificity of the ERF/AP2 domain of *Arabidopsis* DREBs, transcription factors involved in dehydration- and cold-inducible gene expression. Biochem Biophys Res Commun 290(3):998–1009

Sana TR, Fischer S, Wohlgemuth G, Katrekar A, Jung K, Ronald PC, Fiehn O (2010) Metabolomic and transcriptomic analysis of the rice response to the bacterial blight pathogen *Xanthomonas oryzae* pv *oryzae*. Metabolomics 6(3):451–465

Sato M, Tsuda K, Wang L, Coller J, Watanabe Y, Glazebrook J, Katagiri F (2010) Network modeling reveals prevalent negative regulatory relationships between signaling sectors in *Arabidopsis* immune signaling. PLoS Pathog 6:e1001011

Savchenko T, Walley JW, Chehab EW, Xiao Y, Kaspi R, Pye MF, Mohamed ME, Lazarus CM, Bostock RM, Dehesha K (2010) Arachidonic acid: an evolutionarily conserved signaling molecule modulates plant stress signaling networks. Plant Cell 22(10):3193–3205

Schulze-Lefert P, Vogel J (2000) Closing the ranks to attack by powdery mildew. Trends Plant Sci 5(8):343–348

Seki M, Narusaka M, Ishida J, Nanjo T, Fujita M, Oono Y, Kamiya A, Nakajima M, Enju A, Sakurai T, Satou M, Akiyama K, Taji T, Yamaguchi-Shinozaki K, Carninci P, Kawai J, Hayashizaki Y, Shinozaki K (2002) Monitoring the expression profiles of 7000 *Arabidopsis* genes under drought, cold and high salinity stresses using a full-length cDNA microarray. Plant J 31(3):279–292

Seo YS, Chern M, Bartley LE, Han M, Jung KH, Lee I, Walia H, Richter T, Xu X, Cao P, Bai W, Ramanan R, Amonpan F, Arul L, Canlas PE, Ruan R, Park CJ, Chen X, Hwang S, Jeon JS,

Ronald PC (2011) Establishment of a rice stress response interactome. PLoS Genet 7(4): e1002020
Sharma N, Rahman MH, Strelkov S, Thiagarajah M, Bansal VK, Kav NNV (2007) Proteome-level changes in two Brassica napus lines exhibiting differential responses to the fungal pathogen Alternaria brassicae. Plant Science 172(1):95–110
Sobhanian H, Razavizadeh R, Nanjo Y, Ehsanpour AA, Jazii FR, Motamed N, Komatsu S (2010) Proteome analysis of soybean leaves, hypocotyls and roots under salt stress. Proteome Sci 8:19
Spoel SH, Johnson JS, Dong X (2007) Regulation of tradeoffs between plant defenses against pathogens with different lifestyles. Proc Natl Acad Sci USA 104(47):18842–18847
Teige M, Scheikl E, Eulgem T, Doczi R, Ichimura K, Shinozaki K, Dangl JL, Hirt H (2004) The MKK2 pathway mediates cold and salt stress signaling in *Arabidopsis*. Mol Cell 15(1):141–152
Thaler JS, Owen B, Higgins VJ (2004) The role of the jasmonate response in plant susceptibility to diverse pathogens with a range of lifestyles. Plant Physiol 135(1):530–538
Ulm R, Ichimura K, Mizoguchi T, Peck SC, Zhu T, Wang X, Shinozaki K, Paszkowski J (2002) Distinct regulation of salinity and genotoxic stress responses by *Arabidopsis* MAP kinase phosphatase 1. EMBO J 21(23):6483–6493
Van Loon LC, Rep M, Pieterse CM (2006) Significance of inducible defense-related proteins in infected plants. Annu Rev Phytopathol 44:135–162
Walia H, Wilson C, Wahid A, Condamine P, Cui X, Close TJ (2006) Expression analysis of barley (*Hordeum vulgare* L) during salinity stress. Funct Integr Genomics 6(2):143–156
Wang D, Amornsiripanitch N, Dong X (2006) A genomic approach to identify regulatory nodes in the transcriptional network of systemic acquired resistance in plants. PLoS Pathog 2(11):1042–1050
Winfield MO, Lu C, Wilson ID, Coghill JA, Edwards KJ (2010) Plant responses to cold: transcriptome analysis of wheat. Plant Biotechnol J 8(7):749–771
Xu XY, Fan R, Zheng R, Li CM, Yu DY (2011) Proteomic analysis of seed germination under salt stress in soybeans. J Zhejiang Univ Sci B 12(7):507–517
Yamaguchi-Shinozaki K, Shinozaki K (2006) Transcriptional regulatory networks in cellular responses and tolerance to dehydration and cold stresses. Annu Rev Plant Biol 57:781–803
Yan SP, Zhang QY, Tang ZC, Su WA, Sun WN (2006) Comparative proteomic analysis provides new insights into chilling stress responses in rice. Mol Cell Proteomics 5(3):484–496
Yuasa T, Ichimura K, Mizoguchi T, Shinozaki K (2001) Oxidative stress activates ATMPK6, an Arabidopsis homologue of MAP kinase. Plant Cell Physiol 42(9):1012–1016
Zhang L, Tian LH, Zhao JF, Song Y, Zhang CJ, Guo Y (2009) Identification of an apoplastic protein involved in the initial phase of salt stress response in rice root by two-dimensional electrophoresis. Plant Physiol 149(2):916–928
Zhao L, Sun YL, Cui SX, Chen M, Yang HM, Liu HM, Chai TY, Huang F (2011) Cd-induced changes in leaf proteome of the hyperaccumulator plant Phytolacca Americana. Chemosphere 85(1):56–66
Zheng Z, Qamar SA, Chen Z, Mengiste T (2006) *Arabidopsis* WRKY33 transcription factor is required for resistance to necrotrophic fungal pathogens. Plant J 48:592–605
Zhu JK (2002) Salt and drought stress signal transduction in plants. Annu Rev Plant Biol 53:247–273
Zipfel C, Robatzek S, Navarro L, Oakeley EJ, Jones JD, Felix G, Boller T (2004) Bacterial disease resistance in *Arabidopsis* through flagellin perception. Nature 428(6984):764–767
Zou X, Jiang Y, Liu L, Zhang Z, Zheng Y (2010) Identification of transcriptome induced in roots of maize seedlings at the late stage of waterlogging. BMC Plant Biol 10:189

Chapter 3
Signaling by MicroRNAs in Response to Abiotic Stress

Guadalupe Sosa-Valencia, Alejandra A. Covarrubias, and José Luis Reyes

Introduction

As part of a large repertoire of strategies to cope with environmental variations, plants have chosen to include the use a vast array of non-coding RNAs to regulate gene expression. Among these, microRNAs (miRNAs) have been extensively studied in recent years and found to participate in numerous phenomena in plants ranging from metabolic responses to developmental decisions. Thus, it is not surprising that they have also been recruited to participate in pathways selected to counteract the adverse effects of biotic and abiotic stress. In this chapter we will first introduce the general pathways for maturation of microRNAs followed by their mechanisms of action. This overview will provide the context to present examples of microRNA involvement in stress signaling and will provide us with the framework to suggest potential points of regulation by stress signals. We also present recent advances in the field originating from genome-wide analyses and other data suggesting future directions towards a better understanding of the role of microRNAs in modulating plant responses to abiotic stress.

MicroRNAs and Other Small RNAs

MicroRNAs were first identified in animals as result of the characterization of developmental decisions in *Caenorhabditis elegans* (Lee et al. 1993). Almost a decade later, they were shown to be present in animals in large numbers (Lagos-Quintana et al. 2001; Lau et al. 2001; Lee and Ambros 2001), and eventually in

G. Sosa-Valencia • A.A. Covarrubias • J.L. Reyes (✉)
Department of Biología Molecular de Plantas, Instituto de Biotecnología, Universidad Nacional Autónoma de México, Av. Universidad 2001, Cuernavaca, Morelos C.P. 62210, Mexico
e-mail: jlreyes@ibt.unam.mx

M. Sarwat et al. (eds.), *Stress Signaling in Plants: Genomics and Proteomics Perspective, Volume 1*, DOI 10.1007/978-1-4614-6372-6_3,

plants (Llave et al. 2002a; Rhoades et al. 2002). Subsequent studies in Arabidopsis and several plant systems have revealed numerous other small RNAs including small interfering RNAs (siRNAs) which comprise the majority of the small RNA population and have been implicated in various pathways of gene silencing (Chen 2010). Among these, *trans*-acting siRNAs (tasiRNAs), natural antisense siRNAs (nat-siRNAs) or repeat-associated siRNAs (rasiRNAs) represent alternate small RNAs involved in silencing pathways employing additional members of the silencing machinery for their biogenesis and function. While current data on their activities could shed light into the intricacies of microRNA functions, interested readers are encouraged to turn to recent reviews in these subjects (Chen 2010; Law and Jacobsen 2010; Vazquez et al. 2010). Up until now, these small RNAs have not been directly linked to stress responses or stress signaling except for a handful of cases. Thus, these examples will be mentioned later on to mark the potential for other pathways to influence stress responses. In contrast, numerous studies have underscored the contribution of microRNAs to stress responses and we will focus mainly in these RNA molecules.

MicroRNA Biogenesis and Action

Maturation of MicroRNAs

While biogenesis of animal microRNAs is similar to that present in plants, there are certain differences that will be mentioned as we describe the pathway in plants (Fig. 3.1). Plant microRNA genes (*MIR* genes) are in general found as independent transcription units with their own regulatory promoter sequences. Transcription by RNA Polymerase II is the norm and transcripts are in general capped and polyadenylated (Lee et al. 2004; Parizotto et al. 2004). Soon after transcription, the pre-miRNA adopts a characteristic hairpin secondary structure and is sequentially recognized by the cap binding complex components CBP20 and CBP80 (Kim et al. 2008) and DAWDLE (Yu et al. 2008) to be followed by binding and processing by DCL1 (a member of the DICER-LIKE family of proteins) aided by SERRATE (SE) and HYPONASTIC LEAVES 1 (HYL1) (Kurihara and Watanabe 2004; Vazquez et al. 2004; Yang et al. 2006). The product is a double-stranded duplex of 20–24 nts in length with a 5'-phosphate and a two-nucleotide overhang at the 3'-end. This RNA duplex is methylated at the 2'OH position of the last ribose at the 3'-ends of each strand by the HUA-ENHANCER 1 (HEN1) methyltransferase activity (Park et al. 2002; Yu et al. 2005). The RNA duplex is subsequently transported to the cytoplasm via HASTY, a Ran-GTPase homologous to mammalian Exportin 5 (Park et al. 2005), and recruited to an ARGONAUTE 1-containing complex (RISC) where one of the two strands is selected to represent the mature miRNA while the other strand, known as the microRNA* strand is rapidly degraded (Baumberger and Baulcombe 2005; Vaucheret et al. 2004).

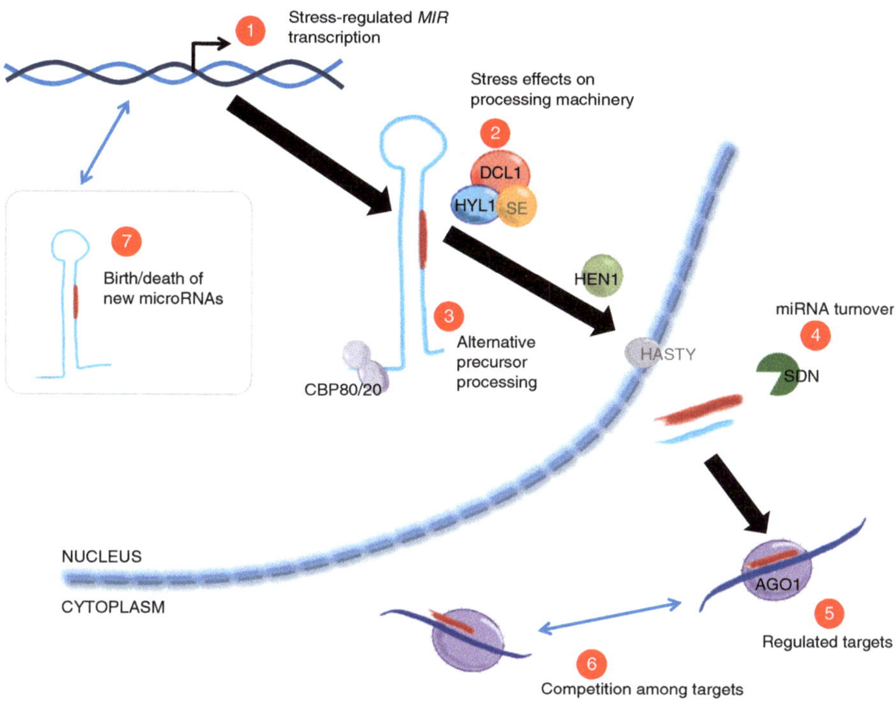

Fig. 3.1 Stress signals affecting microRNA biogenesis and activities. The diagram shows the microRNA biogenesis pathway in plants with selected biogenesis factors that have been found in different studies to be affected by abiotic stress (see main text). Elements influencing the outcome of the pathway discussed in the text are numbered. (*1*): Regulated transcription in response to stress (either repression or induction, as suggested by microarray and high-throughput sequencing experiments); (*2*): mutant analysis has revealed factors in the biogenesis pathway that are required for adequate stress responses (names in *black letters*, factors in *grey* have not been linked to stress); (*3*): alternative processing of microRNAs precursor (sequence variants or additional DCL products) may generate additional functional small RNAs; (*4*): microRNA (and target) turnover could influence the final result of microRNA regulation; (*5*): the MicroRNA-regulated transcript products are proposed to directly contribute to stress responses; (*6*): the presence of multiple targets for the same microRNA could compete for microRNA binding; (*7*): acquisition of novel microRNAs in the genome can reshape how stress responses are modulated over evolutionary time

Although this general biogenesis pathway applies to most conserved plant microRNAs studied so far, there are a few interesting examples deviating from the canonical process, where the differences may be related to precursor features and/or processing, or other associated factor activities that result in the production of additional small RNAs with the properties of microRNAs. In either case, it can be speculated that variations can be due, or at least influenced by, environmental input and by stress signals. For instance, a few reports where high-throughput sequencing has been used to determine the microRNA profile under different growth conditions have found that the otherwise unstable and low abundance microRNA* sequences are enriched up to detectable levels and sometimes even more abundant than the annotated and functional microRNA strand (Devers et al. 2011; Wong et al. 2011). Interestingly, in response to phosphate deficiency, there was an increased accumulation of particular microRNA* sequences in Arabidopsis that did not completely correlate with an increase in the corresponding microRNAs (Pant et al. 2009). These examples raise the possibility of their participation and functionality under stress conditions, and suggest that microRNAs* could participate by regulating expression of other mRNAs, as has been elegantly demonstrated in *Drosophila* (Okamura et al. 2008). Moreover, recent discoveries also originating from small RNA sequencing strategies in different plant species have shown that microRNA precursors possess the potential to generate more that one RNA duplex processed by a DCL protein. First, variants differing in a few nucleotide positions from the canonical mature microRNA sequence, but originating from the same precursor, could potentially target mRNAs with the complementary sequence [for a recent example see (Jeong et al. 2011)]. Alternatively, DCL proteins could sequentially process long microRNA precursors to generate other small RNAs. Selected examples have been described in *Populus trichocarpa*, *Phaseolus vulgaris* and Arabidopsis that could potentially be alternative DCL products with biological activity (Contreras-Cubas et al. 2012; Lu et al. 2008; Zhang et al. 2010).

An interesting example is the miR159 and related miR319 precursor. The majority of the microRNA precursors are recognized from the base of the precursor, where the transition from single to double-stranded RNA is a signal for processing by the DCL1, SE and HYL1 factors (Mateos et al. 2010; Song et al. 2010; Werner et al. 2010). In pre-miR159/319, DCL1 starts cleavage of the precursor from the terminal loop at 20-21nts intervals, until it reaches the position of the mature miR159/319 sequence. In Arabidopsis and *Physcomitrella patens* the intermediate sequences are of extremely low abundance, only detectable by deep sequencing (Addo-Quaye et al. 2009; Bologna et al. 2009). In contrast, other species show higher abundances of one of the equivalent small RNAs, designated as miR159.2, suggesting its functionality. This is consistent with the fact that *P. vulgaris* miR159.2 can be recruited to AGO1-containing complexes and is functional in a heterologous system. Interestingly, the abundance of *Phaseolus* miR159a and miR159.2 did not always correlate under stress conditions (Contreras-Cubas et al. 2012), and possibly in other plants including soybean, rice and maize as well (Li et al. 2011c), suggesting that its abundance is regulated under particular environmental situations (see section *Additional Elements Affecting MicroRNA Activities*).

RISC Activity

Mature microRNAs are recruited to a cytoplasmic multi-protein complex known as RNA-Induced Silencing complex (RISC) composed of several protein factors, but most importantly by the catalytic subunit, a member of the ARGONAUTE (AGO) protein family. *Arabidopsis thaliana* contains ten *AGO* genes, while *Oryza sativa* has eighteen. These diversity has been attributed to specific functions carried out by individual family members (Vaucheret 2008). *In vitro* reconstitution experiments, genetic analysis and deep sequencing of small RNAs specifically associated to AGO proteins, all point to AGO1 as the major AGO protein controlling and executing microRNA activity in plants (Baumberger and Baulcombe 2005; Mi et al. 2008; Vaucheret et al. 2004). A notable exception is miR390, which is bound by AGO7 in the Arabidopsis pathway to initiate processing of TAS3 transcripts, leading to phased-processing of a double-stranded RNA intermediate by DCL4 for tasiRNA production (Montgomery et al. 2008). In the few well-characterized examples available, tasiRNAs are subsequently recruited to AGO-containing complexes to negatively regulate gene expression (Chen 2010), however their possible involvement in pathogen responses has only recently been reported in legumes (Zhai et al. 2011).

A microRNA within the RISC complex recognizes its target mRNA through extensive RNA:RNA base-pairing. This interaction can result in one of two outcomes: AGO1 catalyzes cleavage of the mRNA at the position opposite to nucleotides 10th and 11th of the microRNA as long as these bases are involved in a Watson-Crick base-pairing interaction (Llave et al. 2002b). Alternatively, the mRNA is not degraded but instead the microRNA function is redirected to translational inhibition of its target mRNA. Current evidence indicates that AGO- and microRNA-containing complexes directing translation inhibition are recruited to translating polysomes (Brodersen et al. 2008; Lanet et al. 2009). Although this activity has been observed in selected examples, the extent of its participation during stress responses has not been determined yet.

Due to the extensive base-pairing that has been observed between plant microRNAs and their target transcripts, several studies have widely uncovered the regulatory mechanisms of several microRNAs in the model plant *A. thaliana* using a wide variety of approaches, including genome-scale bioinformatical prediction of target mRNAs (Jones-Rhoades and Bartel 2004) and high throughput sequencing approaches designed to identify mRNAs cleaved by microRNA activity, commonly known as ‘degradome’ or ‘PARE’ (for Parallel Analysis of RNA Ends) (Addo-Quaye et al. 2008; German et al. 2008).

Thus, it would seem that finding the regulated target for a given plant microRNA has become a routine task, with degradome data available for Arabidopsis (Addo-Quaye et al. 2008; German et al. 2008), as well as a variety of plant species (Devers et al. 2011; Li et al. 2010; Pantaleo et al. 2010; Song et al. 2011), which include specific organs, developmental stages, and in our particular case of interest, abiotic stress conditions, such as *Medicago* plants exposed to mercury (Zhou et al. 2012), or *P. euphratica* leaves subjected to drought (Li et al. 2011a), to name some recent reports.

Although these studies provide useful information about those mRNAs subjected to microRNA regulation, however, in addition to the identification of the relevant transcripts and its association to a particular cellular process, it has become increasingly evident that other factors play important roles in modulating the microRNA:target interactions. Features such as the half-life of the microRNA target and the microRNA itself; the competition between one or more coding and non-coding transcripts for miRNA binding; the differences between spatial, temporal and condition-specific expression of both target and microRNA is achieved; or the birth of new miRNAs during evolution and their incorporation into novel regulatory pathways. Although each of these factors could, in principle, be influenced by input from adverse environmental conditions and therefore affect the output of the microRNA pathway, we only know of a few cases, which will be highlighted as we briefly overview these factors.

Additional Elements Affecting MicroRNA Activities

Half-Life of the MicroRNA Target and the MicroRNA Itself

Like other RNA molecules, the microRNA and its mRNA target have a half-life, which is defined by the contribution of different processes, such as the rate of transcription, the rate of processing (i.e. pre-microRNA processing or RNA splicing for mRNAs), or their degradation by exonucleases. In plants, the major 3' to 5' exoribonuclease family for small RNAs is encoded by the *SMALL RNA DEGRADING NUCLEASE* (*SDN*) genes (Ramachandran and Chen 2008). The degradation process is thought to be facilitated by post-transcriptional uridylation of small RNA 3'-ends, possibly mediated by a homolog of MUT68, a *Chlamydomonas reinhardtii* nucleotidyltransferase involved in microRNA and siRNA uridylation (Ibrahim et al. 2010). Interestingly, the levels of Arabidopsis *SDN* transcripts were found to be altered by ABA and drought and even more so by extreme temperatures (Laubinger et al. 2010), suggesting that these stress signals affect the half-life of microRNAs through modulation of the activity of SDN enzymes. In contrast, the XRN2 and XRN3 exoribonucleases in Arabidopsis promote degradation of the loop sequence of microRNA precursor without affecting the mature levels of microRNAs (Gy et al. 2007). Whether lack of these nucleases results in altered responses to stress remains to be addressed.

Similarly, the half-life of the target mRNA may also affect its regulation by microRNAs: it has been shown in animal cells that short-lived mRNAs may be more difficult to process by microRNAs, and conversely a stable target may be more easily recognized and processed by microRNAs, however the relative abundance of the mRNA when compared to that of the microRNA should also be considered (Arvey et al. 2010; Larsson et al. 2010). Ultimately, the microRNA and its targeted transcript should be present in the same cell, at the same time. Finally, in addition to tissue-specificity and developmental stage changes affecting microRNA and/or target mRNA gene co-expression, there is convincing evidence that several small RNAs are transported through the phloem and spread systemically (Yoo et al. 2004), thus changing the distribution of microRNAs and their activity on their targets.

For example, miR399 is induced in response to low-phosphate conditions (see discussion below) and transported from the roots to shoots, however the significance of this movement is not fully understood (Pant et al. 2008).

Competition Between One or More Coding and Non-Coding Transcripts for miRNA Binding

Unlike animal microRNAs that are known each to have several transcripts as targets (Lim et al. 2005), plant microRNAs usually regulate only a few mRNAs, a feature possibly reflecting the extensive microRNA:mRNA base-pairing occurring in plants but contrasting with the limited base-pairing observed in animals (Axtell et al. 2011). The presence of a target mRNA could potentially influence the expression of another transcript already under microRNA control by recruiting the microRNA-loaded RISC and releasing the formerly regulated transcript from its inhibited expression. An example of such regulation in plants occurs in the context of phosphate starvation. miR399 is induced upon phosphate limitation and recognizes the PHO2 mRNA, encoding a E2 ubiquitin conjugase, an important negative regulator of phosphate deprivation responses (Bari et al. 2006). miR399-guided cleavage of PHO2 mRNA allows for proper phosphate limitation responses. To modulate miR399 activity, Arabidopsis and other plants induce the expression of members of the *INDUCED BY PHOSPHATE STARVATION* (IPS) 1 family of RNAs. These transcripts lack an open reading frame, however they contain a sequence partially complementary to miR399, which serves to sequester the available miR399 allowing for accumulation of PHO2 transcripts. IPS1 transcripts, also known as miRNA target mimics effectively compete with PHO2 mRNA for binding of miR399 to modulate its activity and achieve phosphate homeostatic conditions (Franco-Zorrilla et al. 2007). One can envisage that changes in the accumulation patterns of RNA molecules containing microRNA binding sites will affect the regulation of other mRNAs by the same microRNAs. The response to phosphate limitation has acquired this regulatory module and it is expected that other signaling pathways have gained it too. Additionally, we can expect to see competition among different coding target mRNAs for a particular microRNA, effectively establishing communication between mRNAs through a microRNA language, as recently proposed (Salmena et al. 2011).

Birth of New MicroRNAs

The formation of novel *MIR* genes and their subsequent insertion as regulators of target transcripts is a phenomenon that continuously shapes the regulatory mechanisms of plants. Although there are different models to explain the birth of new microRNAs, including a duplication event generating a partially inverted-repeat copy, an interesting model involves the recruitment of miniature inverted repeat elements (MITEs) that become substrates for DCL proteins (Axtell et al. 2011). If the small RNAs generated provide a selective advantage, a novel microRNA could arise. Interestingly, two MITE-derived small RNAs, siR441 and siR446 in rice have been shown to respond to ABA and drought and act as positive regulators of ABA signaling and abiotic stress responses, possibly regulating target mRNAs (Yan et al. 2011). Newly emerged microRNAs may not always possess a regulatory target *ab initio*, thus they may be subjected to selection, and possible extinction if they do not provide an advantage to the plant.

What Is Known About MicroRNAs and Stress?

In principle, given that the microRNA pathway can regulate several processes within the cell, and it can be used to finely regulate gene expression, the effect of stress at multiple levels of this pathway could profoundly influence its outcome. For example, microRNA expression could be specifically regulated by stress to modulate accumulation levels of its downstream mRNA targets, which in turn directly contribute to stress responses. In these cases, defining microRNA accumulation patterns under stress conditions as well as their relevant mRNA targets is essential to understand microRNA contribution to stress responses. Several studies have revealed specific microRNA families involved in stress responses, mainly through the use of high throughput sequencing or microarray analysis (see section *Genome-Wide Analyses* below). Such evidence indicates that microRNAs participate in response to a wide variety of external stimuli including drought, salt, cold, heat and other forms. Although certain microRNA families are repeatedly found (i.e. miR393, miR398, miR169), several studies report novel microRNAs, specifically found in different plant species that could be contributing to stress responses according to their adaptive history.

An additional effect of abiotic stress upon plant processes could be through direct modulation of the factors participating in the microRNA pathways, such as the biogenesis proteins or pathway effectors. Such regulation would influence microRNA activities at the global level and thus would have a distinct and potentially broader consequence on cell processes during stress. Evidence for an effect at this level of microRNA activity has come from the analysis of mutations in factors involved in microRNA metabolism and the corresponding alterations in responses to stress. A brief description of recent advances in these areas will be presented in the following sections.

MicroRNAs Involved in Stress Responses

The specific roles of individual microRNAs during stress responses have been documented, revealing that microRNAs can regulate transcription factors, enzymatic activities or other regulators. The analysis of each of those microRNAs found in Arabidopsis and conserved in other plant species has provided clues as to how plants respond to different stress stimuli. In Arabidopsis seedlings, miR159 was found to accumulate in response to ABA and drought, it controls the levels of two transcription factors, MYB33 and MYB101 that act as positive regulators of plant responses during germination (Reyes and Chua 2007). In contrast to miR159, the levels of miR169 decreased under stress conditions, which allowed for accumulation of its target mRNA encoding NFYA5, a subunit of the trimeric transcription factor Nuclear factor-Y (NFY), and consequently mediating ABA-dependent responses that included stomata closing and a reduction of leaf water loss (Li et al. 2008). Accumulation of miR395 is induced by low-sulfate conditions in the medium (Jones-Rhoades and Bartel 2004), and mediates sulfate homeostasis by cleaving the transcripts encoding the low-affinity sulfur transporter SULTR2;1 and the ATP sulfurylases APS1 and APS4 (Allen et al. 2005; Jones-Rhoades and Bartel 2004). In this case and for miR398, we observe that microRNA regulation is directed towards enzymatic activities and not transcription factors. miR398 regulates the mRNA for the copper-dependent superoxide dismutase CSD1 and cytochrome C oxidase subunit V (Jones-Rhoades and Bartel 2004). It participates in numerous response pathways, including those for oxidative stress, water deficit, salt stress, ultraviolet stress, copper and phosphate deficiency, among others (Zhu et al. 2011). As mentioned earlier, miR399 has been defined as important in phosphate homeostasis in different species (Jones-Rhoades 2011). It controls the levels of PHO2 mRNA, encoding an E2 ubiquitin conjugase, demonstrating that a microRNA can target a regulator different from a transcription factor and even more interesting, that microRNAs can be transported through the phloem to transmit a stress signal (Pant et al. 2008). These examples clearly show the diversity of functions and regulatory circuits in which microRNAs can be involved. The challenge is to understand other relevant microRNA activities to the extent these cases have shown, in plants different from Arabidopsis and under particular stress conditions. A promising way to start is by analyzing the global microRNA expression profile using genome-wide approaches as described next.

Genome-Wide Analyses

Over the last decade, there has been an explosion of microRNA data due to the advent of genome-scale technologies to explore small RNAs by high-throughput sequencing or microarray analysis. In particular, we have seen that stress-related microRNAs have been characterized for diverse plant models. To name a few recent examples, microRNA deep-sequencing data has emerged for *M. truncatula* in response to aluminum toxicity or drought (Chen et al. 2011a; Wang et al. 2011), drought or cold in two *Populus* species (Chen et al. 2011b; Li et al. 2011a), multiple forms of biotic and abiotic stress in soybean (Kulcheski et al. 2011; Li et al. 2011b), among others. This methodology has provided valuable information about how microRNAs are involved in stress responses: which microRNAs are expressed in a specific abiotic/biotic stress condition, and the possibility to identify novel microRNAs as well as other regulatory small RNAs. Due to the large scale of the results obtained, estimations of microRNA differential expression under stress conditions can be inferred as well.

As mentioned earlier, high-throughput sequencing can also be applied to explore the population of mRNA molecules that has been processed by small RNA cleavage. This type of analysis, commonly known as 'Degradome' analysis, provides short sequencing reads that match sites at mRNAs that are prone to degradation and that upon comparison to known or predicted small RNAs can reveal mRNAs under microRNA regulation. This approach provides experimental validation to complement bioinformatical predictions, and has been extended to analyze RNA target processing by tasiRNAs, sites of RNA-dependent DNA methylation (RdDM) implicated in chromatin silencing and even microRNA precursor processing. A recent report in soybean used deep sequencing of small RNAs and degradome analysis to report 26 new miRNAs and 9 miRNAs belonging to conserved microRNAs families and defined 170 transcript targets that could have a function during soybean seed development (Song et al. 2011). Using a similar approach 21 novel microRNAs and 112 mRNA targets were described for *Vitis vinifera* (Pantaleo et al. 2010). In terms of stress responses, a recent study combining deep sequencing of small RNAs and cleaved mRNA targets in maize seedlings exposed to nitrogen deficiency expanded the number of known microRNAs by identifying a total of 99 new loci and confirming 108 target mRNAs (Zhao et al. 2012). Furthermore, responses to drought conditions in leaves of *P. euphratica*, a known stress-resistant woody species, were evaluated through the use of genome-wide strategies: small RNA high-throughput sequencing revealed 58 novel microRNAs (in addition to others already known), while degradome analysis confirmed 47 targets for conserved and novel microRNAs (Li et al. 2011a). Interestingly, this study also used microRNA-specific microarrays to compare with results obtained from deep sequencing and to evaluate the accumulation of microRNAs due to stress conditions.

Small RNA microarrays represent an alternative to high-throughput sequencing to explore the accumulation status of microRNAs. Due to the ability to perform replicates more easily than with sequencing strategies, results obtained with microarrays are more amenable to statistical analysis and can include samples from multiple origins (developmental stages, organs, time points, etc.). A disadvantage is that the evaluation of a global microRNA profile is limited to sequences present in the array and prevents the ability for small RNA discovery. Nevertheless it has been recently used to successfully evaluate microRNA profiles in response to cadmium toxicity (Ding et al. 2011) or during the course of drought at two developmental stages in rice (Zhou et al. 2010), to explore shock drought in *Triticum dicoccoides* leaves and roots (Kantar et al. 2011) or to compare two cotton cultivars differing in their susceptibility to salt stress (Yin et al. 2011), to mention a few recent examples that show how a variety of conditions can be assessed using microarrays. A combination of genome-wide technologies has allowed for a glimpse of the intricate microRNA populations present during stress conditions in plants. The results are now available and will help to determine the contribution of individual microRNAs to plant responses.

Genetic Screens

HYL1 was the first factor involved in the microRNA pathway that was originally isolated as a mutant with a defect in its response to stress. *hyl1* plants are hypersensitive to ABA in addition to other developmental defects, and less sensitivity to auxin and cytokinin (Lu and Fedoroff 2000). Later on, it was recognized that HYL1 participates in precursor processing by aiding DCL1 to correctly recognize cleavage sites together with SE along the precursor secondary structure (Dong et al. 2008). These findings suggested that impairment of microRNA biogenesis might cause a deficiency in a particular microRNA and a consequent disruption of its role in modulating ABA responses. Alternatively, the defect on stress responses could be due to an indirect effect through general failure to accumulate the appropriate amounts of the microRNA population within the cell. In either case, it is expected that mutations in other elements participating in microRNA biogenesis show similar phenotypes.

For example, the *ABA HYPERSENSITIVE 1* (ABH1) gene identified as a mutant in CBP80, the large subunit of the cap binding complex (CBP), showed defects in ABA sensitivity, and reduced wilting upon drought treatment (Hugouvieux et al. 2001). Interestingly, mutations in the CBP20 subunit also turned out to be resistant to drought and ABA hypersensitive during germination (Papp et al. 2004). Because *cbp20* and *cbp80* mutants contained reduced levels of mature microRNA but increased levels of their precursors (Kim et al. 2008), it was postulated that correct recognition of the primary transcript for multiple microRNAs could be impaired in the mutants and in turn result in diminished levels of microRNAs involved in ABA responses, such as miR159 (Reyes and Chua 2007). Another screen recovered *ABA supersensitive during germination* (*absg*) mutants, revealing new alleles for *DCL1* and *HEN1* in Arabidopsis, resulting in hypersensitivity to ABA as well as enhanced sensitivity to drought and salt stress, and increased expression of stress-responsive genes (Zhang et al. 2008). That microRNA biogenesis factors are involved in stress responses is further supported by the finding that the mRNAs encoding *Mt*DCL1 and *Mt*AGO1 increased their accumulation, while microRNAs known to regulate these transcripts, namely miR162 and miR168, decreased their levels in *M. truncatula* roots in response to water deficit (Capitao et al. 2011). Moreover, identification of *DCL* and *AGO* genes in maize and subsequent expression profiling indicated that certain members of these families are affected by osmotic and salt stress (Qian et al. 2011). In contrast to these findings, FBW2 encodes an F-box protein that negatively regulates the levels of Arabidopsis AGO1 protein. The *fbw2* mutant exhibits decreased ABA sensitivity, the opposite phenotype to many microRNA biogenesis mutants, correlating with increased AGO1 abundance (Earley et al. 2010). These results reinforce the idea that the microRNA regulatory pathway is intimately involved in stress responses, and while many examples of specific microRNAs participating at different levels in stress responses have been identified (section *MicroRNAs Involved in Stress Responses*), it is still uncertain what is the largest effect of mutants in this pathway: the absence of certain microRNAs or impairing the overall microRNA levels. While precise and detailed analyses could address this question it is likely to be a combination of both effects what determines the outcome of plant response to adverse conditions.

Future Perspectives

Much has been learned from the study of microRNA expression in the context of stress affecting major plant models, including Arabidopsis, rice, maize, and crops such as legumes, and others. The current evidence has revealed that during a wide variety of adverse conditions the entire landscape of the microRNA population changes according to the condition imposed. In certain cases these changes result in a large variation in microRNA abundance while others are subtler. This scenario should then be reflected in the abundance of those transcripts regulated by the action of microRNAs and possibly other small regulatory RNA molecules, such as tasiRNAs or small RNAs arising from alternate processing of specific precursors. While a large amount of work has been put into identifying the microRNA contribution to this regulatory circuit, much less has been revealed about their mRNA targets. Future work will benefit from large-scale analysis such as degradome or transcriptome analysis to unveil these targets based on the use of a variety of experimental and bioinformatical tools. A subsequent analysis of the changes in mRNA abundance caused by stress conditions should be aimed at integrating the individual effects of microRNA regulation observed into regulatory networks to reveal the effect on cellular pathways and metabolism and how they are ultimately affecting the plant responses to abiotic stress conditions.

Another factor contributing to the outcome of microRNA regulation is the relative abundance of these and other small RNAs within the cell. As we mentioned above, several factors contribute to the final concentration of the small RNA molecules, including, but not limited to, transcription rate, half-life, competition among target mRNAs for microRNA availability, and possibly others yet to be discovered. The cell machineries responsible for these processes might as well be regulated by stress conditions. We exemplified this aspect by highlighting the susceptibility to stress developed by mutants in a few of the microRNA biogenesis factors (section *Genetic Screens*), however other factors involved in regulating microRNA abundance might turn out to be affected by stress and thus in turn alter, directly or indirectly, the responses to those stimuli that affected their function in the first place.

In addition to conserved microRNAs it will be important to analyze less conserved small RNAs that may play important roles in crop species, potentially regulating processes specific to particular plant species. However as it has been mentioned, identification of targets will be essential to place regulatory pathways in these other plants.

Acknowledgements G. S.-V. received a PhD scholarship from CONACyT. Work on microRNAs in the laboratory of the authors is partially supported by grants from DGAPA-PAPIIT (IN-205112) and CONACyT (CB2010-151571) to JLR.

References

Addo-Quaye C, Eshoo TW, Bartel DP, Axtell MJ (2008) Endogenous siRNA and miRNA targets identified by sequencing of the Arabidopsis degradome. Curr Biol 18:758–762

Addo-Quaye C, Snyder JA, Park YB, Li YF, Sunkar R, Axtell MJ (2009) Sliced microRNA targets and precise loop-first processing of MIR319 hairpins revealed by analysis of the Physcomitrella patens degradome. RNA 15:2112–2121

Allen E, Xie Z, Gustafson AM, Carrington JC (2005) MicroRNA-directed phasing during trans-acting siRNA biogenesis in plants. Cell 121:207–221

Arvey A, Larsson E, Sander C, Leslie CS, Marks DS (2010) Target mRNA abundance dilutes microRNA and siRNA activity. Mol Syst Biol 6:363

Axtell MJ, Westholm JO, Lai EC (2011) Vive la difference: biogenesis and evolution of microRNAs in plants and animals. Genome Biol 12:221

Bari R, Datt Pant B, Stitt M, Scheible WR (2006) PHO2, microRNA399, and PHR1 define a phosphate-signaling pathway in plants. Plant Physiol 141:988–999

Baumberger N, Baulcombe DC (2005) Arabidopsis ARGONAUTE1 is an RNA Slicer that selectively recruits microRNAs and short interfering RNAs. Proc Natl Acad Sci USA 102:11928–11933

Bologna NG, Mateos JL, Bresso EG, Palatnik JF (2009) A loop-to-base processing mechanism underlies the biogenesis of plant microRNAs miR319 and miR159. EMBO J 28:3646–3656

Brodersen P, Sakvarelidze-Achard L, Bruun-Rasmussen M, Dunoyer P, Yamamoto YY, Sieburth L, Voinnet O (2008) Widespread translational inhibition by plant miRNAs and siRNAs. Science 320:1185–1190

Capitao C, Paiva JA, Santos DM, Fevereiro P (2011) In Medicago truncatula, water deficit modulates the transcript accumulation of components of small RNA pathways. BMC Plant Biol 11:79

Chen X (2010) Small RNAs - secrets and surprises of the genome. Plant J 61:941–958

Chen L, Wang T, Zhao M, Tian Q, Zhang WH (2011a) Identification of aluminum-responsive microRNAs in Medicago truncatula by genome-wide high-throughput sequencing. Planta 235 (2):375–386

Chen L, Zhang Y, Ren Y, Xu J, Zhang Z, Wang Y (2011b) Genome-wide identification of cold-responsive and new microRNAs in Populus tomentosa by high-throughput sequencing. Biochem Biophys Res Commun 417(2):892–896

Contreras-Cubas C, Rabanal FA, Arenas-Huertero C, Ortiz MA, Covarrubias AA, Reyes JL (2012) The Phaseolus vulgaris miR159a precursor encodes a second differentially expressed microRNA. Plant Mol Biol 80:103–115

Devers EA, Branscheid A, May P, Krajinski F (2011) Stars and symbiosis: microRNA- and microRNA*-mediated transcript cleavage involved in arbuscular mycorrhizal symbiosis. Plant Physiol 156:1990–2010

Ding Y, Chen Z, Zhu C (2011) Microarray-based analysis of cadmium-responsive microRNAs in rice (Oryza sativa). J Exp Bot 62:3563–3573
Dong Z, Han MH, Fedoroff N (2008) The RNA-binding proteins HYL1 and SE promote accurate in vitro processing of pri-miRNA by DCL1. Proc Natl Acad Sci USA 105:9970–9975
Earley K, Smith M, Weber R, Gregory B, Poethig R (2010) An endogenous F-box protein regulates ARGONAUTE1 in Arabidopsis thaliana. Silence 1:15
Franco-Zorrilla JM, Valli A, Todesco M, Mateos I, Puga MI, Rubio-Somoza I, Leyva A, Weigel D, Garcia JA, Paz-Ares J (2007) Target mimicry provides a new mechanism for regulation of microRNA activity. Nat Genet 39:1033–1037
German MA, Pillay M, Jeong DH, Hetawal A, Luo S, Janardhanan P, Kannan V, Rymarquis LA, Nobuta K, German R (2008) Global identification of microRNA-target RNA pairs by parallel analysis of RNA ends. Nat Biotechnol 26:941–946
Gy I, Gasciolli V, Lauressergues D, Morel JB, Gombert J, Proux F, Proux C, Vaucheret H, Mallory AC (2007) Arabidopsis FIERY1, XRN2, and XRN3 are endogenous RNA silencing suppressors. Plant Cell 19:3451–3461
Hugouvieux V, Kwak JM, Schroeder JI (2001) An mRNA cap binding protein, ABH1, modulates early abscisic acid signal transduction in Arabidopsis. Cell 106:477–487
Ibrahim F, Rymarquis LA, Kim EJ, Becker J, Balassa E, Green PJ, Cerutti H (2010) Uridylation of mature miRNAs and siRNAs by the MUT68 nucleotidyltransferase promotes their degradation in Chlamydomonas. Proc Natl Acad Sci USA 107:3906–3911
Jeong DH, Park S, Zhai J, Gurazada SG, De Paoli E, Meyers BC, Green PJ (2011) Massive analysis of rice small RNAs: mechanistic implications of regulated microRNAs and variants for differential target RNA cleavage. Plant Cell 23(12):4185–4207
Jones-Rhoades MW (2011) Conservation and divergence in plant microRNAs. Plant Mol Biol 80(1):3–16
Jones-Rhoades MW, Bartel DP (2004) Computational identification of plant microRNAs and their targets, including a stress-induced miRNA. Mol Cell 14:787–799
Kantar M, Lucas SJ, Budak H (2011) miRNA expression patterns of Triticum dicoccoides in response to shock drought stress. Planta 233:471–484
Kim S, Yang JY, Xu J, Jang IC, Prigge MJ, Chua NH (2008) Two cap-binding proteins CBP20 and CBP80 are involved in processing primary MicroRNAs. Plant Cell Physiol 49:1634–1644
Kulcheski FR, de Oliveira LF, Molina LG, Almerao MP, Rodrigues FA, Marcolino J, Barbosa JF, Stolf-Moreira R, Nepomuceno AL, Marcelino-Guimaraes FC (2011) Identification of novel soybean microRNAs involved in abiotic and biotic stresses. BMC Genomics 12:307
Kurihara Y, Watanabe Y (2004) Arabidopsis micro-RNA biogenesis through Dicer-like 1 protein functions. Proc Natl Acad Sci USA 101:12753–12758
Lagos-Quintana M, Rauhut R, Lendeckel W, Tuschl T (2001) Identification of novel genes coding for small expressed RNAs. Science 294:853–858
Lanet E, Delannoy E, Sormani R, Floris M, Brodersen P, Crete P, Voinnet O, Robaglia C (2009) Biochemical evidence for translational repression by Arabidopsis microRNAs. Plant Cell 21(6):1762–1768
Larsson E, Sander C, Marks D (2010) mRNA turnover rate limits siRNA and microRNA efficacy. Mol Syst Biol 6:433
Lau NC, Lim LP, Weinstein EG, Bartel DP (2001) An abundant class of tiny RNAs with probable regulatory roles in Caenorhabditis elegans. Science 294:858–862
Laubinger S, Zeller G, Henz SR, Buechel S, Sachsenberg T, Wang JW, Ratsch G, Weigel D (2010) Global effects of the small RNA biogenesis machinery on the Arabidopsis thaliana transcriptome. Proc Natl Acad Sci USA 107:17466–17473
Law JA, Jacobsen SE (2010) Establishing, maintaining and modifying DNA methylation patterns in plants and animals. Nat Rev Genet 11:204–220
Lee RC, Ambros V (2001) An extensive class of small RNAs in Caenorhabditis elegans. Science 294:862–864

Lee RC, Feinbaum RL, Ambros V (1993) The C. elegans heterochronic gene lin-4 encodes small RNAs with antisense complementarity to lin-14. Cell 75:843–854

Lee Y, Kim M, Han J, Yeom KH, Lee S, Baek SH, Kim VN (2004) MicroRNA genes are transcribed by RNA polymerase II. EMBO J 23:4051–4060

Li WX, Oono Y, Zhu J, He XJ, Wu JM, Iida K, Lu XY, Cui X, Jin H, Zhu JK (2008) The Arabidopsis NFYA5 transcription factor is regulated transcriptionally and posttranscriptionally to promote drought resistance. Plant Cell 20:2238–2251

Li YF, Zheng Y, Addo-Quaye C, Zhang L, Saini A, Jagadeeswaran G, Axtell MJ, Zhang W, Sunkar R (2010) Transcriptome-wide identification of microRNA targets in rice. Plant J 62:742–759

Li B, Qin Y, Duan H, Yin W, Xia X (2011a) Genome-wide characterization of new and drought stress responsive microRNAs in Populus euphratica. J Exp Bot 62:3765–3779

Li H, Dong Y, Yin H, Wang N, Yang J, Liu X, Wang Y, Wu J, Li X (2011b) Characterization of the stress associated microRNAs in Glycine max by deep sequencing. BMC Plant Biol 11:170

Li Y, Li C, Ding G, Jin Y (2011c) Evolution of MIR159/319 microRNA genes and their post-transcriptional regulatory link to siRNA pathways. BMC Evol Biol 11:122

Lim LP, Lau NC, Garrett-Engele P, Grimson A, Schelter JM, Castle J, Bartel DP, Linsley PS, Johnson JM (2005) Microarray analysis shows that some microRNAs downregulate large numbers of target mRNAs. Nature 433(7027):769–773

Llave C, Kasschau KD, Rector MA, Carrington JC (2002a) Endogenous and silencing-associated small RNAs in plants. Plant Cell 14:1605–1619

Llave C, Xie Z, Kasschau KD, Carrington JC (2002b) Cleavage of scarecrow-like mRNA targets directed by a class of Arabidopsis miRNA. Science 297:2053–2056

Lu C, Fedoroff N (2000) A mutation in the Arabidopsis HYL1 gene encoding a dsRNA binding protein affects responses to abscisic acid, auxin, and cytokinin. Plant Cell 12:2351–2366

Lu S, Sun YH, Chiang VL (2008) Stress-responsive microRNAs in Populus. Plant J 55:131–151

Mateos JL, Bologna NG, Chorostecki U, Palatnik JF (2010) Identification of microRNA processing determinants by random mutagenesis of Arabidopsis MIR172a precursor. Curr Biol 20:49–54

Mi S, Cai T, Hu Y, Chen Y, Hodges E, Ni F, Wu L, Li S, Zhou H, Long C (2008) Sorting of small RNAs into Arabidopsis argonaute complexes is directed by the 5' terminal nucleotide. Cell 133:116–127

Montgomery TA, Howell MD, Cuperus JT, Li D, Hansen JE, Alexander AL, Chapman EJ, Fahlgren N, Allen E, Carrington JC (2008) Specificity of ARGONAUTE7-miR390 interaction and dual functionality in TAS3 trans-acting siRNA formation. Cell 133:128–141

Okamura K, Phillips MD, Tyler DM, Duan H, Chou YT, Lai EC (2008) The regulatory activity of microRNA* species has substantial influence on microRNA and 3' UTR evolution. Nat Struct Mol Biol 15:354–363

Pant BD, Buhtz A, Kehr J, Scheible WR (2008) MicroRNA399 is a long-distance signal for the regulation of plant phosphate homeostasis. Plant J 53:731–738

Pant BD, Musialak-Lange M, Nuc P, May P, Buhtz A, Kehr J, Walther D, Scheible WR (2009) Identification of nutrient-responsive Arabidopsis and rapeseed microRNAs by comprehensive real-time polymerase chain reaction profiling and small RNA sequencing. Plant Physiol 150:1541–1555

Pantaleo V, Szittya G, Moxon S, Miozzi L, Moulton V, Dalmay T, Burgyan J (2010) Identification of grapevine microRNAs and their targets using high-throughput sequencing and degradome analysis. Plant J 62:960–976

Papp I, Mur LA, Dalmadi A, Dulai S, Koncz C (2004) A mutation in the cap binding protein 20 gene confers drought tolerance to Arabidopsis. Plant Mol Biol 55:679–686

Parizotto EA, Dunoyer P, Rahm N, Himber C, Voinnet O (2004) In vivo investigation of the transcription, processing, endonucleolytic activity, and functional relevance of the spatial distribution of a plant miRNA. Genes Dev 18:2237–2242

Park W, Li J, Song R, Messing J, Chen X (2002) CARPEL FACTORY, a Dicer homolog, and HEN1, a novel protein, act in microRNA metabolism in Arabidopsis thaliana. Curr Biol 12:1484–1495

Park MY, Wu G, Gonzalez-Sulser A, Vaucheret H, Poethig RS (2005) Nuclear processing and export of microRNAs in Arabidopsis. Proc Natl Acad Sci USA 102:3691–3696

Qian Y, Cheng Y, Cheng X, Jiang H, Zhu S, Cheng B (2011) Identification and characterization of Dicer-like, Argonaute and RNA-dependent RNA polymerase gene families in maize. Plant Cell Rep 30:1347–1363

Ramachandran V, Chen X (2008) Degradation of microRNAs by a family of exoribonucleases in Arabidopsis. Science 321:1490–1492

Reyes JL, Chua NH (2007) ABA induction of miR159 controls transcript levels of two MYB factors during Arabidopsis seed germination. Plant J 49:592–606

Rhoades MW, Reinhart BJ, Lim LP, Burge CB, Bartel B, Bartel DP (2002) Prediction of plant microRNA targets. Cell 110:513–520

Salmena L, Poliseno L, Tay Y, Kats L, Pandolfi PP (2011) A ceRNA hypothesis: the Rosetta Stone of a hidden RNA language? Cell 146:353–358

Song L, Axtell MJ, Fedoroff NV (2010) RNA secondary structural determinants of miRNA precursor processing in Arabidopsis. Curr Biol 20:37–41

Song QX, Liu YF, Hu XY, Zhang WK, Ma B, Chen SY, Zhang JS (2011) Identification of miRNAs and their target genes in developing soybean seeds by deep sequencing. BMC Plant Biol 11:5

Vaucheret H (2008) Plant ARGONAUTES. Trends Plant Sci 13:350–358

Vaucheret H, Vazquez F, Crete P, Bartel DP (2004) The action of ARGONAUTE1 in the miRNA pathway and its regulation by the miRNA pathway are crucial for plant development. Genes Dev 8:1187–1197

Vazquez F, Gasciolli V, Crete P, Vaucheret H (2004) The nuclear dsRNA binding protein HYL1 is required for microRNA accumulation and plant development, but not posttranscriptional transgene silencing. Curr Biol 14:346–351

Vazquez F, Legrand S, Windels D (2010) The biosynthetic pathways and biological scopes of plant small RNAs. Trends Plant Sci 15:337–345

Wang T, Chen L, Zhao M, Tian Q, Zhang WH (2011) Identification of drought-responsive microRNAs in Medicago truncatula by genome-wide high-throughput sequencing. BMC Genomics 12:367

Werner S, Wollmann H, Schneeberger K, Weigel D (2010) Structure determinants for accurate processing of miR172a in Arabidopsis thaliana. Curr Biol 20:42–48

Wong CE, Zhao YT, Wang XJ, Croft L, Wang ZH, Haerizadeh F, Mattick JS, Singh MB, Carroll BJ, Bhalla PL (2011) MicroRNAs in the shoot apical meristem of soybean. J Exp Bot 62:2495–2506

Yan Y, Zhang Y, Yang K, Sun Z, Fu Y, Chen X, Fang R (2011) Small RNAs from MITE-derived stem-loop precursors regulate abscisic acid signaling and abiotic stress responses in rice. Plant J 65:820–828

Yang L, Liu Z, Lu F, Dong A, Huang H (2006) SERRATE is a novel nuclear regulator in primary microRNA processing in Arabidopsis. Plant J 47:841–850

Yin Z, Li Y, Yu J, Liu Y, Li C, Han X, Shen F (2011) Difference in miRNA expression profiles between two cotton cultivars with distinct salt sensitivity. Mol Biol Rep 39(4):4961–4970

Yoo BC, Kragler F, Varkonyi-Gasic E, Haywood V, Archer-Evans S, Lee YM, Lough TJ, Lucas WJ (2004) A systemic small RNA signaling system in plants. Plant Cell 16:1979–2000

Yu B, Yang Z, Li J, Minakhina S, Yang M, Padgett RW, Steward R, Chen X (2005) Methylation as a crucial step in plant microRNA biogenesis. Science 307:932–935

Yu B, Bi L, Zheng B, Ji L, Chevalier D, Agarwal M, Ramachandran V, Li W, Lagrange T, Walker JC (2008) The FHA domain proteins DAWDLE in Arabidopsis and SNIP1 in humans act in small RNA biogenesis. Proc Natl Acad Sci USA 105:10073–10078

Zhai J, Jeong DH, De Paoli E, Park S, Rosen BD, Li Y, Gonzalez AJ, Yan Z, Kitto SL, Grusak MA (2011) MicroRNAs as master regulators of the plant NB-LRR defense gene family via the production of phased, trans-acting siRNAs. Genes Dev 25:2540–2553

Zhang JF, Yuan LJ, Shao Y, Du W, Yan DW, Lu YT (2008) The disturbance of small RNA pathways enhanced abscisic acid response and multiple stress responses in Arabidopsis. Plant Cell Environ 31:562–574

Zhang W, Gao S, Zhou X, Xia J, Chellappan P, Zhang X, Jin H (2010) Multiple distinct small RNAs originate from the same microRNA precursors. Genome Biol 11:R81

Zhao M, Tai H, Sun S, Zhang F, Xu Y, Li WX (2012) Cloning and characterization of maize miRNAs involved in responses to nitrogen deficiency. PLoS One 7:e29669

Zhou L, Liu Y, Liu Z, Kong D, Duan M, Luo L (2010) Genome-wide identification and analysis of drought-responsive microRNAs in Oryza sativa. J Expt Bot 61:4157–4168

Zhou ZS, Zeng HQ, Liu ZP, Yang ZM (2012) Genome-wide identification of Medicago truncatula microRNAs and their targets reveals their differential regulation by heavy metal. Plant Cell Environ 35:86–99

Zhu C, Ding Y, Liu H (2011) MiR398 and plant stress responses. Physiol Plant 143:1–9

Chapter 4
Signal Transduction and Regulatory Networks in Plant-Pathogen Interaction: A Proteomics Perspective

M.Z. Abdin, Mather Ali Khan, Athar Ali, Pravej Alam, Altaf Ahmad, and Maryam Sarwat

Introduction

Plant diseases are amongst the major limiting factors of agricultural crop production worldwide. Depending upon the time of infection and severity of the disease, they can cause average yield losses of about 10–90 %. Plant disease directly or indirectly affects the life of human, as it may cause famine, mass migration and even death. For example, Irish potato famine of 1845–1846 killed hundreds of thousands of people. This event initiated a large scale migration. Within decade that follows the population of Ireland dropped from 8 million to 4 million (Ristaino 2002). Plant diseases significantly influence world economy, as crop plants make up large proportion of the world's economy, and in many countries constitute main sustenance for humans. According to one estimate, plant diseases could cost the US alone $33 billion per year (Maor and Shirasu 2005). To meet the ever increasing food demands of the rapidly increasing population, crop production will need to increase by 50 % by 2025 (Khush 2001). Currently, worldwide crop losses due to diseases are estimated to exceed $140 billion (Shani et al. 2006). Although application of fungicides and pesticides has helped in controlling plant diseases, chemical control is economically costly as well as environmentally undesirable. Keeping in view of the global food scarcity, there is, hence, an urgent need to develop crop plants with increased biotic stress tolerance so as to meet the global

M.Z. Abdin (✉) • M.A. Khan • A. Ali • P. Alam
Department of Biotechnology, Centre for Transgenic Plant Development,
Jamia Hamdard, New Delhi 110062, India
e-mail: mzabdin@rediffmail.com

A. Ahmad
Department of Botany, Faculty of Science, Jamia Hamdard, New Delhi, India

M. Sarwat
Pharmaceutical Biotechnology, Amity Institute of Pharmacy,
Amity University, NOIDA, India

M. Sarwat et al. (eds.), *Stress Signaling in Plants: Genomics and Proteomics Perspective, Volume 1*, DOI 10.1007/978-1-4614-6372-6_4,

food demands. A detailed study of the molecular interactions between crops plants and their pathogens would, therefore, be of primary importance for devising new strategies based on plants self defense mechanisms to develop crops with increased disease tolerance for sustainable agricultural production.

When a plant and a pathogen come into contact, close communications occur between the two organisms (Hammond-Kosack and Jones 2000). Pathogen activities focus on colonization of the host and utilization of its resources, while plants are adapted to detect the presence of pathogens and to respond accordingly with antimicrobial defenses and other stress responses. During the long process of host-pathogen coevolution, plants have developed various elaborate mechanisms to ward off pathogen attack. Whereas some of these defense mechanisms are preformed and provide physical and chemical barriers to hinder pathogen infection, others are induced only after pathogen attack (Yang et al. 1997). Intercellular detection of the pathogen activates the first line of defense, termed innate or basal resistance, which involves recognition of evolutionarily conserved and essential features of pathogens such as bacterial lipopolysaccharides (LPS) or fungal chitin. These Extracellular signals are perceived by host cells through plasma membrane (PM) receptors that transduce the signals to an intracellular signal transduction cascade. It ends in the activation of transcription of the appropriate set of genes, which results into alteration or modification of cellular metabolism, accumulation of barrier forming substances (thickening of cell walls) and production of anti microbial compounds. In most cases, the transduction of signal relies on post translational modifications of the signaling proteins and the generation of so called secondary messenger molecules.

Among all post-translational modifications, phosphorylation has been studied most intensively (Pawson and Scott 2005), which can lead to changes in conformation, protein–protein interaction and protein activity. In eukaryotic cells, protein phosphorylation occurs predominantly on serine, threonine and tyrosine residues, but has also been described to occur on aspartate and histidine residues. The regulatory mechanisms of plant–pathogen interaction are extremely complex and dynamic, and the ongoing interactions between the pathogen and the plant are difficult to monitor with more traditional genetic and biochemical methods. The two approaches that are most promising for understanding the full network of the responses are microarray and proteomic analyses. Both permit a global analysis of cellular regulation while the microarray is restricted to the analysis of gene expression. The proteomics follows the accumulation and modification of proteins directly responsible for final cellular responses. Recent advancement in liquid chromatography–tandem mass spectrometry (LC–MS/MS) has greatly improved the throughput and sensitivity of protein measurements. In order to efficiently describe the status of phosphorylated molecules, a variety of enrichment strategies for phosphorylated peptides have been developed. The most commonly used are based on affinity purification of phosphoryl-group containing peptides and include immobilized metal affinity chromatography (IMAC), strong cation exchange chromatography (SCX) and metal oxide affinity chromatography (MOAC) (Mithoe and Menke 2011).

Over the last few years, significant progress has been made in understanding the signaling processes involved in plant-pathogen interactions. In this chapter, we focus on signaling pathways involved in plant defense against pathogens and the role of proteomics technology in understanding the underlying mechanisms.

Plant Signal Transduction in Plant Defense Against Pathogens

Plants have integrated signaling system that mediate the perception and responses to the hormones, nutrients, environmental and stresses that govern plant growth and development. Interactions between plants and pathogens induce a series of plant defense responses (Hammond-Kosack and Jones 1996). The rate at which the plant cell mobilise its defenses often determines whether it survives or succumbs to the attack. Therefore, highly sensitive perception systems for either pathogen derived (exogenous) or plant-derived (endogenous) elicitors are the key to successful plant pathogen defense. The sensing of stress signals and their transduction into appropriate responses is crucial for the adaptation and survival of plants. Plant receptors are instrumental for signal recognition and initiation of an intracellular signal transduction cascade mediating activation of multifaceted defense reactions, both in host and non-host incompatible plant pathogen interactions.

The current knowledge of plant signal transduction pathways has come from the identification of the sensors and receptors that perceive the signal, the transcription factors and target genes that coordinate the response (Hammond-Kosack and Jones 1996). The activation of defense responses in plants is initiated by host recognition of pathogen-encoded molecules called elicitors (e.g., microbial proteins, small peptides, and oligosaccharides, etc.). The term 'elicitor', originally coined for compounds that induce accumulation of antimicrobial phytoalexins in plants, is now commonly applied to agents stimulating any type of defense response (Ebel and Scheel 1997). Elicitors of diverse chemical nature and from a variety of different plant pathogenic microbes have been characterized and shown to trigger defense responses in intact plants or cultured plant cells. These elicitors include (poly) peptides, glycoproteins, lipids and oligosaccharides. Binding of the elicitor ligand to its receptor initiates a signal transduction cascade that may involve protein phosphorylation, ion fluxes, reactive oxygen species (ROS), nitric oxide (NO) and other signaling events.

Ion Fluxes and Reactive Oxygen Species

The earliest reactions of plant cells include changes in plasma membrane permeability leading to calcium and proton influx and potassium and chloride efflux (McDowell and Dangl 2000). Various fungal and bacterial elicitors have been reported to trigger fluxes of H^+, K^+, Cl^-, and Ca^{2+} across the plasma membrane (Atkinson et al. 1996; Mathieu et al. 1991; Bach et al. 1993; Kuchitsu et al. 1993; Popham et al. 1995). In suspension cells of parsley, a transient influx of Ca^{2+} and H^+ and an efflux of K^+ and Cl^- are initiated within two to five minutes after the addition of a fungal oligopeptide elicitor (Hahlbrock et al. 1995). Ion fluxes subsequently induce extracellular production of reactive oxygen intermediates, such as superoxide (O^{2-}), hydrogen peroxide (H_2O_2) and hydroxyl free radical (OH^-), known as oxidative burst, catalyzed by a plasma membrane-located NADPH oxidase and/or apoplastic-localized peroxidases (Somssich and Hahlbrock 1998). Oxidative burst is a central component of plants defense machinery (Alvarez et al. 1998). The generation of ROS is likely dependent on the activation of a plasma membrane NADPH oxidase similar to that present in mammalian phagocytes. Using the mammalian system as a model, homologues of the large gp91-phox protein of NADPH oxidase have been cloned from several plant species including *Arabidopsis* and rice (Desikan et al. 1998; Torres et al. 1998). Eight such sequences have been identified in the *Arabidopsis* genome (The Arabidopsis Genome Initiative 2000).

Interestingly, plant homologues contain calcium-binding EF-hand regions, suggesting that calcium may be important in the regulation of their activity (Desikan et al. 1997). Immunological evidence points to the presence of the NADPH oxidase cytoplasmic peptides too, in several species of plants including *Arabidopsis* (Desikan et al. 1996), tobacco (*Nicotiana tabacum*), tomato (*Lycopersicon esculentum*), soybean (*Glycine max*) and cotton (*Gossypium hirsutum*) (Dwyer et al. 1996; Tenhaken et al. 1995; Xing et al. 1997).

Cell wall peroxidases have also been reported to be involved in ROS generation following pathogen challenge (Bolwell et al. 1995). For example, in French bean (*Phaseolus vulgaris*) ROS production in response to a cell wall elicitor from *Colletotrichum lindemuthianum* was dependent on an exocellular peroxidase, and *Arabidopsis* plants expressing an antisense peroxidase from French bean exhibit enhanced disease susceptibility (Bolwell 1999). Therefore, it can be seen that there are several potential sources of ROS in plant tissues, and future research should aim to elucidate the role of distinct sources of ROS not only in plant defense, but also in response to a variety of abiotic stresses.

Superoxide anion and H_2O_2 generated during the oxidative burst play multiple roles in plant defense responses. During a hypersensitive response (HR), a highly localized accumulation of H_2O_2 was found in the lettuce cell walls adjacent to invading bacteria (Bestwick et al. 1997). In addition, constitutive expression of an H_2O_2 generating glucose oxidase in the transgenic potato was shown to confer enhanced resistance to the bacterial pathogen, *Erwinia carotovora* pv. *carotovora* and the fungal pathogen, *Phytophthora infestans* (Wu et al. 1995). H_2O_2 was also demonstrated to have direct antimicrobial activity (Peng and Kuc 1992) and to contribute to cell wall reinforcement by stimulating lignification and crosslinking of cell wall hydroxyproline-rich glycoproteins (Bradley et al. 1992; Brisson et al. 1994). Furthermore, superoxide anion and H_2O_2 may also act as secondary messengers to induce plant defense-related genes (Levine et al. 1994; Green and Fluhr 1995) and hypersensitive host cell death (Doke 1983a; Doke 1983b; Doke and Ohashi 1988; Levine et al. 1994). Elicitor-stimulated superoxide anion from the oxidative burst was also shown to be essential in triggering defense gene activation and phytoalexin synthesis in parsley (Jabs et al. 1997).

Cellular Generation of Nitric Oxide

Nitric Oxide (NO) like ROS, is an important signaling molecule that is rapidly generated after recognition of pathogens (Perchepied et al. 2010). It is a small gaseous radical with broad spectrum of regulatory functions in lateral root development, germination, leaf expansion, stomatal closure, flowering, hormonal signaling, defense against biotic and abiotic stresses and cell death (Leitner et al. 2009). The sources of NO synthesis in plants include nitrate reductase (NR) dependent NO formation, oxidation of arginine to citrulline by NO Synthase (NOS) like activity, and a non-enzymatic NO generation system in the apoplast (Leitner et al. 2009). Although a number of studies had demonstrated NOS-like activity in plants, no gene or protein that has a sequence similar to known mammalian-type NOS has been establish in plants (Leitner et al. 2009). Guo et al. (2003) reported a NOS-like enzyme activity from *Arabidopsis thaliana* (At NOS1) with a sequence similar to a protein that has been implicated in NO synthesis in the snail *Helix pomatia.* Recently, the only postulated plant NOS (AtNOA1/RIF1) has been shown to have no NOS activity (Moreau et al. 2008). Instead, it is the chloroplast-targeted GTPase essential for proper ribosome assembly (Moreau et al. 2008). Mutation in this gene leads to reduced NO accumulation, probably because of its rapid reaction with the elevated amounts of ROS observed in the Atnoa1 mutant (Moreau et al. 2008). *Arabidopsis* mutant *noa*1, however, is still useful for its phenotype, which shows reduced levels of NO in plant growth, fertility, hormonal signaling, salt tolerance, and plant-pathogen responses. Knocking out or down NOA1 expression provides a powerful tool to analyze NO function (Asai et al. 2008). So, the identification and characterization of NO-producing enzymes in plants, other than NR, still remains a challenging tasks for plant biologists.

Recently, Perchepied et al. (2010) reported that NO production in *Arabidopsis* leaf was significantly reduced by the mammalian NO synthase (NOS) inhibitor L-NAME (37% inhibition). They further demonstrated that like ROS, NO is an early-induced signal during the interaction between *Arabidopsis* and *S. sclerotiorum.* In order to genetically determine the role of the signaling molecules during the interaction, mutants altered in their production were tested for their response to *S. sclerotiorum* and found that NO synthesis was strongly reduced in *noa1* mutant (83% inhibition) and also in *nia1 nia2* mutant (62% inhibition). Therefore, these results demonstrate that NO might have an important role in disease resistance to *S. sclerotiorum.* To further analyze the role of ROS and NO in the activation of defenses to *S. sclerotiorum*, Perchepied et al. (2010) also demonstrated that expression of *PR1*, *PDF1-2*, and *ABI1* was found to be abolished or strongly reduced compared with the wild type in double mutant *nia1 nia2*, whereas *VSP1* expression was delayed by 24 h. Similarly, in the *rboh-D rboh-F* double mutant, *PR1* expression was abolished, *PDF1-2* and *VSP1* expression was delayed, and *ABI1* expression was upregulated. These results suggest a differential regulatory effect of NO and ROS on the different defense pathways (Perchepied et al. 2010).

G Proteins

G Proteins are one of the most important components of signaling system. There are two types of G proteins, monomeric and heterotrimeric. The heterotrimeric G proteins contain α, β and γ subunits, and the monomeric small G proteins appear to be similar to free α subunits, operating without the $\beta\gamma$ heterodimer. The Gα subunit has both GTP-binding and GTPase activity and acts as a molecular switch for signal transduction. G proteins have been implicated in plant defense, however much remains to be explored especially in case of Arabidopsis and rice. The genome of the model plant *Arabidopsis thaliana* contains one prototypical Gα (GPA1), one Gβ (AGB1), and two identified Gγ (AGG1 and AGG2) subunits (Jones and Assmann 2004) and one RGS protein, AtRGS1 (Chen et al. 2003). Two Gα subunits (PGA1 and PGA2) are reported from pea (Marsh and Kaufman 1999).

A variety of evidence suggest that heterotrimeric GTP-binding proteins are involved in transferring elicitor signals from the receptor to calcium channels that activate downstream reactions, such as the oxidative burst and phytoalexin accumulation (Gelli et al. 1997; Xing et al. 1997). Llorente et al. (2005) reported that the ERECTA receptor-like kinase and G proteins are required for resistance to the necrotrophic fungus *Plectosphaerella cucumerina* in *Arabidopsis*. In cultured soybean cells, mastoparan, a G protein-activating peptide, was found to stimulate calcium influx, increases in cytosolic calcium levels and production of reactive oxygen species in the absence of elicitor (Chandra and Low 1997). Ectopic expression of the cholera toxin A1 subunit inhibiting GTPase activity of G proteins in tobacco plants resulted in high salicylate levels, constitutive expression of PR proteins and enhanced pathogen resistance (Beffa et al. 1995). Recently, Trusov et al. (2006) found in *Arabidopsis* that Gβ deficient mutants are more susceptible to infection with *A. brassicicola* and *F. oxysporum* when compared to wild type (Col-0), while Gα-deficient mutants are less susceptible to the disease than wild type. They also reported that the G$\beta\gamma$ subunit is an integral component and a positive regulator of the JA-signaling cascade.

Phospholipid Derived Molecules

Phospholipids are emerging as important secondary messengers in plant defense signaling. Recent research has begun to reveal the signals produced by the enzymes phospholipase C (PLC), phospholipase D (PLD) and phospholipase A2 (PLA_2), and their putative downstream targets (Laxalt and Munnik 2002). Upon perception of the invading pathogen, several phospholipid hydrolyzing enzymes are activated that contribute to the establishment of an appropriate defense response. These include activation of G proteins based signaling leading to the production of oxylipins and jasmonates, as well as the potent second messenger, phosphatidic acid (PA) (Canonne et al. 2011).

Three PAT-PLA genes (AAF98368, AAF98369, AAF98370) found in tobacco, were strongly induced in response to microbial infections or elicitors before the accumulation of jasmonic acid in the infiltrated zone, but poorly induced in response to wounding (Dhondt et al. 2002), indicating that further studies are required. Additionally, PLA2 activity was found to be involved in the wound-activated cascade leading to the production of aldehydes by generating C20 fatty acids in diatoms (Pohnert 2002). PLA activation has been reported in elicitor-treated cultured parsley cells. Virus infection also activates PLA (Ryu 2004). The role of PLA in cell elongation, auxin signaling, shoot gravitropism, pollen maturation, anther dehiscence and flower opening are well established, but its role in plant defense is still need to be revealed.

The involvement of PLC in stress signaling has been indicated in a number of studies. The genes encoding PI-PLC were found to be induced to a significant extent under environmental stresses (Tuteja and Sopory 2008). Van der Luit et al. (2000) reported that tomato cells responded to general elicitors, such as xylanase, flagellin or chitotetraose, by rapidly (i.e. within minutes) and transiently producing phosphatidic acid (PA). Role of PLC in Avr-induced disease resistance has also been implicated. In this case PLC/diacylglycerol kinase (PLC/DGK) mediated production of PA was found to be involved in disease resistance signaling (Andersson et al. 2006).

PLD activity has been associated to a variety of stress responses in plants. Activation of PLD activity during plant defense was first described in rice, after infection by *Xanthomonas oryzae* (Young et al. 1996). In *Arabidopsis*, expression of the α, β and γ class of *PLD* genes is induced after infiltration by both virulent and avirulent strains of *P. syringae* (de Torres Zabela et al. 2002). The effects of PLD activation during plant-pathogen interactions are varied. Indeed, PA has been shown to induce ROS production1 and activate defense-related or ethylene-responsive genes PLDs also participate in salicylic acid-dependent signaling (Canonne et al. 2011). Five different tomato PLDs have been cloned and only one of them, PLDβ1, is specifically upregulated in response to xylanase. Silencing of this gene in tomato cell suspensions resulted in the loss of the xylanase-PLD response, indicating that PLDβ1 generates PA in response to xylanase treatments (Laxalt and Munnik 2002).

Protein Kinase

Protein phosphorylation plays an important role in plant responses to pathogen attack. Signaling systems which involve phosphorylation, and can lead directly to altered gene expression pattern in cells, are the MAPK (Miotgen Activated Protein Kinase) pathways (Hancock et al. 2002). The MAPK cascades are highly conserved modules in all eukaryotes (Pitzschke et al. 2009). In plants, MAPK pathways are involved in the regulation of development, growth and programmed cell death in responses to a diversity of environmental stimuli including cold, heat, reactive oxygen species, UV, drought and pathogen attack (Colcombet and Hirt 2008). These cascades are minimally composed of a MAPKKK (MAPK kinase kinase), a MAPKK (MAPK kinase) and a MAPK. The *Arabidopsis* genome contains about 110 genes coding for putative MAPK pathway components: 20 MAPKs, 10 MAPKKs and more than 80 MAPKKKs (MAPK Group 2002). In plants, pathogen challenge along with cold, drought and phytohormones may lead to the activation of MAPK cascades, resulting in the modulation of nuclear gene expression (Hirt 1997). Exogenous hydrogen peroxide can lead to the activation of MAP kinases (Kovtun et al. 2000; Samuel et al. 2000). In *Arabidopsis* suspension cultures and leaves, hydrogen peroxide treatment activates AtMPK6 (Desikan et al. 2001). One of the potential targets of NO in cells is also the MAPK cascade. MAPK activation by NO has been reported in *Arabidopsis* (Clarke et al. 2000) and tobacco (Durner and Klessig 2000). Activation of such MAPK cascades is likely to lead to alteration in gene expression profiles. In *Arabidopsis*, MPK3, MPK4 and MPK6 are all activated by bacterial and fungal elicitors (Desikan et al. 2001; Nuhse et al. 2000). The flagellin derived peptide, flg22 triggers a rapid and strong activation of MPK3, MPK4 and MPK6 (Droillard et al. 2004). MPK4 andMPK6 are also activated by harpin proteins, which are encoded by hrp (hypersensitive response and pathogenicity) genes in many plant pathogenic bacteria. This activation is followed by the induction of pathogenesis-related (PR) genes (Desikan et al. 2001), encoding for proteins with antimicrobial activities. Similarly, various NLPs (necrosis and ethylene-inducing peptide1-like proteins) trigger MAPK activation and induce defence responses (Qutob et al. 2006).

Salicylic Acid, Jasmonic Acid and Ethylene

Most of the inducible, defense-related genes are regulated by signal pathways involving one or more of the three regulators jasmonate, ethylene and salicylic acid (Sticher et al. 1997; Reymond and Farmer 1998; Ananieva and Ananiev 1999).

SA levels increase in plant tissue following pathogen infection, and exogenous application of SA results in enhanced resistance to a broad range of pathogens (Kunkel and Brooks 2002). Genetic studies have shown that SA is required for the rapid activation of defense responses that are mediated by several resistance genes, for the induction of local defenses that contain the growth of virulent pathogens, and for the establishment of systemic acquired resistance (SAR) (Ryals et al. 1996). Several studies have also demonstrated that when SA accumulation is prevented, resistance is compromised. Transgenic tobacco and *Arabidopsis* plants unable to accumulate SA because of the expression of the *Pseudomonas putida nahG* gene encoding salicylate hydroxylase, exhibit poor induction of *PR* genes after pathogen infection and fail to develop SAR (Gaffney et al. 1993; Delaney et al. 1994). The signal transduction pathway downstream of SA leads to the expression of a number of PR genes, such as PR-1 and β-1,3-glucanase (Ryals et al. 1996). Activation of R-gene-mediated defense signaling induces SA synthesis and downstream defense responses. Significantly, the application of SA activates the expression of R genes of the toll-interleukin-2 receptor (TIR)–nucleotide-binding site (NBS)–leucine-rich repeat (LRR) type (Shirano et al. 2002). Similarly, SA activates expression of RPW8, which confers resistance to the powdery mildew pathogen (Xiao et al. 2003).

SA also activates expression of the EDS1 gene, which is required for SA accumulation and resistance conferred by these R-gene-activated pathways (Feys et al. 2001). *Arabidopsis* mutants that are impaired in SA responsiveness, such as npr1 (nonexpressor of PR) or are defective in pathogen-induced SA accumulation, such as eds1 (enhanced disease susceptibility 1), eds5 (enhanced disease susceptibility 5), sid2 (isochorishmate synthase) and pad4 (phytoalexin deficient 4), exhibit enhanced susceptibility to pathogen infection and show impaired PR gene expression (Venugopal et al. 2009).

Jasmonates are produced from the major plant plasma membrane lipid, linolenic acid via the octadecanoid biosynthetic pathway (Yang et al. 1997). First indications for a role of jasmonates in the regulation of gene expression were obtained by Parthier and co-workers who observed the accumulation of jasmonate inducible proteins (JIPs) in senescing barley leaves (Weidhase et al. 1987; Mueller-Uri et al. 1988). The rapid accumulation of jasmonate has been observed in many cultured plant cells in response to various elicitor treatments (Ebel and Scheel 1997; Gundlach et al. 1992). In suspension-cultured rice cells, an *N*-acetylchitohepatose elicitor induces the synthesis of the phytoalexin, momilactone A, which is preceded by transient accumulation of jasmonate (Nojiri et al. 1996). *A. thaliana* mutants that are impaired in JA production (e.g. *fatty acid desaturase fad3/fad7/fad8* triple mutants) or perception (e.g. *coronatine insensitive1* [*coi1*] and *jasmonic acid resistant1* [*jar1*]) exhibit enhanced susceptibility to a variety of pathogens, including the fungal pathogens *Alternaria brassicicola*, *Botrytis cinerea*, and *Pythium sp.*, and the bacterial pathogen *Erwinia carotovora* (Thomma et al. 1998; Stintzi et al. 2001; Vijayan et al. 1998; Staswick et al. 1998; Norman-Setterblad et al. 2000). Perchepied et al. (2010) demonstrated that the JA-insensitive *coi1-1 arabidopsis* mutant was highly susceptible to *S. sclerotiorum*, thus indicating that JA is a major signal for activation of defenses against this fungus. In the same study, however, *jar1-1*, a jasmonate-resistant mutant shown to exhibit enhanced sensitivity to the fungal necrotroph *Pythium irregulare*, was not affected for responsiveness to *S. sclerotiorum*. Wounded tissue rapidly activates JA biosynthesis, and increased JA triggers the SKP1/Cullin/F-box E3 ubiquitin ligase complex containing the F-box subunit CORONATINE INSENSITIVE1 (SCFCOI1) to degrade the repressors of JA signaling – the JASMONATE-ZIM (JAZ) family proteins – by the ubiquitin/26S-proteasome pathway (Chini et al. 2007; Yoshida et al. 2009). In addition to their local synthesis and action, JAs also move systemically via vascular strands to transmit wound signals to distal tissues (Thorpe et al. 2007).

Ethylene (ET) plays a critical role in the activation of plant defenses against different biotic stresses through its participation in a complex signaling network that includes jasmonic acid (JA), salicylic acid (SA), and abscisic acid (ABA) (Adie et al. 2007). ET has been implicated in both local and systemic defense responses to *A. brassicicola* through its regulation of GLIP1. This secreted lipase has antifungal properties and is induced by ET but not by SA or JA (Oh et al. 2005). It is believed that crosstalk between ethylene and JA pathways enables plants to optimize their defense strategies more efficiently and economically (Baldwin 1998). It has also been reported that ethylene and jasmonate pathways converge in the transcriptional activation of ETHYLENE RESPONSE FACTOR1 (ERF1), which encodes a transcription factor that regulates the expression of pathogen response genes that prevent disease progression. The expression of *ERF1* can be activated rapidly by ethylene or jasmonate and can be activated synergistically by both hormones (Lorenzo et al. 2003). In another study with ET-insensitive (Tetr) tobacco plants, it was reported that ET is essential for the onset of SA-dependent SAR that is triggered upon infection by tobacco mosaic virus (Verberne et al. 2003). Moreover, ET was shown to enhance the response of *Arabidopsis* to SA, resulting in a potentiated expression of the SA-responsive marker gene *PR-1* (Lawton et al. 1994; De Vos et al. 2006). This synergistic effect of ET on SA-induced *PR-1* expression was blocked in the ET-insensitive mutant *ein2* (De Vos et al. 2006), which indicates that the modulation of the SA pathway by ET is EIN2 dependent and thus functions through the ET signaling pathway. Therefore, the SA, JA and ET response pathways serve as the backbone of the induced defense signaling network in plants.

Plant-Pathogen Interactions and Proteomics

The common approach utilized for proteomics based experiments comprises two-dimensional gel electrophoresis protein profile followed by MS analysis of differential expressed spots (MALDI-TOF or MS-MS) and identification by DNA, EST or protein database searching using specific algorithms (i.e. MASCOT, phenyx and OMSAA). In brief, the workflow of a standard proteomics experiment includes all or most of the following steps: experimental design, sampling, sample preparation, protein extraction/fractionation/purification, labeling/modification, separation, MS analysis, protein identification, and statistical analysis of data and validation. The most appropriate protocol to be used depends on and must be optimized for the biological system and type of tissue/cells, as well as the objectives of the research (descriptive, comparative, Post Translational Modifications, interactions, targeted Proteomics) (Jorrin-Novo et al. 2009). There are number of technical advances available and constantly evolving particularly for sample preparation, gel free proteomics, protein identification and data analysis, but this will not be the focus of this chapter. Those interested in further reading can refer to the reviews (Ong et al. 2003; Chen and Harmon 2006; Domon and Aebersold 2006; Rossignol et al. 2006; Everberg et al. 2008; Carpentier et al. 2008; Chen 2008; Jorrin-Novo et al. 2009). Herein, we address the challenges in proteomics and phosphoproteomics studies of plant pathogen interactions.

Comparative Proteomics

The aim of most of the proteomics studies was to compare the plant response to infection by the pathogen and to identify and characterize common and specific changes in protein expression patterns. Geddes et al. (2008), using two-dimensional Electrophoresis (2-DE) coupled with LC-MS/MS, identified differentially expressed proteins in *Fusarium* head blight-resistant and *Fusarium* head blight-susceptible barley genotypes under infected and uninfected conditions. In this study, approximately 600 protein spots were resolved in the pH range of 4–7 in the 2-DE gels. A total of 16 different acidic proteins associated with resistance mechanisms against *Fusarium* head blight were identified, out of which 12 proteins were associated with oxidative burst response and 4 proteins were associated with PR-Proteins. Takemoto et al. (1997) reported that chitinase (PR3) and osmotin (PR5) were associated with the actin cytoskeleton that is involved in cytoplasmic aggregation in the early stages of the hypersensitive response (HR) between *Phytophthora infestans* and potato. *Fusarium* head blight, caused mainly by *F. graminearum*, is one of the most destructive diseases of wheat. The interaction between *F. graminearum and wheat* has been investigated by Zhou and his coworkers in 2006. They found that 33 plant proteins were expressed in response to *F. graminearum* in wheat spikes. These proteins were divided into two groups, each related to defense response or metabolism. The authors suggested that several of these proteins were directly involved in mounting the plant defense against infection by protecting against the oxidative burst inside the plant cell. Such a burst can be caused in plant cells by invading fungus.

Proteomics analysis was carried out to study the compatible and incompatible interactions between rice and bacteria, *Xanthomonas oryzae* pv. *oryzae (Xoo)* (Mahmood et al. 2006). In this study, four different defense-related proteins were identified, namely thaumatin-like protein (PR5), PBZ, SOD, and peroxiredoxin. Overexpression of PR5 in transgenic rice plants enhanced the resistance of rice to *Rhizoctonia solani*, the causal organism of sheath blight of rice (Datta et al. 1999). Wei et al. (2009) used isobaric tag- based methodology for relative peptide quantification (iTRAQ) coupled with multidimensional liquid chromatography and tandem mass spectrometry to study the response of rice to brown plant hopper (BPH) attack. In this study, three proteins involved in JA biosynthesis were induced in rice in response to infestation by the BPH: cytochrome P450, AOC 4 and alpha-DOX2. Alpha-DOX2 is a dioxygenase that catalyzes the synthesis of 13-hydroperoxylinolenic acid from linolenic acid in JA biosynthesis. Koeduka et al. (2005) reported that alpha-DOX can be induced by blight bacteria infection, and both oxidative and heavy metal stresses, through the jasmonate signaling pathway in the leaves of rice seedling. AOC4 catalyzes the stereospecific cyclization of an unstable allene oxide to (9S, 13S)-12-oxo- (10, 15Z)-phytodienoic acid, and experiments with the JA deficient *Arabidopsis* mutant opr3 indicate that AOC is the preferential target in the regulation of JA biosynthetic capacity (Stenzel et al. 2003). Since the BPH is a phloem-feeding insect, AOC may have a role in systemic defense signaling. Liang et al. (2008), using 2-DE identified 9 proteins related with defense responses. Out of these two proteins were identifies as methionine adenosyltransferase (MAT) involved in ethylene biosynthesis, and JA-responsive proteins (JR1) MAT catalyzes the synthesis of the ethylene precursor, S-adenosylmethionine (AdoMet) and plays an important role in mediating the cross talk between ethylene and NO signaling pathways (Lindermayr et al. 2006). JA has crucial role in regulating many plant processes including mediation of resistance to pathogens (Creelman and Mullet 1997). JA-responsive (JR) genes, including JR1, have been demonstrated to be induced by wounding (Leon et al. 1998). Oh et al. (2005) started with a proteomic comparison of the proteins secreted by *Arabidopsis* cultured cells in the presence of salicylic acid (SA). Thirteen different proteins that responded to the SA treatment were identified by MALDI-ToF MS. One of them was GDSL LIPASE 1, or GLIP 1, a SA-induced protein. Upon further characterization, it was found to play a role in the defense against the necrotrophic fungus *Alternaria brassicola*. In another study, a proteomics analysis was carried out to understand the molecular mechanism of interaction between *Fusarium graminearum* and *Triticum aestivum*. About 1,380 protein spots were resolved on 2-D gels stained with Sypro Ruby. In total, 41 proteins were detected which are differentially regulated due to *F. graminearum* infection, and were analyzed with LC-MS/MS for their identification. The proteins involved in the antioxidant and jasmonic acid signaling pathways, pathogenesis-related response, amino acid synthesis and nitrogen metabolism were up-regulated, while those related to photosynthesis were less abundant following *F. graminearum* infection (Zhou et al. 2006).

Beet necrotic yellow vein virus (BNYVV) is a devastating sugar beet pathogen. Resistance is limited and resistance-breaking isolates are becoming problematic. Larson et al. (2008) studied the differential sugar beet protein expression induced by BNYVV- with multidimensional liquid chromatography. Of more than 1,000 protein peaks detected in root extracts, 7.4 and 11% were affected by BNYVV in the resistant and susceptible genotypes, respectively. Using tandem MALDI-TOF-MS, 65 proteins were identified in this study. Proteomic data suggest involvement of systemic resistance components in Rz1-mediated resistance and phytohormones in symptom development. Several proteins affected by BNYVV are classically associated with plant defense, suggesting inducible resistance may contribute to viral disease suppression. These include pathogenesis-related proteins, such as chitinase, protease, glucanase, peroxidase and defensin. Interestingly, induction of these proteins was not always limited to the resistant genotypes. Some oxidative enzymes, which are also known to contribute to plant defense, appear to have similar timing dependent expression. Polyphenol oxidase, a protein responsible for physical barrier development, and toxic compound and ROS production, is more highly and rapidly expressed in the resistant genotype when compared with expression patterns from the susceptible genotype.

A study on rice proteomics was performed to analyse the protein profile after *Magnaporthe grisea* infection, and was conducted using infected leaf blades fertilized with various levels of nitrogen (Konishi et al. 2001). Rice plants grown with high levels of nitrogen nutrient are more susceptible to infection by the blast fungus (Long et al. 2000). Though, leaf proteins revealed some minor changes when plants grown under different levels of nitrogen were compared, this study failed to establish any direct correlation between nitrogen application and disease resistance (Rakwal and Agrawal 2003). Twelve proteins, including the rice thaumatin-like protein (TLP) (PR-5), were identified with accumulation changes at different levels of nitrogen. Another study of the same interaction was performed by Kim et al. (2003) using rice suspension cultured cells. In this study, twelve proteins were identified, including the rice pathogenesis-related protein class 10 (OsPR-10), isoflavone reductase-like protein (PBZ1), glucosidase and putative receptor-like protein kinase (RLK), which had not been reported previously in suspension-cultured rice cells. The authors, followed with another proteome study using rice leaves, identified eight proteins newly induced or with increased expression (Kim et al. 2004). The identified proteins belonged to several groups of PR proteins, and included two RLKs, two b-1,3-glucanases (Glu1, Glu2), TLP, peroxidase (POX 22.3), PBZ1 and OsPR-10. Lee et al. (2006) investigated rice sheath leaves after infection with this fungus, *Rhizoctonia solani* and the results revealed six proteins whose relative abundance varied significantly in the resistant and susceptible lines, and 11 additional proteins which were identified in abundance in response of the resistant line only. These proteins have been reported previously to be involved in antifungal activity, signal transduction, energy metabolism, photosynthesis, protein folding and degradation, and antioxidation, indicating a common pathway for both stress and non-stress plant functions.

Using 2-DE, the root protein profiles of *M. truncatula* were analysed after *Aphanomyces euteiches* pathogen infection (Colditz et al. 2004). The majority of the induced proteins belonged to the PR-10 family, whereas others corresponded to putative cell wall proteins and enzymes of the phenylpropanoid–isoflavonoid pathway. Another study focused on *Zea mays* embryos in response to the fungus, *Fusarium verticillioides* (Campo et al. 2004). The proteins identified included PR proteins, antioxidant enzymes and proteins involved in protein biosynthesis, folding and stabilization.

Phosphoproteomics

Protein phosphorylation plays a key role in signal transduction in plant during defense responses. The importance of phosphorylation in plant basal defense responses is exemplified by the FLS22activated MAP kinase cascade. The phosphorylated proteins are primarily involved in the early steps of signal transduction pathways as demonstrated by Lecourieux-Ouaked et al. (2000) in tobacco by using cryptogein, an elicitor of defense reactions. Kinases are implicated in direct interactions with R protein signaling complexes (RPS5/PBS1) and also in the modification of the key effector protein RIN4. Additional kinase activities are necessary for downstream signaling events. Proteomics not only monitors the steady state level of proteins but also co- and post-translational modifications of proteins. These include not only kinases and phosphatases but also their substrates (Xing et al. 2002).

Lecourieux-Ouaked et al. (2000), using 2-DE, tested the in vivo phosphorylation status of proteins after cryptogein, staurosporine (a kinase inhibitor), and calyculin A (a phosphatase inhibitor) treatments in tobacco cells. Out of about 100 phospholabelled polypeptides, 19 showed increased ^{32}P incorporation after cryptogein treatment and 12 of these depended upon calcium influx. Staurosporine inhibited the phosphorylation induced by cryptogein whereas calyculin A activated the phosphorylation of 18 polypeptides indicating that the phosphorylation of these proteins were activated by certain protein kinases and inhibited by certain phosphatases. These results demonstrate the power of phosphoproteomics to identify key proteins.

In suspension-cultured cells of *Arabidopsis*, Peck et al. (2000) used ^{32}P pulse-label method in conjunction with 2-DE and MS to identify proteins that are rapidly phosphorylated in response to bacterial and fungal elicitors. One of these proteins, AtPhos43 was found to be phosphorylated within minutes after treatment with flagellin. They also found that phosphorylation of AtPhos43 after flagellin treatment was dependent on FLS2, a receptor-like kinase involved in flagellin perception (Gomez-Gomez and Boller 2000). It has also been found that this protein was phosphorylated in response to both bacterial and fungal elicitors, and related proteins are phosphorylated in other monocot and dicot species (Peck 2003; Peck et al. (2000)).

Jones et al. (2006) describe the application of differential mass tags (iTRAQ) to provide relative quantification of phosphorylated peptides during the early stages in plant pathogen interaction the model plant *Arabidopsis thaliana* after challenging with three strains of *Pseudomonas syringae* pv. *tomato* DC3000 (DC3000). The HR induced by the *avrRpm1*/RPM1 interaction was compared with basal resistance through examination of responses to the *hrpA* mutant of DC3000. They identified five proteins which showed reproducible differences between a mock-inoculated control and different bacterial challenges 3 hours post inoculation (hpi), thus identifying proteins. Four of the five proteins, a dehydrin, a putative p23 co-chaperone, heat shock protein 81 and a plastidassociated protein (PAP)/fibrillin, are known to be phosphorylated or have potential phosphorylation sites. One another protein, the large subunit of Rubisco, showed a significant difference between tissue undergoing the hypersensitive response and a basal defence response. This novel study shows the application of iTRAQ to plant–pathogen interactions and the challenge of examining phosphoproteins from intact green leaf tissue, rather than the more commonly used cell culture system.

Conclusion and Future Perspectives

In the post genomic era, proteomics has emerged as an indispensable tool for understanding signaling mechanisms of plant against pathogen, its potential impact in plant pathology, and the study of plant-pathogen interaction. Previously, a limited number of genes involved in infection process had been identified using conventional molecular genetics and biochemical methods. With the advent of proteomics technology, a number of proteins involved in plant defenses have been identified. Phosphoproteomics will continue to play a major role in identifying post-translational modifications and, therefore, have an additional benefit of identifying signaling components that may not be revealed by transcriptome analysis alone. Currently, the major challenges for the plant phosphoproteomics are to identify the relevant phosphorylation sites from the vast majority of phosphopeptides. A high throughput technical advancement, therefore, will be an important development that will help to identify these relevant phosphorylation sites in proteins of interest. Another much needed improvement, highly desired in phosphoproteomics study, is the development of improved and novel enrichment strategies for phosphorylated peptides. In future, the integration of proteomics with genomics, transcriptomics, and metabolomics will play a major role in understanding the plant biology and will uncover many unexpected links within the signaling networks in plants. The continued proteomics advances in unrevealing the molecular mechanisms will lead to a better understanding of plant-pathogen interactions, which may ultimately contribute to the development of novel disease tolerant varieties of agriculturally and economically important crops.

References

Adie B, Chico JM, Rubio-Somoza I, Solano R (2007) Modulation of plant defenses by ethylene. J Plant Growth Regul 26:160–177

Alvarez ME, Pennel RI, Meijer PJ, Ishikawa A, Dixon RA, Lamb C (1998) Reactive oxygen intermediates mediate a systemic signal network in the establishment of plant immunity. Cell 92:773–784

Ananieva KI, Ananiev ED (1999) Effect of methyl ester of jasmonic acid and benzylaminopurine on growth and protein profile of excised cotyledons of Cucurbita pepo (zucchini). Biol Plant 42:549–557

Andersson MX, Kourtchenko O, Dangl JL, Mackey D, Elerstrom M (2006) Phopspholipase--dependent signaling during the AvrRpm1 and AvrRpt2—Induced disease resistance response in Arabidopsis thaliana. Plant J 47:947–949

Asai S, Ohta K, Yoshioka H (2008) MAPK signaling regulates nitric oxide and NADPH oxidase-dependent oxidative bursts in Nicotiana benthamiana. Plant Cell 20:1390–1406

Atkinson MM, Midland SL, Sims JJ, Keen NT (1996) Syringolide 1 triggers Ca + influx, K+ efflux, and extracellular alkalization in soybean cells carrying the disease resistance gene Rpg4. Plant Physiol 112:297–302

Bach M, Schnitzler JP, Ulrich H (1993) Elicitor-induced changes in Ca+ influx, K+ efflux, and 4-hydroxybenzoic acid synthesis in protoplasts of Daucus carota. Plant Physiol 103:407–412

Baldwin IT (1998) Jasmonate-induced responses are costly but benefit plants under attack in native populations. Proc Natl Acad Sci USA 95:8113–8118

Beffa R, Szell M, Meuwly P, Pay A, Vogeli-Lange R, Métraux JP, Neuhaus G, Meins F, Nagy J, Nagy F (1995) Cholera toxin elevates pathogen resistance and induces pathogenesis-related gene expression in tobacco. EMBO J 14:5753–5761

Bestwick CS, Brown IR, Bennett MHR, Mansfield JW (1997) Localization of hydrogen peroxide accumulation during the hypersensitive reaction of lettuce cells to Pseudomonas syringae pv. phaseolicola. Plant Cell 9:209–221

Bolwell GP (1999) Role of active oxygen species and NO in plant defence. Curr Opin Plant Biol 2:287–294
Bolwell GP, Butt VS, Davies DR, Zimmerlin A (1995) The origin of the oxidative burst in plants. Free Rad Res 23:517–532
Bradley DJ, Kjellbom P, Lamb CJ (1992) Elicitor- and wound-induced oxidative cross-linking of a proline-rich plant cell wall protein: A novel, rapid defense response. Cell 70:21–30
Brisson LF, Tenhaken R, Lamb C (1994) Function of oxidative cross-linking of cell wall structural proteins in plant disease resistance. Plant Cell 6:1703–1712
Campo S, Carrascal M, Coca M, Abian J, San SB (2004) The defence response of germinating maize embryos against fungal infection: a proteomics approach. Proteomics 4:383–396
Canonne J, Froidure-Nicolas S, Rivas S (2011) Phospholipases in action during plant defense signaling. Plant Signal Behav 6:13–18
Carpentier SC, Panis B, Vertommen A, Swennen R, Sergeant K, Renaut J, Laukens K, Witters E, Samyn B, Devreese B (2008) Proteome analysis of non model plants: a challenging but powerful approach. Mass Spectrom Rev 27:354–377
Chandra S, Low PS (1997) Measurement of Ca2+ fluxes during elicitation of the oxidative burst in aequorin-transformed tobacco cells. J Biol Chem 272:28274–28280
Chen CH (2008) Review of a current role of mass spectrometry for proteome research. Anal Chim Acta 624:16–36
Chen S, Harmon AC (2006) Advances in plant proteomics. Proteomics 6:5504–5516
Chen JG, Willard FS, Huang J, Liang J, Chasse SA, Jones AM, Siderovski DP (2003) A seven-transmembrane RGS protein that modulates plant cell proliferation. Science 301:1728–1731
Chini A, Fonseca S, Fernández G, Adie B, Chico JM, Lorenzo O, García-Casado G, López-Vidriero I, Lozano FM, Ponce MR, Micol JL, Solano R (2007) The JAZ family of repressors is the missing link in jasmonate signaling. Nature 448:666–671
Clarke A, Desikan R, Hurst RD, Hancock JT, Neill SJ (2000) NO way back: nitric oxide and programmed cell death in Arabidopsis thaliana suspension cultures. Plant J 24:667–677
Colcombet J, Hirt H (2008) Arabidopsis MAPKs: a complex signaling network involved in multiple biological processes. Biochem J 413:217–226
Colditz F, Nyamsuren O, Niehaus K, Eubel H, Braun HP, Krajinski F (2004) Proteomic approach: identification of Medicago truncatula proteins induced in roots after infection with the pathogenic oomycete Aphanomyces euteiches. Plant Mol Biol 55:109–120
Creelman RA, Mullet JE (1997) Biosynthesis and action of jasmonates in plants. Annu Rev Plant Physiol Plant Mol Biol 48:355–381
Datta K, Velazhahan R, Oliva N, Ona L, Mew T, Khush GS, Muthukrishnan S, Datta SK (1999) Over-expression of the cloned rice thaumatin-like protein (PR-5) gene in transgenic rice plants enhances environmental friendly resistance to Rhizoctonia solani causing sheath blight disease. Theor Appl Genet 98:1138–1145
de Torres Zabela M, Fernandez-Delmond I, Niittyla T, Sanchez P, Grant M (2002) Differential expression of genes encoding Arabidopsis phospholipases after challenge with virulent or avirulent Pseudomonas isolates. Mol Plant Microbe Interact 15:808–816
De Vos M, Van Zaanen W, Koornneef A, Korzelius JP, Dicke M, Van Loon LC, Pieterse CM (2006) Herbivore-induced resistance against microbial pathogens in Arabidopsis. Plant Physiol 142:352–363
Delaney TP, Uknes S, Vernoij B, Friedrich L, Weyman K, Negrotto D, Gaffney T, Gut-Rella M, Kessmann H, Ward E, Ryals J (1994) A central role of salicylic acid in plant disease resistance. Science 266:1247–1250
Desikan R, Hancock JT, Coffey MJ, Neill SJ (1996) Generation of active oxygen in elicited cells of Arabidopsis thaliana is mediated by a NADPH oxidase-like enzyme. FEBS Lett 382:213–217
Desikan R, Neill SJ, Hancock JT (1997) Generation of active oxygen in Arabidopsis thaliana. Phyton 37:65–70
Desikan R, Burnett EC, Hancock JT, Neill SJ (1998) Harpin and hydrogen peroxide induce the expression of a homologue of gp91-phox in Arabidopsis thaliana suspension cultures. J Exp Bot 49:1767–1771

Desikan R, Hancock JT, Ichimura K, Shinozaki K, Neill SJ (2001) Harpin induces activation of the Arabidopsis MAP kinasesATMPK4 and ATMPK6. Plant Physiol 126:1579–1587
Dhondt S, Gouzerh G, Müller A, Legrand M, Heitz T (2002) Spatio-temporal expression of patatin-like lipid acyl hydrolases and accumulation of jasmonates in elicitor-treated tobacco leaves are not affected by endogenous levels of salicylic acid. Plant J 32:749–762
Doke N (1983a) Generation of superoxide anion by potato tuber protoplasts during the hypersensitive response to hyphal wall components of Phytophthora infestans and specific inhibition of the reaction by suppressors of hypersensitivity. Physiol Plant Pathol 23:359–367
Doke N (1983b) Involvement of superoxide anion generation in the hypersensitive response of potato tuber tissues to infection with an incompatible race of Phytophthora infestan and to the hyphal wall components. Physiol Plant Pathol 23:345–357
Doke N, Ohashi Y (1988) Involvement of an O2- generating system in the induction of necrotic lesions on tobacco leaves infected with tobacco mosaic virus. Physiol Mol Plant Pathol 32:163–175
Domon B, Aebersold R (2006) Mass spectrometry and protein analysis. Science 312:212–217
Droillard MJ, Boudsocq M, Barbier-Brygoo H, Lauriere C (2004) Involvement of MPK4 in osmotic stress response pathways in cell suspensions and plantlets of Arabidopsis thaliana: activation by hypoosmolarity and negative role in hyperosmolarity tolerance. FEBS Lett 574:42–48
Durner D, Klessig DF (2000) Differential induction of tobacco MAP kinases by the defense signals nitric oxide, salicylic acid, ethylene, and jasmonic acid. Mol Plant Microbe Interact 13:347–351
Dwyer SC, Legendre L, Low PS, Leto TL (1996) Plant and human neutrophil oxidative burst complexes contain immunologically related proteins. Biochem Biophys Acta 1289:231–237
Ebel J, Scheel D (1997) Signals in host–parasite interactions. In: Carroll GC, Tudzynski P (eds) The mycota, vol V, Plant relationships, Part A. Springer, Berlin, pp 85–105
Everberg H, Gustavasson N, Tjerned F (2008) Enrichment of membrane proteins by partitioning in detergent/polymer aqueous two-phase systems. Met Mol Biol 424:403–412
Feys BJ, Moisan LJ, Newman MA, Parker JE (2001) Direct interaction between the Arabidopsis disease resistance signaling proteins, EDS1 and PAD4. EMBO J 20:5400–5411
Gaffney T, Friedrich L, Vernooij B, Negrotto D, Nye G, Uknes S, Ward E, Kessmann H, Ryals J (1993) Requirement of salicylic acid for the induction of systemic acquired resistance. Science 26:754–756
Geddes J, Eudes F, Laroche A, Selinger B (2008) Differential expression of proteins in response to the interaction between the pathogen Fusarium graminearum and its host, Hordeum vulgare. Proteomics 8:545–554
Gelli A, Higgins VJ, Blumwald E (1997) Activation of plant plasma membrane Ca2+−permeable channels by race-specific fungal elicitors. Plant Physiol 113:269–279
Gomez-Gomez L, Boller T (2000) FLS2: a LRR receptor-like kinase involved in recognition of the flagellin elicitor in Arabidopsis. Mol Cell 5:1003–1011
Green R, Fluhr R (1995) UV-B-induced PR-1 accumulation is mediated by active oxygen species. Plant Cell 7:203–212
Gundlach H, Muller MJ, Kutchan TM, Zenk MH (1992) Jasmonic acid is a signal transducer in elicitor-induced plant cell cultures. Proc Natl Acad Sci USA 89:2389–2393
Guo FQ, Okamoto M, Crawford NM (2003) Identification of a plant nitric oxide synthase gene involved in hormonal signaling. Science 302:100–103
Hahlbrock K, Scheel D, Logemenn E, Nrnberger T, Papniske M, Reinold S, Sacks WR, Schmelzer E (1995) Oligopeptide elicited defense gene activation in cultured parsley cells. Proc Natl Acad Sci USA 92:4150–4157
Hammond-Kosack KE, Jones JDG (1996) Resistance gene dependent plant defense responses. Plant Cell 8:1773–1791

Hammond-Kosack KE, Jones JDG (2000) Responses to plant pathogens. In: Buchanan BB, Gruissem W, Jones RL (eds) Biochemistry and Molecular Biology of Plants. American Society of Plant Physiology, Maryland, pp 1102–1156

Hancock JT, Desikan R, Clarke A, Hurst RD, Neill SJ (2002) Cell signaling following plant/pathogen interactions involves the generation of reactive oxygen and reactive nitrogen species. Plant Physiol Biochem 40:611–617

Hirt H (1997) Multiple roles of MAP kinases in plant signal transduction. Trends Plant Sci 2:11–15

Jabs T, Tschope M, Colling C, Hahlbrock K, Scheel D (1997) Elicitor-stimulated ion fluxes and O^{2-} from the oxidative burst are essential components in triggering defense gene. Proc Natl Acad Sci USA 94(9):4800–4805

Jones AM, Assmann SM (2004) Plants: the latest model system for G-protein research. EMBO Reports 5:572–578

Jones AME, Bennett MH, Mansfield JW, Grant M (2006) Analysis of the defence phosphoproteome of Arabidopsis thaliana using differential mass tagging. Proteomics 6:4155–4165

Jorrin-Novo JV, Maldonado AM, Echevarria-Zomeno S, Valledor L, Castillejo MA, Curto M, Valero J, Sghaier B, Donoso G, Redondo I (2009) Plant proteomics update (2007–2008): second-generation proteomic techniques, an appropriate experimental design, and data analysis to fulfil MIAPE standards, increase plant proteome coverage and expand biological knowledge. J Proteomics 72:285–314

Khush GS (2001) Challenges for meeting the global food and nutrient needs in the new millennium. Proc Nutr Soc 60:15–26

Kim ST, Cho KS, Yu S, Kim SG, Hong JC, Han CD, Bae DW, Nam MH, Kang KY (2003) Proteomic analysis of differentially expressed proteins induced by rice blast fungus and elicitor in suspension-cultured rice cells. Proteomics 3:2368–2378

Kim ST, Kim SG, Hwang DH, Kang SY, Kim HJ, Lee BH, Lee JJ, Kang KY (2004) Proteomic analysis of pathogen-responsive proteins from rice leaves induced by rice blast fungus, Magnaporthe grisea. Proteomics 4:3569–3578

Koeduka T, Matsui K, Hasegawa M, Akakabe Y, Kajiwara T (2005) Rice fatty acid alpha-dioxygenase is induced by pathogen attack and heavy metal stress: activation through jasmonate signaling. J Plant Physiol 162:912–920

Konishi H, Ishiguro K, Komatsu S (2001) A proteomics approach towards understanding blast fungus infection of rice grown under different levels of nitrogen fertilization. Proteomics 1:1162–1171

Kovtun Y, Chiu WL, Tena G, Sheen J (2000) Functional analysis of oxidative stress-activated mitogen-activated protein kinase cascade in plants. Proc Natl Acad Sci 97:2940–2945

Kuchitsu K, Kikuyama M, Shibuya N (1993) N-acetylchitooligosaccharides, biotic elicitors for phytoalexin production, induce transient membrane depolarization in suspension-cultured rice cells. Protoplasma 174:79–81

Kunkel BN, Brooks DM (2002) Cross talk between signaling pathways in pathogen defense. Curr Opin Plant Biol 5:325–331

Larson RL, Wintermantel WM, Hill A, Fortis L, Nunez A (2008) Proteome changes in sugar beet in response to Beet necrotic yellow vein virus. Physiol Mol Plant Pathol 72:62–72

Lawton KA, Potter SL, Uknes S, Ryals J (1994) Acquired resistance signal transduction in Arabidopsis is ethylene independent. Plant Cell 6:581–588

Laxalt AM, Munnik T (2002) Phospholipid signaling in plant defence. Curr Opin Plant Biol 5:332–338

Lecourieux-Ouaked F, Pugin A, Lebrun-Garcia A (2000) Phosphoproteins involved in the signal transduction of cryptogein, an elicitor of defense reactions in tobacco. Mol Plant Microbe Interact 13:821–829

Lee J, Bricker TM, Lefevre M, Pinson SRM, Oard JH (2006) Proteomic and genetic approaches to identify defence-related proteins in rice challenged with the fungal pathogen Rhizoctonia solani. Mol Plant Pathol 7:405–416

Leitner M, Vandelle E, Gaupels F, Bellin D, Delledonne M (2009) NO signals in the haze: Nitric oxide signaling in plant defence. Curr Opin Plant Biol 12:451–458

Leon J, Rojo E, Titarenko E, Sanchez-Serrano JJ (1998) Jasmonic acid-dependent and independent wound signal transducyión pathways are differentially regulated by Ca2+/calmodulin in Arabidopsis thaliana. Mol Gen Genet 258:412–419

Levine A, Tenhaken R, Dixon R, Lamb C (1994) H2O2 from the oxidative burst orchestrates the plant hypersensitive disease resistance response. Cell 79:583–595

Liang Y, Srivastava S, Rahman MH, Strelkov SE, Kav NNV (2008) Proteome changes in leaves of Brassica napus L. as a result of Sclerotinia sclerotiorum challenge. J Agri Food Chem 56:1963–1976

Lindermayr C, Saalbach G, Bahnweg G, Durner J (2006) Differential inhibition of Arabidopsis methionine adenosyltransferases by protein S-nitrosylation. J Biol Chem 281:4285–4291

Llorente F, Alonso-Blanco C, Sanchez-Rodriguez C, Jorda L, Molinà A (2005) ERECTA receptor-like kinase and heterotrimeric G protein from Arabidopsis are required for resistance to the necrotrophic fungus Plectosphaerella cucumerina. Plant J 43:165–180

Long DH, Lee FN, TeBeest DO (2000) Effect of nitrogen fertilization on disease progress of rice blast on susceptible and resistant cultivars. Plant Dis 84:403–409

Lorenzo O, Piqueras R, Sánchez-Serrano JJ, Solano R (2003) ETHYLENE RESPONSE FACTOR1 integrates signals from ethylene and jasmonate pathways in plant defense. Plant Cell 15:165–178

Mahmood T, Jan A, Kakishima M, Komatsu S (2006) Proteomic analysis of bacterial-blight defense responsive proteins in rice leaf blades. Proteomics 6:6053–6065

Maor R, Shirasu K (2005) The arms race continues: battle strategies between plants and fungal pathogens. Curr Opin Microbiol 8:399–404

MAPK Group (2002) Mitogen-activated protein kinase cascades in plants: a new nomenclature. Trends Plant Sci 7:301–308

Marsh JF, Kaufman LS (1999) Cloning and characterization of PGA1 and PGA2: Two G protein α-subunits from pea that promote growth in the yeast Saccharomyces cerevisiae. Plant J 19:237–247

Mathieu Y, Kurkjian A, Xia H, Guern J, Spiro M, O'Neill M, Albersheim P, Darvill A (1991) Membrane responses induced by oligogalacturonides in suspension-cultured tobacco cells. Plant J 1:333–343

McDowell JM, Dangl JL (2000) Signal transduction in the plant immune response. Trends Biochem Sci 25:79–82

Mithoe SC, Menke FLH (2011) Phosphoproteomics perspective on plant signal transduction and tyrosine phosphorylation. Phytochem 72:997–1006

Moreau M, Lee GI, Wang Y, Crane BR, Klessig DF (2008) AtNOS/AtNOA1 is a functional Arabidopsis thaliana cGTPase and not a nitric-oxide synthase. J Biol Chem 283:32957–32967

Mueller-Uri F, Parthier B, Nover L (1988) Jasmonate-induced alteration of gene expression in barley leaf segments analyzed by in-vivo and in-vitro protein synthesis. Planta 176:241–247

Nojiri H, Sugimori M, Yamane H, Nishimura Y, Yamada A, Shibuya N, Kodama O, Murofushi N, Omori T (1996) Involvement of jasmonic acid in elicitor-induced phytoalexin production in suspension-cultured rice cells. Plant Physiol 110:387–392

Norman-Setterblad C, Vidal S, Palva ET (2000) Interacting signal pathways control defense gene expression in Arabidopsis in response to cell wall-degrading enzymes from Erwinia carotovora. Mol Plant Microbe Interact 13:430–438

Nuhse TS, Peck SC, Hirt H, Boller T (2000) Microbial elicitors induce activation and dual phosphorylation of the Arabidopsis thaliana MAPK 6. J Biol Chem 275:7521–7526

Oh IS, Park AR, Bae MS, Kwon SJ, Kim YS, Lee JE, Kang NY, Lee S, Cheong H, Park OK (2005) Secretome analysis reveals an Arabidopsis lipase involved in defense against Alternaria brassicicola. Plant Cell 17:2832–2847

Ong S, Foster LJ, Mann M (2003) Mass spectrometric-based approaches in quantitative proteomics. Methods 29:124–130

Pawson T, Scott JD (2005) Protein phosphorylation in signaling–50 years and counting. Trends Biochem Sci 30:286–290

Peck SC (2003) Early phosphorylation events in biotic stress. Curr Opin Plant Biol 6:334–338

Peck SC, Nuhse TS, Hess D, Iglesias A, Meins F, Boller T (2000) Directed proteomics identifies a plant specific protein rapidly phosphorylated in response to bacterial and fungal elicitors. Plant Cell 13:1467–1475

Peng M, Kuc J (1992) Peroxidase-generated hydrogen peroxide as a source of antifungal activity in vitro and on tobacco leaf discs. Phytopathol 82:696–699

Perchepied L, Balague C, Rioum C, Claudel-Renard C, Riviere N, Grezes-Besset B, Roby D (2010) Nitric oxide participates in the complex interplay of defense-related signaling pathways controlling disease resistance to Sclerotinia sclerotiorum in Arabidopsis thaliana. Mol Plant Microbe Interact 23:846–860

Pitzschke A, Schikora A, Hirt H (2009) MAPK cascade signaling networks in plant defence. Curr Opin Plant Biol 12:1–6

Pohnert G (2002) Phospholipase A2 activity triggers the woundactivated chemical defense in the diatom Thalassiosira rotula. Plant Physiol 129:103–111

Popham P, Pike S, Novacky A (1995) The effect of harpin from Erwinia amylovora on the plasmalemma of suspesioncultured tobacco cells. Physiol Mol Plant Pathol 47:39–50

Qutob D, Kemmerling B, Brunner F, Kufner I, Engelhardt S, Gust AA, Luberacki B, Seitz HU, Stahl D, Rauhut T, Glawischnig E, Schweene G, Lacombe B, Watanabe N, Lam E, Schlichting R, Scheel D, Nau K, Dodt G, Hubert G, Gijzen M, Nürnberger T (2006) Phytotoxicity and innate immune responses induced by Nep1-like proteins. Plant Cell 18:3721–3744

Rakwal R, Agrawal GK (2003) Rice proteomics: current status and future perspectives. Electrophoresis 24:3378–3389

Reymond P, Farmer EE (1998) Jasmonate and salicylate as global signals for defense gene expression. Curr Opin Plant Biol 1:404–441

Ristaino JB (2002) Tracking historic migrations of the Irish potato famine pathogen, Phytophthora infestan. Microbes Infect 4:1369–1377

Rossignol M, Peltier JB, Mock HP, Matros A, Maldonado AM, Jorrin JV (2006) Plant proteome analysis: a 2004–2006 update. Proteomics 6:5529–5548

Ryals JA, Neuenschwander UH, Willits MG, Molina A, Steiner H, Hunt MD (1996) Systemic acquired resistance. Plant Cell 8:1809–1819

Ryu SB (2004) Phospholipid-derived signaling mediated by phospholipase A in plants. Trends Plant Sci 9:229–235

Samuel MA, Miles GP, Ellis BE (2000) Ozone treatment rapidly activates MAP kinase signaling in plants. Plant J 22:367–376

Shani BB, Suleiman ML, Mohammed US (2006) Performance evaluation of an improved animal drawn ground Metered shrouded disc (gmsd) sprayer. J Agri Engg Tech 14:4–11

Shirano Y, Kachroo P, Shah J, Klessig DF (2002) A gain-of-function mutation in an Arabidopsis toll interleukin 1 receptor-nucleotide binding site-leucine-rich repeat type R gene triggers defense responses and results in enhanced disease resistance. Plant Cell 14:3149–3162

Somssich I, Hahlbrock K (1998) Pathogen defense in plants - a paradigm of biological complexity. Trends Plant Sci 3:86–90

Staswick PE, Yuen GY, Lehman CC (1998) Jasmonate signaling mutants of Arabidopsis are susceptible to the soil fungus Pythium irregulare. Plant J 15:747–754

Stenzel I, Hause B, Miersch O, Kurz T, Maucher H, Weichert H, Ziegler J, Feussner I, Wasternack C (2003) Jasmonate biosynthesis and the allene oxide cyclase family of Arabidopsis thaliana. Plant Mol Biol 51:895–911

Sticher L, Mauch-Mani B, Metraux JP (1997) Systemic acquired resistance. Ann Rev Phytopathol 35:235–270

Stintzi A, Weber H, Reymond P, Browse J, Farmer EE (2001) Plant defense in the absence of jasmonic acid: the role of cyclopentenones. Proc Natl Acad Sci USA 98:12837–12842

Takemoto D, Furuse K, Doke N, Kawakita K (1997) Identification of chitinase and osmotin-like protein as actin-binding proteins in suspension-cultured potato cells. Plant Cell Physiol 38:441–448

Tenhaken R, Levine A, Brisson LF, Dixon RA, Lamb C (1995) Function of the oxidative burst in hypersensitive disease resistance. Proc Natl Acad Sci USA 92:4158–4163

The Arabidopsis Genome Initiative (2000) Analysis of the genome sequence of the flowering plant Arabidopsis thaliana. Nature 408:796–815

Thomma BPHJ, Eggermont K, Penninckx IAMA, Mauch-Mani B, Vogelsang R, Cammue BPA, Broekaert WF (1998) Separate jasmonate dependent and salicylate-dependent defense pathways in Arabidopsis are essential for resistance to distinct microbial pathogens. Proc Natl Acad Sci USA 95:15107–15111

Thorpe MR, Ferrieri AP, Herth MM, Ferrieri RA (2007) 11Cimaging: methyl jasmonate moves in both phloem and xylem, promotes transport of jasmonate, and of photoassimilate even after proton transport is decoupled. Planta 226:541–551

Torres MA, Onuchi H, Hamada S, Machida C, Hammond-Kosack KE, Jones JDG (1998) Six Arabidopsis thaliana homologues of the human respiratory burst oxidase (gp91phox). Plant J 14:365–370

Trusov Y, Rookes JE, Chakravorty D, Armour D, Schenk PM, Botella JR (2006) Heterotrimeric G proteins facilitate Arabidopsis resistance to necrotrophic pathogens and are involved in jasmonate signaling. Plant Physiol 140:210–220

Tuteja N, Sopory SK (2008) Plant signaling in stress. Plant Signal Behav 3:79–86

Van der Luit AH, Piatti T, Van Doorn A, Musgrave A, Felix G, Boller T, Munnik T (2000) Elicitation of suspension-cultured tomato cells triggers the formation of phosphatidic acid and diacylglycerol pyrophosphate. Plant Physiol 123:1507–1515

Venugopal SC, Jeong RD, Mandal MK, Zhu S, Chandra-Shekara AC, Xia Y, Hersh M, Stromberg AJ, Navarre D, Kachroo A, Kachroo P (2009) Enhanced disease susceptibility 1 and salicylic acid act redundantly to regulate resistance gene-mediated signaling. PLoS Genet 5. doi:10.1371/journal.pgen.1000545

Verberne MC, Hoekstra J, Bol JF, Linthorst HJM (2003) Signaling of systemic acquired resistance in tobacco depends on ethylene perception. Plant J 35:27–32

Vijayan P, Shockey J, Levesque CA, Cook RJ, Browse J (1998) A role for jasmonate in pathogen defense of Arabidopsis. Proc Natl Acad Sci USA 95:7209–7214

Wei Z, Hu W, Lin Q, Cheng X, Tong M, Zhu L, Chen R, He G (2009) Understanding rice plant resistance to the Brown Planthopper (Nilaparvata lugens): A proteomic approach. Proteomics 9:2798–2808

Weidhase RA, Kramell HM, Lehmann J, Liebisch HW, Lerbs W, Parthier B (1987) Methyl jasmonate-induced changes in the polypeptide pattern of senescing barley leaf segments. Plant Sci 51:177–186

Wu G, Shortt BJ, Lawrence EB, Levine EB, Fitzsimmons KC, Shah DM (1995) Disease resistance conferred by expression of a gene encoding H2O2-generating glucose oxidase in transgenic potato plants. Plant Cell 7:1357–1368

Xiao S, Brown S, Patrick E, Brearley C, Turner JG (2003) Enhanced transcription of the Arabidopsis disease resistance genes RPW8.1 and RPW8.2 via a salicylic acid-dependent amplification circuit is required for hypersensitive cell death. Plant Cell 15:33–45

Xing T, Higgins VJ, Blumwald E (1997) Race-specific elicitors of Cladosporium fulvum promote translocation of cytosolic components of NADPH oxidase to the plasma membrane of tomato cells. Plant Cell 9:249–259

Xing T, Ouellet T, Miki BL (2002) Towards genomic and proteomic studies of protein phosphorylation in plant-pathogen interactions. Trends Plant Sci 7:224–230

Yang Y, Shah J, Klessig DF (1997) Signal perception and transduction in plant defense responses. Genes Dev 11:1621–1639

Yoshida Y, Sano R, Wada T, Takabayashi J, Okada K (2009) Jasmonic acid control of GLABRA3 links inducible defense and trichome patterning in Arabidopsis. Development 136:1039–1048

Young SA, Wang X, Leach JE (1996) Changes in the plasma membrane distribution of rice phospholipase D during resistant interactions with Xanthomonas oryzae pv.oryzae. Plant Cell 8:1079–1090

Zhou W, Eudes F, Laroche A (2006) Identification of differentially regulated proteins in response to a compatible interaction between the pathogen Fusarium graminearum and its host, Triticum aestivum. Proteomics 6:4599–4609

Chapter 5
Auxin Genes and Auxin Responsive Factors in Signaling During Leaf Senescence

Maryam Sarwat, Preeti Rathore, Gowher Nabi, M.Z. Abdin, and Altaf Ahmad

Introduction

Auxin is a plant hormone whose main functions are- regulating cell division, cell expansion, differentiation, flowering, lateral root formation and tropic responses (Davies 2004). First auxin i.e. IAA was discovered in 1930s and since then, IAA has become synonymous with auxins.

Auxins have been the oldest fields of study in plant research and one of the earliest auxin effect noted was phototropism. Another effect was observed in tissue culture, where auxins promote rooting and shoot formation from undifferentiated callus (Skoog and Miller 1957) and cultured callus (Krikorian 1995) respectively.

Various endogenous compounds with auxin like activity have been discovered like IAA(Indole-3-Acetic Acid), 4-Cl-IAA (chlorinated form of IAA), PAA(Phenyl Acetic Acid), IBA (Indole-3-Butyric Acid). Two synthetic plant growth regulators have been described NAA(1-Naphthalacetic acid) and 2,4,-D related compounds eg. 2,4-Dichloro Phenoxy Butyric Acid(2,4-DB).

M. Sarwat (✉)
Pharmaceutical Biotechnology, Amity Institute of Pharmacy, Amity University, NOIDA 201303, India
e-mail: maryam21_7@yahoo.com; msarwat@amity.edu

P. Rathore
Pharmaceutical Biotechnology, Amity Institute of Pharmacy, NOIDA, India

G. Nabi
Genetics and Molecular Biology, Jazan University, College of Applied Medical Sciences, Jazan, Kingdom of Saudi Arabia

M.Z. Abdin
Centre for Transgenic Plant Development, Department of Biotechnology, Faculty of Science, Jamia Hamdard, New Delhi, India

A. Ahmad
Department of Botany, Faculty of Science, Jamia Hamdard, New Delhi, India

M. Sarwat et al. (eds.), *Stress Signaling in Plants: Genomics and Proteomics Perspective, Volume 1*, DOI 10.1007/978-1-4614-6372-6_5,

The biological function of auxin requires strict coordination of three complex processes: auxin metabolism, auxin transport, and auxin signaling. It is imperative to have an understanding about auxin biosynthetic pathway.

Auxin Biosynthetic Pathway

The biosynthesis of auxin in plants occurs via several interconnecting pathways and tryptophan (Trp) is the main precursor for IAA (Woodward et al. 2005; Zhao 2010). Although workers have proposed a Trp-independent pathway, its genetic basis is still unclear (Strader and Bartel 2008). The four proposed pathways for IAA biosynthesis from Trp are:

1. The YUCCA (YUC) pathway: a common IAA biosynthetic pathway where tryptophan is converted to tryptamine and produces auxin.
2. The indole-3-pyruvic acid (IPA) pathway: which is considered as a major IAA biosynthetic pathway in Arabidopsis and converts tryptophan to indole-3-pyruvic acid.
3. The indole-3-acetamide (IAM) pathway: exists widely in plants, which converts tryptophan to indole-3-acetamide
4. The indole-3-acetaldoxime (IAOx) pathway (previously called the CYP79B pathway): which was previously known as CYP79B pathway and converts tryptophan to indole-3-acetaldoxime. This last pathway is active only in those plants that have CYP79B family members to convert Trp to IAOx.

The pathway most prominent in plants is the IAM pathway. It has been very recently completely defined, resulting in the main IAA biosynthetic pathway in Arabidopsis (Mashiguchi et al. 2011; Won et al. 2011). While, the most widely studied pathway is the YUC pathway. The genes of the YUC pathway have been ubiquitously present in various plant species (Gallavotti et al. 2008). For example the SPARSE INFLORESCENCE 1 (SPI1) of maize, plays critical roles in vegetative and reproductive development (Gallavotti et al. 2008).

YUC family plays very important role as they encode flavin monooxygenase-like proteins that catalyze a rate-limiting step in IAA biosynthesis (Zhao et al. 2001). 11 YUC genes have been reported in Arabidopsis (Cheng et al. 2006; 2007). YUC catalyzes the conversion of tryptamine (TAM) to N- hydroxy - TAM (HTAM) in vitro (Zhao et al. 2001; Kim et al. 2007).

$$\text{TAM (Tryptamine)} \xrightarrow{\text{YUC}} \text{(HTAM) N-Hydroxy-TAM}$$

Studies on Arabidopsis mutants have helped in the better understanding of IAA biosynthetic pathways. YUCCA members (YUCCA1 and YUCCA6) have overlapping functions in localised auxin synthesis, which is important for the development of various plant organs. Kim et al. (2011) reported over expression of YUCCA6 causes delay in leaf senescence and produces phenotypes resulting in curled rosette leaves and long hypocotyls, which are characteristic of auxin overproduction.

The three Arabidopsis mutants shade avoidance 3, weak ethylene insensitive 8 (wei8), and transport inhibitor response 2—in which the TRYPTOPHAN AMINOTRANSFERASE OF ARABIDOPSIS 1 (TAA1) gene is disrupted (Tao et al. 2008). TAA1 is an important enzyme as it mediates the conversion of Trp to IPA in the first step of the IPA pathway. TAA1 plays critical roles in embryogenesis, flower development, seedling growth, vascular patterning, lateral root formation, tropism, shade avoidance, and temperature-dependent hypocotyl elongation (Tao et al. 2008). Two TAA1-related proteins TAR1 and TAR2 have been reported in Arabidopsis. Hence, the double-KO mutants of TAA1 and TAR2 genes, *wei8 tar2*, have significantly reduced IAA production and therefore exibit severe growth defects (Stepanova et al. 2008).

YUC and IPA pathways independently produce IAA and studies have shown that YUC and TAA families probably have same IAA biosynthetic pathway, which is suggested by the similarities between TAA deficient and YUC deficient mutants.

Auxin Transport

Auxin is produced in many tissues but it is synthesized in large amounts in the shoot apical meristem (Ljung et al. 2001). It is transported to distant sites for normal developments of later al root (Bhalerao et al. 2002) vasculature (Mattsson et al. 1999), embryonic axis (Friml et al. 2003), tropism (Friml et al. 2002) and phyllotaxis (Reinhardt et al. 2003). IAA is transported basipetally in shoots (Lomax et al. 1995) and suppresses lateral shoot growth (Thimann and Skoog 1934). Both acropetal (Scott and Wilkins 1968) and basipetal (Davies and Mitchell 1972) transport occurs in roots. Proteins of the PIN family are involved in polar auxin transport from one cell to the other (Bosco et al. 2012). *In silico* approaches have shown PIN- LIKES proteins (PILS) which can facilitate auxin transport by determining auxin sensitivity to cells (Barbez et al. 2012).

The influx of IAA into cells is mediated by AUX1 and other closely related proteins (Bennett et al. 1996). AUX1 is a transmembrane protein, localized asymmetrically in certain cells, facilitating directional auxin transport (Swarup et al. 2004).

The efflux of IAA is controlled by a multigene family (including EIR1/AGR1/PIN2. One of its members is PIN1 which is asymmetrically localized in the cell (Gälweiler et al. 1998).

In addition to PIN proteins, certain MULTI DRUG RESISTANCE-like (MDR) proteins play important role in polar auxin transport in arabidopsis (Noh et al. 2001), maize and *Sorghum bicolor* (Multani et al. 2003).

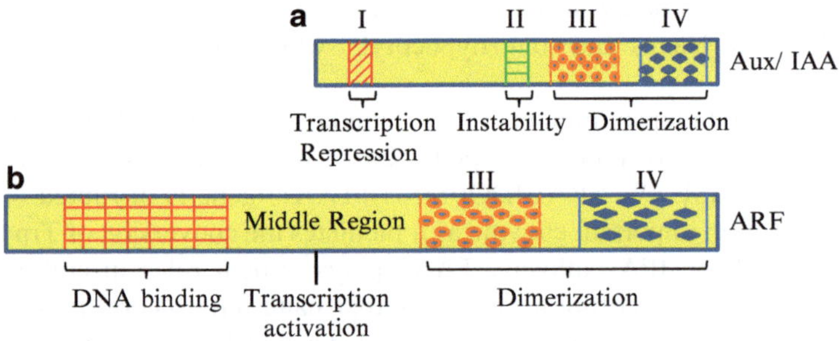

Fig. 5.1 Structure of (**a**) Aux/IAA and (**b**) ARF proteins. Conserved domains present in most Aux/IAA or ARF proteins. ARF domains III and IV are similar to the corresponding domains of Aux/IAA proteins. The middle region of most of the ARF proteins can activate transcription

Auxin-Induced Transcripts

Auxin induces accumulation of 3 families of transcripts: 1. SAUR's (Small Auxin Up RNA's) - SAUR are the auxin early responsive genes and their transcript accumulates just after an auxin exposure in Arabidopsis. SAUR transcripts are reported from soybean (Walker and Key 1982), Arabidopsis (Gil et al. 1994) maize (Knauss et al. 2003) and other plant species. These transcripts have a very short half-life. Kant and Rothstein (2009) have shown a rice SAUR 39 gene which negatively regulates auxin synthesis and transport. Constitutive expression of this gene causes low auxin levels and thus causes reduced growth and seed yield in rice plants. Polar auxin transport is also reduced, resulting in less chlorophyll in rice leaves.

GH-3 related transcripts – GH3 transcript accumulation is induced by auxin (Hagen et al. 1984). GH3 members are auxin inducible genes. Increased expression of GH3 family leads to increase in senescence.

The first auxin responsive genes are Aux/IAA. They encode 25- to 35-kDa proteins that are short-lived and localized to the nucleus (Hagen and Guilfoyle 2001). They are encoded by a large gene family in *Arabidopsis thaliana* with 29 members (Abel et al. 1995; Remington et al. 2004).

Most of them contain four highly conserved domains (I to IV) (Reed 2001) (Fig. 5.1a). Each domain has its own characteristic functional properties. Domain I is responsible for the transcriptional repressing activity of the proteins (Tiwari et al. 2004). Domain II plays a role in destabilizing Aux/IAA proteins and may be a target for ubiquitination (Colon-Carmona et al. 2000; Ouellet et al. 2001). Domains III and IV serve for homodimerization and heterodimerization with other *Aux/IAA* gene family members as well as for heterodimerization with the Auxin Response Factors (ARFs) (Kim et al. 1997; Ulmasov et al. 1999b). Domain III is part of a motif predicted to resemble the amphipathic fold found in the ribbon multimerization and DNA binding domains of Arc and MetJ repressor proteins (Abel et al. 1994). The predicted motif has been shown to play a role in dimerization/multimerization of Aux/IAA proteins and in heterodimerization among Aux/IAA and auxin response factor (ARF) proteins (Ouellet et al. 2001).

The genes with auxin-induced expression (including SAUR, GH3, Aux/IAA genes), have a common sequence (TGTCTC) in upstream regulatory regions. Such regions form Auxin Responsive Element (AuxRE) and confer auxin-induced gene expression (Ulmasov et al. 1995, 1997).

Role of Auxin in Senescence

Senescence is controlled by plethora of plant hormones (Sarwat et al. 2013). Some of them that induce senescence are ethylene, abscisic acid and jasmonates; other hormones like auxin, cytokinin and gibberellins suppress senescence (Lim et al. 2003).

Auxin Interactions with Other Hormones

Auxin interacts with several other hormones and it regulates or is regulated by these hormones. The most common interaction is between Auxin and cytokinins, which have an inversely proportional relationship. Auxin treatment inhibits biosynthesis of cytokinin (Eklöf et al. 2000; Nordstrom et al. 2004). Another important and interesting relation is that of Auxin and Ethylene. Although, auxin exposure stimulates the production of ethylene, but, ethylene inhibits auxin transport (Suttle 1988). Similarly, auxin induces gibberellin production (Wolbang et al. 2004). Auxin works in synergy with Brassinosteroids (BRs). Induction of similar transcripts was reported by independent application of these two hormones (Goda et al. 2004). Abscisic acid causes decrease in auxin level and inhibits lateral root formation (DeSmet et al. 2003).

1. Auxin-cytokinins → in-vitro → induction of root and shoot development

   ↓

   In-vivo → inverse relationship, auxin treatment can inhibit cytokinin biosynthesis
2. Auxin-Ethylene → Exogenous auxin application stimulates ethylene production

   ↓

   Ethylene inhibits lateral and basipetal auxin transport
3. Auxin-Gibberelic Acid → auxin necessary for production of gibberellins
4. Auxin-Abscissic Acid → Exposure to abscisic acid decreases free IAA levels.

The role of auxin in repression of transcription of some genes whose expression is correlated with senescence and/or abscission has been demonstrated long ago. Shoji et al. (1951) reported a gradient of auxin levels between the leaf blade and the stalk in bean leaves. Addicott et al. (1955) also suggests importance of auxin gradients in triggering the onset of senescence.

Auxin Gradient Theory

Auxin gradient theory of abscission regulation was given in 1955. Abscission is a type of senescence of leaves, fruits and flowers, which increases or decreases due to many factors like- light intensity, anaesthetics, auxins, photoperiod, mineral nutrients, mechanical injury, temperature, carbohydrates, temperature, diseases, insects, water, carbon dioxide and oxygen.

Shoji et al. (1951) investigated auxins in beans and found that concentration of auxins in leaflets (distal to the leaflet abscission zone) was 3 times the concentration in leaf stalks (proximal to abscission zone), but before the leaflets abscised, auxin concentration in the leaflets fell but remained same in the stalks, therefore suggesting that auxin gradient across the abscission zone is a factor in abscission regulation. These results were confirmed with cotton.

Relationship of auxins and other factors affecting abscission were also studied. Oxygen increases abscission and is required for auxin inactivation. Chemical defoliants cause a rapid decrease in leaf auxin. Ethylene increases abscission and decreases auxin in some species. Injury due to disease/insects that may decrease auxin, e.g. *Omphalia* (fungus) defoliates coffee by producing auxin inactivating enzyme and zinc, whose deficiency causes abscission and decreases auxin.

According to these studies, Auxin is a principal endogenous regulator of abscission and its gradient across abscission zone regulates the onset and rate of abscission. Auxin signaling involves two receptor systems, Auxin Binding Protein 1 (ABP1) and Transport Inhibitor Resistant 1 (TIR 1). This type of system helps plants to strengthen the tissue autonomy (Scherer 2011).

Signaling During Senescence

Recent studies have revealed a complex network of signaling pathways involving various signaling factors. The exact role of the genes involved in senescence is not clear because of the complex interactions of these pathways within the signaling network (van der Graaff et al. 2006). The microarray-based expression profiling and suppression subtractive hybridisation revealed hundreds of genes changing their expression during developmentally-regulated leaf senescence in Arabidopsis or when senescence was artificially induced through prolonged dark incubation or leaf detachment (van der Graaff et al. 2006). Some of these genes are down-regulated, such as those encoding photosynthetic proteins, termed as senescence down-regulated genes (SDGs), while other genes are up-regulated, referred to as senescence-associated genes (SAGs).

The genes which encode transcription factors (TFs) are also an important constituent of senescence associated expression clusters. The auxin response factor (ARF) family of transcription factors also falls in this category and regulate many responses to auxin. These proteins bind to auxin response elements (5-TGTCTC-3) in the promoters of auxin regulated genes and either activate or repress transcription of these genes (Ulmasov et al. 1997).

The ARF proteins are encoded by a large gene family in Arabidopsis (23 members). The N-terminal region consists of a B3 like DNA binding domain which helps an ARF to bind with the auxin-responsive *cis* -acting elements (*AuxREs*) found in the promoter region of auxin-responsive genes (Ulmasov et al. 1999a). The Aux/IAA proteins regulate auxin-gene expression through interaction with the ARF proteins. The III and IV domins of ARFs are similar to those found in the C terminus of Aux/IAAs (Fig. 5.1b). The amino acid composition of the middle region (MRs) between the DNA binding domain and domains III/IV determines whether an ARF protein functions as an activator or repressor (Ulmasov et al. 1999b; Tiwari et al. 2003). Some examples include ARFs containing Glutamine rich MRs which function as activators of Auxin Responsive Gene Expression. These include MP/ARF5, involved in embryo patterning and vascular formation; NPH4/ARF7, involved in phototropism and gravitropism; ARF19, which acts redundantly with NPH4/ ARF7 and control leaf expansion and lateral root growth. Another example is Proline/ Serine rich MRs which repress auxin responsive gene expression and these include ARF1 and ARF2.

ARFs (Auxin Response Factors) Mediate Auxin-Induced Changes in Gene Expression

After identification of AuxREs, ARF1 was isolated which is founding member of AuxRE binding protein family. ARFs can bind to AuxREs as homodimers, dimer with other ARFs or dimers with repressive Aux/IAA proteins. Mutations in several ARF genes of Arabidopsis led to various developmental defects like:

ETTIN/ARF3 (ETT) - floral abnormalities
MONOPTEROS/ARF5 (MP) - aberrant seedling morphology
NPH4/TIR5/MSG1/ARF7- deficient shoot phototropism

The mutants of ARF genes show gene-specific defects in plant development and exibit developmental phenotypes. This shows complex level of interaction between ARFs and AuxREs.

Aux/IAA Proteins Repress ARF Function

Interaction between Aux/IAA proteins and ARF proteins occur via C-terminal domains III and IV, which are conserved between Aux/IAA and ARF proteins. It is intriguing to know that auxin induced expression of some Aux/IAA genes function to repress auxin signaling. But after auxin exposure

Aux/IAA protein levels fall → increased transcription of Aux/IAA genes
↓
ensures transient auxin expression

Gain-of-function Aux/IAA mutations reduce auxin sensitivity in root elongation assays, confer auxin related developmental defects:

Gain of Function Mutants

1. *axr2/iaa7, axr3/iaa17*- altered gravitropism, apical dominance
2. *iaa28, slr/iaa14* - lateral root defects
3. *shy2/iaa3* - photomorphogenic defects
4. *msg2/iaa19* - hypocotyl tropism defects

Loss of Function Mutants

5. *shy2/iaa3*- large cotyledons, short hypocotyls

In an observation, 8 of 15 arabidopsis IAA biosynthetic genes (putative) are significantly regulated, which suggests that leaf senescence is associated with changes in IAA synthesis.

Accumulation of Auxin transcript activates self sustaining feed-back system by controlling auxin/ IAA inhibitors that prevents ARF transcription factors from regulating the target genes. Recently, Lau et al. (2011) have shown the ARF protein MONOPTEROS (MP) controls its own expression and the expression of its Aux/ IAA inhibitor BODELNOS (BDL).

Gene Mutations and Their Effects

1. ARF3/ETT—Affects gynoecium patterning
2. ARF7/NPH4/MSG1/TIR5—Impaired hypocotyl response to blue light and differential growth responses
3. ARF5/MP—Affects vascular strand formation and initiation of body axis in early embryo
4. ARF1, ARF2—Delay Rosette leaf senescence, flowering time affected

Each ARF protein is thought to play a central role in various auxin-mediated developmental processes.

An intensive study of ARF2 gene shows that ARF2 gene has 15 exons and encodes a 95.7 kDa polypeptide (version 5.0 of Arabidopsis genome annotation, Okushima et al. 2005a). Insertions of ARF2 gene show that all three insertions are

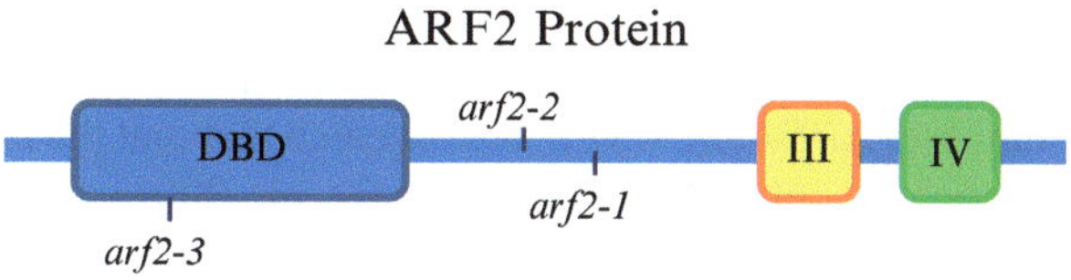

Fig. 5.2 Site of T-DNA insertions in the ARF2 protein showing the corresponding mutants

located downstream of DNA binding domain of ARF2. The *arf2-6* allele has been used for detailed phenotypic analysis as it has most upstream T-DNA insertion among the three insertions.

ARF1 and ARF2 mRNA, each present in roots, rosette and cauline leaves and flowers. ARF2 functions independently of ethylene and cytokinin response pathways. Studies revealed ARF2 to function independent of cytokinin and it has antagonising effect on apical hook formation. In etiolated seedlings ethylene promotes the degradation of ARF2 protein (Li et al. 2004).

ARF2 affects many aspects of senescence, whether induced by age, darkness, hormones, or oxidative stress. Second, phenotypic analyses of T-DNA insertion lines for ARF family genes have revealed that ARF2 plays a major role in controlling leaf senescence and that ARF1 acts only in a partially redundant manner (Ellis et al. 2005). Figure 5.2 shows the site of T-DNA insertion mutations in ARF 2 proteins. Third, microarray analysis has shown that ARF 2 transcripts increase in senescing leaves when induced by developmental ageing or darkness (Buchanan-Wollaston et al. 2005; Ellis et al. 2005). The ARF7 and ARF19 genes are also induced in senescing leaves (Lin and Wu 2004). However, mutations in these genes do not alter the leaf senescence phenotype, although they enhance the delay in senescence conferred by the arf2 mutation (Ellis et al. 2005).

Of all the ARF genes in *Arabidopsis thaliana*, ARF2 is most similar to ARF1 (Remington et al. 2004). The T-DNA insertion lines of the two genes (*arf1, arf2, arf1-4, arf2-8*) and the double stranded RNA interference (dsRNAi) lines of ARF2 are studied to throw light on the functions of the two genes. However, the T-DNA insertion lines of ARF-1 (*arf1*) exhibit no developmental defects in senescence, flowering time, abscission of floral organs, fertility, auxin-mediated lateral root initiation, auxin-inhibited root elongation, hypocotyl elongation in response to different light regimes, gravitropism, phototropism or shoot branching (Ellis et al. 2005), but they enhance late flowering, floral organ abscission and stamen elongation phenotypes of *arf2-8* andWs-0 *dsARF2* plants, and also cause delayed leaf senescence in Ws0 *dsARF2* (Ellis et al. 2005) plants. *arf1* mutations also enhance the effects of *arf2* mutations on apical hook formation (Li et al. 2004).

Gene expression analyses of ARF1 and ARF2 show that they have distinct functions. *arf1-4* mutant plants had increased stamens and in the abscission zone at the base of floral organs, whereas *arf2* and *dsARF2* plants do not show any abnormal phenotype of these organs.

ARF1 cause repression of auxin-induced gene expression in transient assays (Ulmasov et al. 1999a) and in planta, whereas ARF2 does not exhibit any such function.

It shows that ARF2 function involves additional mechanisms, and ARF2 may not conform to the canonical auxin response model. Through phylogenetic studies of *ARF1* and *ARF2,* it's revealed that that the two genes diverged prior to the monocot dicot split (Remington et al. 2004) and thus evolved distinguished functions.

As stated above ARF2 appears to function at multiple stages in the *Arabidopsis* life cycle. Lim et al. (2010) reported *arf2* plants to have delayed rosette leaf senescence, induction of flowering, stamen elongation, floral organ abscission and silique ripening. While, other workers have found that *arf2* mutant plants have delayed apical hook opening, enlarged seeds, stems and cotyledons, and elongated hypocotyls under red light (Li et al. 2004; Okushima et al. 2005b; Schruff et al. 2006).

Most probably, ARF2 acts through distinct mechanisms in different tissues and/ or at different developmental stages. Consistent with this idea, ARF2 is a target of ethylene signaling in etiolated seedlings (Li et al. 2004), but appears to act independently of ethylene in senescing leaves and flowers.

Activating ARFs Also Affects Senescence and Abscission

ARF1, ARF2 and other ARFs like, NPH4/ARF7 and ARF19 are all present in the same tissues and might interact together in the same cells. ARF2 might facilitate recognition of promoters by ARFs. ARFs reported to have different specificities for different promoters (Tiwari et al. 2003). In other words, different ARFs may target different promoters and thus affect different aspects of senescence.

As stated above senescence is closely related to abscission. Studies of Arabidopsis mutants exhibiting delayed senescence and abscission gave further input in this regard. Studies on transgenic plants overexpressing MADS transcription factor gene *AGL15* also show delayed flowering, floral organ abscission and fruit ripening (Fernandez et al. 2000; Fang and Fernandez 2002). AGL15 can play a role in maintaining plants in a juvenile state. Contrarily, ARF2 accelerates them. Thus, ARF2 have an antagonizing effect. Other studies further confirm that *ARF2* and *AGL15* do not regulate the expression of one another.

It was observed that the *ore14/ arf2* mutant is highly sensitive to auxin, as assessed by the inhibition of hypocotyl growth. Thus, ARF2 acts as transcription repressor of auxin-responsive gene expression (Tiwari et al. 2003; Lim et al. 2010). These results suggest that ARF2 mediates auxin signaling in a temporal and spatial manner.

Conclusion

The most prominent auxin responsive genes are Aux/IAA and ARFs. Studies revealed their involvement in auxin signaling during senescence. As, auxin is involved in various aspects of plant development, its role in senescence is very important, and so is the auxin gradient theory of leaf ageing. The various factors involved in auxin signaling are more promising and constitute the latest area of research. Most important of these are ARFs. The ARF1 and ARF2 are reported to have overlapping and distinct functions. The Arabidopsis T-DNA insertion mutants have paved our way to understand this complex and interesting phenomenon.

References

Abel S, Nguyen MD, Chow W, Theologis A (1995) ACS4, a primary indoleacetic acid-responsive gene encoding 1-aminocyclopropane1-carboxylate synthase in Arabidopsis thaliana. J Biol Chem 270:19093–19099

Addicott FT, Lynch RS, Carns HR (1955) Auxin gradient theory of abscission regulation. Science 121:644

Barbez E, Kubeš M, Rolčík J, Béziat C, Pěnčík A, Wang B, Rosquete MR, Zhu J, Dobrev PI, Lee Y, Zažímalovà E, Petrášek J, Geisler M, Friml J, Kleine-Vehn J (2012) A novel putative auxin carrier family regulates intracellular auxin homeostasis in plants. Nature 485:119–122

Bennett MJ, Marchant A, Green HG, May ST, Ward SP, Millner PA, Walker AR, Schultz B, Feldmann KA (1996) Arabidopsis AUX1 gene: a permease-like regulator of root gravitropism. Science 273:948–950

Bhalerao RP, Eklöf J, Ljung K, Marchant A, Bennett M, Sandberg G (2002) Shoot-derived auxin is essential for early lateral root emergence in Arabidopsis seedlings. Plant J 29:325–332

Bosco CD, Dovzhenko A, Liu X, Woerner N, Rensch T, Eismann M, Eimer S, Hegermann J, Paponov IA, Ruperti B, Heberle-Bors E, Touraev A, Cohen JD, Palme K (2012) The endoplasmic reticulum localized PIN8 is a pollen-specific auxin carrier involved in intracellular auxin homeostasis. Plant J (in press). doi:10.1111/j.1365-313X.2012.05037.x

Cheng Y, Dai X, Zhao Y (2006) Auxin biosynthesis by the YUCCA flavin monooxygenases controls the formation of floral organs and vascular tissues in Arabidopsis. Genes Dev 20:1790–1799

Cheng Y, Dai X, Zhao Y (2007) Auxin synthesized by the YUCCA flavin monooxygenases is essential for embryogenesis and leaf formation in Arabidopsis. Plant Cell 19:2430–2439

Davies PJ, Mitchell EK (1972) Transport of indoleacetic acid in intact roots of Phaseolus coccineus. Planta 105:139–154

DeSmet I, Signora L, Beeckman T, Inzé D, Foyer CH, Zhang H (2003) An abscisic acid-sensitive checkpoint in lateral root development of Arabidopsis. Plant J 33:543–555

Eklöf S, Astot C, Sitbon F, Moritz T, Olsson O, Sandberg G (2000) Transgenic tobacco plants co-expressing Agrobacterium iaa and ipt genes have wild-type hormone levels but display both auxin- and cytokinin-overproducing phenotypes. Plant J 23:279–284

Ellis CM, Nagpal P, Young JC, Hagen G, Guilfoyle TJ, Reed JW (2005) AUXIN RESPONSE FACTOR1 and AUXIN RESPONSE FACTOR2 regulate senescence and floral organ abscission in Arabidopsis thaliana. Development 132:4563–4574

Fang SC, Fernandez DE (2002) Effect of regulated overexpression of the MADS domain factor AGL15 on flower senescence and fruit maturation. Plant Physiol 130:78–89

Friml J, Wísniewska J, Benková E, Mendgen K, Palme K (2002) Lateral relocation of auxin efflux regulator PIN3 mediates tropism in Arabidopsis. Nature 415:806–809

Friml J, Vieten A, Sauer M, Weijers D, Schwarz H, Hamann T, Offringa R, Juurgens G (2003) Efflux-dependent auxin gradients establish the apical-basal axis of Arabidopsis. Nature 426:147–153

Gallavotti A, Barazesh S, Malcomber S, Hall D, Jackson D, Schmidt RJ, McSteen P (2008) Sparse inflorescence1 encodes a monocot-specific YUCCAlike gene required for vegetative and reproductive development in maize. Proc Natl Acad Sci USA 105:15196–15201

Gälweiler L, Guan C, Muuller A, Wisman E, Mendgen K, Yephremov A, Palme K (1998) Regulation of polar auxin transport by AtPIN1 in Arabidopsis vascular tissue. Science 282:2226–2230

Gil P, Liu Y, Orbović V, Verkamp E, Poff KL, Green PJ (1994) Characterization of the auxin-inducible SAUR-AC1 gene for use as a molecular genetic tool in Arabidopsis. Plant Physiol 104:777–784

Goda H, Sawa S, Asami T, Fujioka S, Shimada Y, Yoshida S (2004) Comprehensive comparison of auxin-regulated and brassinosteroid-regulated genes in Arabidopsis. Plant Physiol 134:1555–1573

Kant S, Rothstein S (2009) Auxin-responsive SAUR39 gene modulates auxin level in rice. Plant Signal Behav 4:1174–1175

Kim JI, Murphy AS, Baek D, Lee SW, Yun DJ, Bressan RA, Narasimhan ML (2011) YUCCA6 over expression demonstrates auxin function in delaying leaf senescence in Arabidopsis thaliana. J Exp Bot 62:3981–3992

Knauss S, Rohrmeier T, Lehle L (2003) The auxin-induced maize gene ZmSAUR2 encodes a short-lived nuclear protein expressed in elongating tissues. J Biol Chem 278:23936–23943

Lau S, De Smet I, Kolb M, Meinhardt H, Jürgens G (2011) Auxin triggers a genetic switch. Nat Cell Biol 13:611–615

Lehmann T, Hoffmann M, Hentrich M, Pollmann S (2010) Indole-3-acetamide-dependent auxin biosynthesis: A widely distributed way of indole-3-acetic acid production? Eur J Cell Biol 89:895–905

Li H, Johnson P, Stepanova A, Alonso JM, Ecker JR (2004) Convergence of signaling pathways in the control of differential cell growth in Arabidopsis. Dev Cell 7:193–204

Lim PO, Woo HR, Nam HG (2003) Molecular genetics of leaf senescence in Arabidopsis. Trends Plant Sci 8:272–278

Lim PO, Lee IC, Kim J, Kim HJ, Ryu JS, Woo HR, Nam HG (2010) Auxin response factor 2 (ARF2) plays a major role in regulating auxin-mediated leaf longevity. J Exp Bot 61:1419–1430

Ljung K, Bhalerao RP, Sandberg G (2001) Sites and homeostatic control of auxin biosynthesis in Arabidopsis during vegetative growth. Plant J 28:465–474

Lomax TL, Muday GK, Rubery PH (1995) Auxin transport. In: Davies PJ (ed) Plant hormones. Kluwer, Dordrecht, pp 509–530

Mashiguchi K, Tanaka K, Sakai T, Sugawara S, Kawaide H, Natsume M, Hanada A, Yaeno T, Shirasu K, Yao H, Mcsteen P, Zhao Y, Hayashi KI, Kamiya Y, Kasahara H (2011) The main auxin biosynthesis pathway in Arabidopsis. Proc Natl Acad Sci USA 108:18512–18517

Multani DS, Briggs SP, Chamberlin MA, Blakeslee JJ, Murphy AS, Johal GS (2003) Loss of an MDR transporter in compact stalks of maize br2 and sorghum dw3 mutants. Science 302:81–84

Noh B, Murphy AS, Spalding EP (2001) Multidrug resistance-like genes of Arabidopsis required for auxin transport and auxin-mediated development. Plant Cell 13:2441–2454

Nordstrom A, Tarkowski P, Tarkowska D, Norbaek R, Astot C, Dolezal K, Sandberg G (2004) Auxin regulation of cytokinin biosynthesis in Arabidopsis thaliana: a factor of potential importance for auxin-cytokinin-regulated development. Proc Natl Acad Sci USA 101:8039–8044

Okushima Y, Mitina I, Quach HL, Theologis A (2005a) AUXIN RESPONSE FACTOR2 (ARF2): a pleiotropic developmental regulator. Plant J 43:29–46

Okushima Y, Overvoorde PJ, Arima K, Alonso JM, Chan A, Chang C, Ecker JR, Hughes B, Lui A, Nguyen D, Onodera C, Quach H, Smith A, Yu G, Theologis A (2005b) functional genomic analysis of the auxin response factor gene family members in Arabidopsis thaliana: unique and overlapping functions of ARF7 and ARF19. Plant Cell 17:444–463

Ouellet F, Overvoorde PJ, Theologis A (2001) IAA17/AXR3: biochemical insight into an auxin mutant phenotype. Plant Cell 13:829–841

Reinhardt D, Pesce ER, Stieger P, Mandel T, Baltensperger K, Bennett M, Traas J, Friml J, Kuhlemeier C (2003) Regulation of phyllotaxis by polar auxin transport. Nature 426:255–260

Remington DL, Vision TJ, Guilfoyle TJ, Reed JW (2004) Contrasting modes of diversification in the Aux/IAA and ARF gene families. Plant Physiol 135:1738–1752

Sarwat M, Naqvi AR, Ahmad P, Ashraf M, Akram NA (2013) Phytohormones and MicroRNAs as Sensors and Regulators of Leaf Senescence: Assigning Macro Roles to Small Molecules. Biotechnology Advances. doi: 10.1016/j.biotechadv.2013.02.003

Scherer GF (2011) AUXIN-BINDING-PROTEIN1, the second auxin receptor: what is the significance of a two-receptor concept in plant signal transduction? J Exp Bot 62:3339–3357

Schruff MC, Spielman M, Tiwari S, Adams S, Fenby N, Scott RJ (2006) The AUXIN RESPONSE FACTOR 2 gene of Arabid opsis links auxin signaling, cell division, and the size of seeds and other organs. Development 133:251–261

Scott TK, Wilkins MB (1968) Auxin transport in roots II. Polar flux of IAA in Zea roots. Planta 83:323–334

Shoji K, Addicott FT, Swets WA (1951) Auxin in response to leaf blade abscission. Plant Physiol 26:189–191

Stepanova AN, Robertson-Hoyt J, Yun J, Benavente LM, Xie D-Y, Doležal K, Schlereth A, Jürgens G, Alonso JM (2008) TAA1-mediated auxin biosynthesis is essential for hormone crosstalk and plant development. Cell 133:177–191

Strader LC, Bartel B (2008) A new path to auxin. Nat Chem Biol 4:337–339

Suttle JC (1988) Effect of ethylene treatment on polar IAA transport, net IAA uptake and specific binding of N-1-naphthylphthalamic acid in tissues and microsomes isolated from etiolated pea epicotyls. Plant Physiol 88:795–799

Swarup R, Kargul J, Marchant A, Zadik D, Rahman A, Mills R, Yemm A, May S, Williams L, Millner P et al (2004) Structurefunction analysis of the presumptive Arabidopsis auxin permease AUX1. Plant Cell 16:3069–3083

Tao Y, Ferrer J-L, Ljung K, Pojer F, Hong F, Long JA, Li L, Moreno JE, Bowman ME, Ivans LJ, Cheng Y, Lim J, Zhao Y, Ballaré CL, Sandberg G, Noel JP, Chory J (2008) Rapid synthesis of auxin via a new tryptophan-dependent pathway is required for shade avoidance in plants. Cell 133:164–176

Thimann KV, Skoog F (1934) On the inhibition of bud development and other functions o f growth substance in Vicia faba. Proc Royal Soc Lond Biol Sci 114:317–339

Tiwari SB, Hagen G, Guilfoyle TJ (2003) The roles of auxin response factor domains in auxin-responsive transcription. Plant Cell 15:533–543

Tiwari SB, Hagen G, Guilfoyle TJ (2004) Aux/IAA proteins contain a potent transcriptional repression domain. Plant Cell 16:533–543

Ulmasov T, Hagen G, Guilfoyle TJ (1997) ARF1, a transcription factor that binds auxin response elements. Science 276:1865–1868

Ulmasov T, Hagen G, Guilfoyle TJ (1999a) Activation and repression of transcription by auxin-response factors. Proc Natl Acad Sci USA 96:5844–5849

Ulmasov T, Hagen G, Guilfoyle TJ (1999b) Dimerization and DNA binding of auxin response factors. Plant J 19:309–319

Van der Graaff E, Schwacke R, Schneider A, Desimone M, Flugge UI, Kunze R (2006) Transcription analysis of Arabidopsis membrane transporters and hormone pathways during developmental and induced leaf senescence. Plant Physiol 141:776–792

Walker JC, Key JL (1982) Isolation of cloned cDNAs to auxin-responsive poly(A)+ RNAs of elongating soybean hypocotyl. Proc Natl Acad Sci USA 79:7185–7189

Wolbang CM, Chandler PM, Smith JJ, Ross JJ (2004) Auxin from the developing inflorescence is required for the biosynthesis o f active gibberellins in barley stems. Plant Physiol 134:769–776

Woodward C, Bemis SM, Hill EJ, Sawa S, Koshiba T, Torii KU (2005) Interaction of auxin and ERECTA in elaborating Arabidopsis inflorescence architecture revealed by the activation tagging of a new member of the YUCCA family putative flavin monooxygenases. Plant Physiol 139:192–203

Zhao Y (2010) Auxin biosynthesis and its role in plant development. Annu Rev Plant Biol 61:49–64

Zhao Y, Christensen SK, Fankhauser C, Cashman JR, Cohen JD, Weigel D, Chory J (2001) A role of flavin monooxygenase-like enzymes in auxin biosynthesis. Science 291:306–309

Chapter 6
CBF-Dependent Cold Stress Signaling Relevant Post Translational Modifications

Prakriti Kashyap and Renu Deswal

Abbreviations

CBF c-repeat binding factor
ICE Inducer of CBF expression
PTM Post translational modifications

Introduction

Plants acquire tolerance to stress environment by reprogramming metabolism and gene expression (Mazzucotelli et al. 2008). The eukaryotic transcriptome is highly dynamic and changes in response to the environment. The changes in gene expression are controlled by a large array of transcriptional regulators, some of which are repressors while others are activators (Spoel et al. 2010). Temporal activation or repression of specific genes is accomplished via a plethora of transcriptional regulators. In past two decades, efforts had been made to better understand the transcriptional changes induced by the environmental changes and many important advances have been made. This has led to the identification of signaling proteins and transcription factors which regulate gene expression. But transcriptional knowledge is not enough to completely explore the complicated mechanism of

P. Kashyap • R. Deswal (✉)
Molecular Physiology and Proteomics Laboratory, Department of Botany, University of Delhi, Delhi 110007, India
e-mail: rdeswal@botany.du.ac.in

M. Sarwat et al. (eds.), *Stress Signaling in Plants: Genomics and Proteomics Perspective, Volume 1*, DOI 10.1007/978-1-4614-6372-6_6,

signal transduction. Recent advances in proteomics and metabolomics have provided chances to integrate gene expression with proteins and metabolites. These studies give us a hint that post transcriptional and post translational mechanisms are involved in abiotic stress response (Mazzucotelli et al. 2008). The post-translational modifications affect the function of transcription regulators by affecting their localization, conformation and stability.

Environment variables are major determinants of plant growth and development. Low temperature is an environmental factor that severely affects plant growth and development, and limits crop distribution and yield. To acquire tolerance plants undergo various biochemical and physiological changes. Most of these changes are due to up- and down- regulation of hundreds of genes in stress conditions including low temperature. Among these regulons the most studied and best characterized cold responsive regulon is the CBF regulon driven by ICE and CBF transcription factors. CBFs have been described to regulate the expression of around 12 % of Arabidopsis cold inducible genes (Fowler and Thomashow 2002). Like other regulons, the study of this regulon is also incomplete without the information about post translational modifications. Out of about 350 post translational modifications known till date, phosphorylation, S-nitrosylation, ubiquitylation and sumoylation are major post translational modifications associated with cold stress signaling. All these play important role in CBF dependent signaling either directly or indirectly.

Ubiquitination

Gene transcription and ubiquitin mediated proteolysis are two contrasting processes. Transcription leads to protein synthesis via translation whereas proteolysis causes the death of a protein by degrading it. During gene regulation, transcriptional activation has its major role but it is not sufficient as turning off the transcription also has its essential role. This turning off is done by repressors and proteolysis. The proteolysis is mediated by a complex structure referred to as proteasome (Glickman and Ciechanover 2002). In eukayotic cells the proteosome is found in the cytosol and the nucleus. It has a large mass with a sedimentation coefficient of 26S. 26S proteasomes are ATP-driven, multisubunit proteolytic machines that preferentially degrade proteins tagged with polyubiquitin chains (Hershko et al. 1982; Pickart and Rose 1985; Hershko and Ciechanover 1998; Voges et al. 1999; Elsasser and Finley 2005). The 26S proteasome comprises a 20S barrel-shaped catalytic core as well as 19S regulatory complexes at both ends. Immunoprecipitation and DNA microarrays have revealed that proteasome are associated with many gene loci (Auld et al. 2006). Moreover, yeast transcriptome when treated with proteasome inhibitor showed upregulation of a large set of genes further verifying the role of proteasome in gene regulation (Fleming et al. 2002; Bhaumik and Malik 2008).

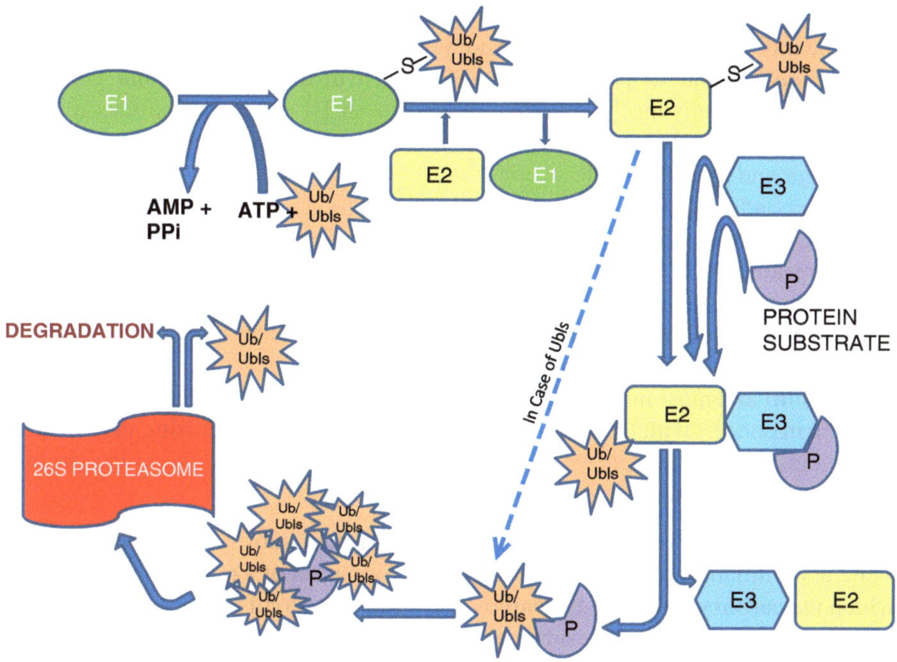

Fig. 6.1 Pathway for Ubiquitin (Ub) and ubiquitin like proteins (Ubls) mediated proteasomal degradation of protein substrates

Degradation of proteins via ubiquitin mediated proteolysis is highly specific in its action. Ubiquitination is covalent addition of ubiquitin to the selected target proteins (Weissman 2001). The addition of multi-ubiquitin chain to proteins targets these proteins to undergo intracellular degradation through the 26S proteasome. Recent structural studies of polyubiquitylated model substrates showed that conjugation of a tetra-ubiquitin chain represents the minimum signal for efficient proteasomal targeting (Thrower et al. 2000).

Ubiquitin-Proteasome Pathway

Ubiquitin is a small protein of 76 amino acids present only in eukaryotes. Moreover, ubiquitin is very conserved among eukaryotic organisms. Ubiquitin itself does not degrade proteins. Role of ubiquitin is to tag the proteins for degradation and degradation occurs through 26S proteasome. Ubiquitin mediated proteolysis is an ATP dependent process and is energetically expensive. Of course, ATP is not needed to degrade proteins but is needed to specifically target the proteins to be degraded. Conjugation of ubiquitin to protein occurs in a three step cascade mechanism involving three enzyme activities (Myung et al. 2001) (Fig. 6.1).

1. The first enzyme in this cascade is E1, the ubiquitin activating enzyme. As the name indicates, it activates the carboxyl group of Gly-76 of ubiquitin in an ATP dependent manner. In this step, ATP gets hydrolyzed to PPi and generates ubiquitinyl adenylate intermediate bound to an E1 enzyme. The Cys residue, an active site of E1 gets covalently bound to ubiquitin via a high energy thioester linkage and releases AMP.
2. The second enzyme is E2, the ubiquitin carrier proteins or ubiquitin conjugating enzyme. After activation, ubiquitin is transferred to a thiol group of an active site Cys residue of E2 by transacylation reaction. E2 transfers ubiquitin directly to a protein substrate or it can also work in concert with E3 to form an amide isopeptide bond between the carboxyl group of Gly-76 of ubiquitin and protein substrate's internal lysine residue.
3. The third and most important enzyme is E3, ubiquitin protein ligases. E3 ligases work in concert with E2. E3 is suggested to have role in recognizing the substrate proteins and E2 recognizes the substrate protein to transfer the ubiquitin to it because it is ligated to E3.

After the formation of mono-ubiquitinated protein substrate, the repetition of same ubiquitination cascade further cause the formation of a polyubiquitin chain in which the carboxyl group of the carboxy terminal Gly-76 of ubiquitin is covalently linked to an internal Lys residue of ubiquitin that is already linked to protein substrate. This mono- or polyubiquitylation favours the recognition of ubiquitylated protein substrates by specific receptors within the 26S proteasome or within adaptor proteins associated with proteasome. Moreover, these mono- or polyubiquitylations via Ub lys residue regulates several biological processes like subcellular localization, chromatin structure, signal transduction, DNA damage repair and protein synthesis. The E3 enzymes and E3 multi protein complexes are highly specific to corresponding protein substrates and E2 enzymes. This specificity of E3 ligases is the reason for the specificity of ubiquitination. The different combinations of E2 and E3 enzymes allow selective tagging of proteins and make them susceptible to degradation. Different from E2 and E3, E1 enzyme family is highly conserved.

Ubiquitination of ICE1

The important task of recognition of specific target proteins to be degraded is accomplished by E3 ligases. They recognize specific motifs in their substrates. Some substrates carry constitutively active degradation signals whereas some undergo some sort of modification of motif like phosphorylation. Some members of such E3 ligases family are N-end rule family, Hect (homologous to E6-AP carboxy terminus) domain family and Ring finger domain family (Myung et al. 2001).

Although more than 200 functionally distinct proteins in eukaryotes are found to contain the RING finger protein sequence motif and its variants, until recently no specific function(s) had been identified (Myung et al. 2001). The RING finger protein, high expression of osmotically responsive gene (HOS1), an E3 ligase is known to participate in the modification of ICE1 in Arabidopsis post translationally and negatively regulates the CBF regulon and cold responses. Yeast two hybrid assays done brought us the information that HOS1 physically interacts with ICE1 (Dong et al. 2006). It has been found that cold induces the degradation of ICE for attenuation of cold response and this proteasome mediated degradation of ICE1 requires HOS1. Ubiquitination assays confirm that HOS1 mediates the ubiquitination of ICE1 both in vitro and *in vivo*. Overexpression of HOS1 in *Arabidopsis thaliana* represses the expression of CBFs (CBF1, CBF2 and CBF3) and their downstream genes (COR15, COR47, RD29A) which results in decreased cold tolerance and sensitivity to freezing stress. At the same time, increased cold tolerance in *hos1* mutants (lost the function of HOS1) confirmed the role of HOS1 in repressing the CBF regulon. Recent studies to evaluate the role of serine/threonine residues in regulation of ICE1 suggest that serine 403 is a key residue for attenuation of cold stress response by HOS1-mediated degradation of ICE1 (Miura et al. 2011). The substitution of serine 403 by alanine enhanced the transactivational activity of ICE1 in Arabidopsis protoplasts. The transgenic plants overexpressing ICE1 (S403A) showed the increased expression of cold induced genes, such as CBF3/DREB1A, COR 47 and KIN, resulting in enhanced freezing tolerance. Moreover, the serine to alanine substitution increased the stability of ICE1 against cold induced proteasome degradation. Ubiquitylation assays revealed that the serine 403 is not the main target residue for ubiquitylation by HOS1, but it inhibits the polyubiquitylation of ICE1 *in vivo* as ICE1 (S403A) protein showed reduced polyubiquitylation mediated by E3 ligase HOS1 *in vivo* but could be polyubiquitylation in vitro. Thus, ICE1 (S403A) may facilitate protein stability which enhanced expression of CBF3/DREB1A and its regulon genes (Miura et al. 2011).

Sumoylation

The ubiquitin modifying the protein substrates post translationally and targeting them to proteasome mediated degradation is most widely studied (Konstantinova et al. 2008). But over the past several years, the existence of some other ubiquitin like proteins has gained the attention. These ubiquitin like proteins (Ubls) have been identified as post translational modifiers in yeasts, plants and metazoans (Hochstrasser 2009). Many biological functions have been associated with these like several cellular and developmental processes and other molecular level changes occurring in response to hormonal stimuli, host–pathogen interaction-related stimuli and environmental stimuli. The Ubls are present in wide range of size from 72 to 186 amino acids and share a much conserved three dimensional structure among eukaryotes and highly similar to that of ubiquitin (Hochstrasser 2009; Kerscher et al. 2006). Recent studies have evidenced at least four functional Ubl subfamilies in plants: SUMO (small ubiquitin-related modifier), RUB [related to ubiquitin, an ortholog of mammalian NEDD8 (neural precursor cell expressed, developmentally down-regulated 8)], and ATG8 and ATG12 (autophagy 8 and 12). These Ubls have a core Ubfold three-dimensional structure and are conserved in eukaryotes (Downes and Vierstra 2005). Among these the SUMO proteins are associated with the broadest range of functions and the largest number of known substrates. Sumoylation is a post-translational modification of protein substrates by covalent conjugation of the SUMO peptide (Hay 2005).

Sumoylation Pathway

Sumoylation share same biochemical steps involved with ubiquitination. Like ubiquitination, conjugation in sumoylation also occurs in three biochemical steps utilizing activating enzymes (E1), conjugating enzymes (E2) and E3 ligases (Fig. 6.1).

1. E1 enzyme- Sumoylation involves two heteromeric SUMO- specific E1 enzymes (SAE1 and SAE2). These enzymes catalyze thioesterification between a catalytic cysteine residue in SAE2 and SUMO. This reaction requires energy and hence takes place in an ATP dependent manner (Geiss-Fridelander and Melchior 2007).
2. E2 enzymes- These are conjugating enzymes which facilitate the transfer of SUMO to protein substrate. SUMO is transferred to SUMO E2 conjugating enzyme (SCE1) through a thioester linkage and then finally to protein substrate. (Geiss-Fridelander and Melchior 2007.) E2 can directly sumoylate the substrates through covalent linkage to the ε amino group of the lysine residue in conserved sumoylation motif CKxE/D (C: Large hydrophobic amino acids, K: lysine, x: any amino acid, E/D: acidic amino acids) (Bernier-Villamor et al. 2002; Melchior et al. 2003).
3. E3 enzymes- These are ligases (SIZ1). In sumoylation E3 ligases act on many different proteins. There are several reports that sumoylation can occur in vitro without involving E3 ligases. Athough biological significance of E3 independent sumoylation is not determined yet (Geiss-Fridelander and Melchior 2007).

The E2 and E3 enzymes of the sumoylation machinery act on many different proteins different from most of the ubiquitination pathways in which E3 ligases specifically recognize target proteins. Phosphorylation of substrates is expected to play an important role in regulation of sumoylation as it seems to be regulated specifically at target level (Gao and Karim 2005). Sumoylation is a reversible modification in which the protein substrates get deconjugated by SUMO specific proteases (Mazzucotelli et al. 2008). Plant SUMO-specific proteases have both peptidase activity for processing pre-SUMO and isopeptidase activity to recycle SUMO from substrates (Miura et al. 2007).

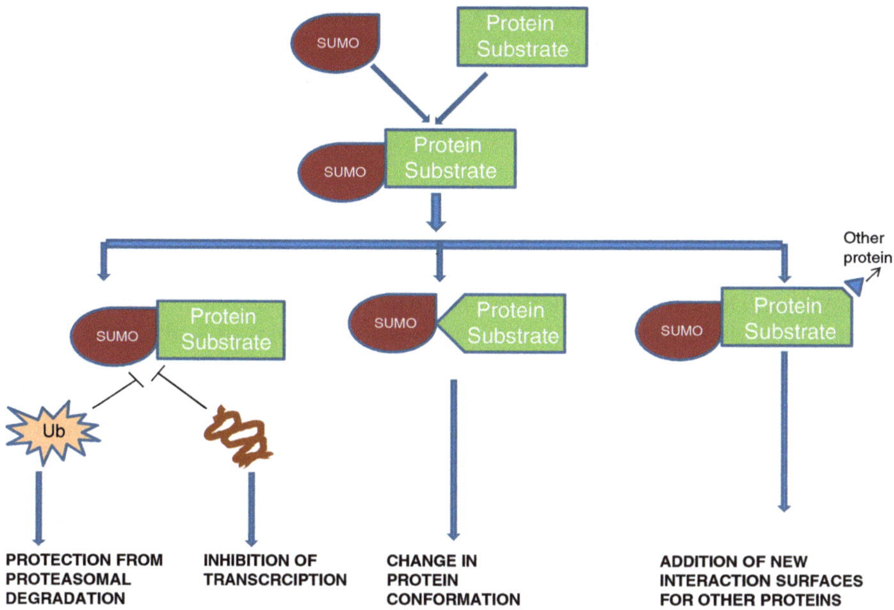

Fig. 6.2 Effect of Sumoylation on protein substrates

Post translational modification by sumoylation affects transcriptional activity by various mechanisms (Gill 2005) (Fig. 6.2):

(a) Inhibition of the interaction of sumoylated substrate with other proteins or other binding factors like DNA.

    As sumoylation occurs at lysine residue, it can compete with other post-translational modifications for lysine. It has been reported that sumoylation can block ubiquitylation at a particular lysine residue and protect the protein substrate from proteasomal degradation.

(b) Change in the conformation of proteins.

    Sumoylation might cause conformational changes in protein substrates as reported for thymine DNA glycosylase (TDG), a DNA repair enzyme. However, significance of the conformational changes has not been determined yet.

(c) Addition of new interaction surfaces on the protein substrates.

    After the covalent addition of SUMO polypeptide, transcription factors gets modified and this modification has been shown to increase the interaction of these with proteins that had little or no affinity for the respective transcription factors otherwise.

    All these changes are not mutually exclusive. SUMO conjugation can both promote and impair transcription by favouring nuclear import of transcription factors or further sending them in sub nuclear domains having repressive environment respectively. SUMO can also influence the promoter binding ability of transcription factors affecting gene regulation. Whatever may be the approach, activating or repressing the transcription, sumoylation has established its key role in gene regulation in response to environmental stimuli.

Sumoylation of ICE1

Sumoylation/Desumoylation has been shown to have pivotal role in plant responses to biotic and abiotic stress. In genomic expression analysis of Arabidopsis, out of 17,000 drought induced genes identified, 300 were found to be upregulated by SIZ1 SUMO E3 ligase (Catala et al. 2007). SIZ1 is a controller of low temperature adaptation in plants. SIZ1 is known to affect the function of transcription factors. Recent studies have established that SIZ1 participates in cold acclimation by regulating the activity of ICE1 and CBF3/DREB1A expression in *Arabidopsis thaliana* (Miura et al. 2007). The *siz1* mutants which impair SIZ1 ligase function showed freezing sensitivity revealing that SIZ1 regulates freezing and chilling tolerances. The *siz1* mutation did not affect the ICE1 which means that SIZ1 is not a transcriptional regulator of ICE1 but could regulate its activity post translationally. SIZ1 is necessary for sumoylation of ICE1 in vitro but substitution of lysine at position 393 with (K393R) blocked SIZ1 mediated sumoylation in vitro. This tells us that lysine at position 393 is the principal site for sumoylation in Arabidopsis. Sumoylation of ICE1 at K393 reduces polyubiquitylation of ICE1, protecting it from proteasome mediated degradation and increasing its stability. This leads to enhanced expression of CBF3/DREB1A resulting in freezing tolerance. Sumoylated ICE1 is less polyubiquitylated than unsumoylated ICE1 but K393 is not the principal residue for ubiquitination. Furthermore, sumoylated ICE1 represses MYB15, which is a negative regulator of CBF3/DREB1A and confers freezing tolerance (Miura et al. 2007).

Phosphorylation

When we talk of post translational modification, the most investigated modification that one can think of is phosphorylation. Phosphorylation is most important and diverse post translational modification that regulates the biological function of proteins. It affects enzymatic activity, subcellular localization, interaction with other proteins and half life (Bentem et al. 2006). Proteins gets phosphorylated and dephosphorylated and this interchange is responsible for regulation of biological function of protein. This regulation of proteins is known as phosphoregulation. During phosphorylation, protein kinase phosphorylates the protein substrate by addition of covalent bound phosphate group. Approximately 1,000 genes in Arabidopsis genome are predicted to encode protein kinases (The Arabidopsis

Genome Initiative 2000). The serine, threonine and tyrosine amino acid residues are targeted for phosphorylation. The enzymes that dephosphorylate proteins are phosphatases. All phosphorylation via protein kinases are mediated by divalent cations like Mn^{2+}, Mg^{2+} which stabilizes the high-energy bonds of the donor molecule. The donors are ATP or ATP like derivatives. Phosphorylation of a protein helps it to interact with other proteins having recognition domains in them to identify the phosphorylated residues, the serine, threonine or tyrosine. This sequential phosphorylation regulates for many signaling pathways.

Phosphorylation in Cold Stress

Cold stress in plants accompanies alteration of gene expression and many biochemical and physiological functions (Levitt 1980; Nishida and Murata 1996). It has been shown that during cold stress, the calcium is exported from apoplast to cytosol (Monroy and Dhindsa 1995). Moreover, cold specific phosphorylation helps in cold acclimation (Monroy and Dhindsa 1995). Calcium is a well-known second messenger and calcium influx is one of the essential factors in signaling pathways during different stress conditions. In Arabidopsis and alfalfa, this influx of calcium has been reported (Knight et al. 1991; Trewavas and Gilroy 1991). The inhibitor of cold induced phosphorylation, calcium chelators and calcium channel blockers prevents the phosphorylation and cold induced gene expression. In alfalfa, cold induced influx of calcium is inhibited by calcium chelators such as BAPTA (1,2-bis (o-aminophenoxy)ethane N, N, N', N'– tetracetic acid) and calcium channel blockers that results in decrease in the expression of a cold inducible cas 15 gene and ultimately decreased cold tolerance (Monroy and Dhindsa 1995). Inhibition of alfalfa cas 15 induction by the protein kinase inhibitor staurosporine and enhanced expression at 25 °C by protein phosphatase inhibitor okadaic acid has been reported by Monroy et al. (1998). Moreover, they also showed that low temperature causes a calcium influx dependent rapid and dramatic decrease in protein phosphatase 2A activity. This decrease in phosphatase activity leads to the phosphorylation of proteins involved in cold stress signaling cascades resulting in cold acclimation (Monroy et al. 1998). Another interesting protein kinase responsible for inducing the expression of cold regulated genes and inferring cold tolerance to plants is mitogen activated protein (MAP) kinase (Jonak et al. 1996). There is specificity in activation of MAP kinases in response to cold. Two MAP kinases in alfalfa MMK2 and MMK3 are not activated by low temperature whereas p44 MMK4 is activated within 10 min upon exposure to low temperature. Interestingly, the transcript levels for p44MMK4 increase within 20 min upon exposure to cold but the amount p44MMK protein does not change (Jonak et al. 1996). It has been reported that in arabidopsis, the expression of MAP kinase and MAP kinase kinase kinase genes is increased in response to low temperature (Mizoguchi et al. 1996). Moreover, there are evidences that calcium dependent protein kinases (CDPKs) in Arabidopsis and alfalfa accumulate in response to low temperatures. All these kinases are known to have role in cold acclimation (Thomashow 1999).

Phosphorylation of transcription factors have also found its significance in proteasome mediated degradation of transcription factors during signaling cascades. Many unstable transcription factors with the fate to be degraded contain a conserved phosphodegron motif, a sequence containing phosphorylatable serine residues. This phosphorylation acts as a signal to induce ubiquitin mediated degradation.

Phosphorylation and CBF Pathway

Though there are no direct evidences of CBF getting phosphorylated but there are some suggestions. These suggestions are based on the fact that phosphorylation is enhanced after cold induced calcium influx. This enhanced phosphorylation during cold increase the induction of cold stress related genes and confers cold tolerance to plants. Moreover ICE1, an upstream regulator functioning in CBF signaling cascade is degraded via ubiquitin mediated proteasomal degradation upon exposure to cold stress (Dong et al. 2006). ICE1 also gets sumoylated as described before. Phophorylation tag the protein substrates to get recognized by E3 ligases in ubiquitin pathways for degradation through 26S proteasome. So, there are suggestions that ICE1 gets modified by this post translational modification before it gets ubiquitinated or sumoylated but there are no direct reports or evidences for this. Phosphorylation is most common post translational modification and CBF signaling pathway is most common and most studied cold regulon. So these may be related but their relationship is yet to be experimentally proved.

S-Nitrosylation

Nitric oxide (NO) is a ubiquitous diatomic gas involved in physiological functions in plant and animals (Besson-Bard et al. 2008). In plants, S- nitrosylation has emerged as an important post-translational modification affecting protein function. It has been established that thiol-disulphide bonding controls transcription regulator conformation. Reactive oxygen species (ROS) and reactive nitrogen species (RNS) can act as signal molecules in plants and regulates protein function by reversible oxidative modifications of cysteine residues (Hess et al. 2005). These reactive species are also capable of causing serious injury to cells by altering the protein functioning. The different reversible thiol modifications, including disulphide bonding, S-nitrosylation (covalent attachment of nitric oxide (NO)), S-glutathionylation (disulphide attachment of glutathione) and S-hydroxylation are known to affect the binding affinities and therefore the activity of transcription factor, e.g. a bacterial transcription factor (OxyR) (Kim et al. 2002). The unregulated disulphide linkage can cause misfolding of proteins, altering their conformation and thus physiological

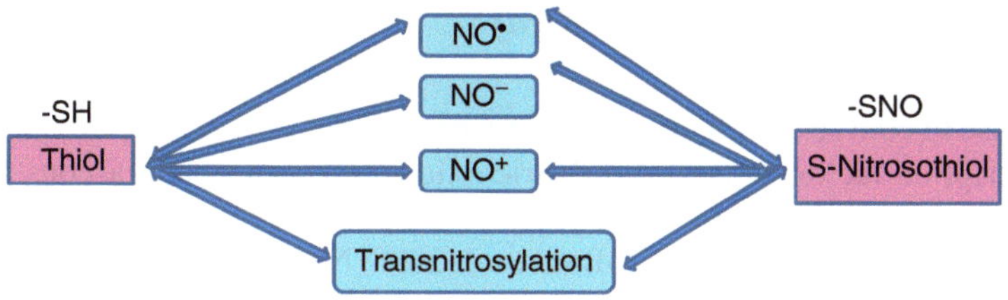

Fig. 6.3 Nitrosylation of thiol group

functions which can prove to be fatal. The thiol- disulphide linkages are controlled by reversible action of oxidative and reductive reactions regulated by both enzymatic and non-enzymatic systems (Spoel et al. 2010). In plants, S- nitrosylation has a well known impact on disulphide bond formation.

S-nitrosylation is the reaction of NO with the sulfur atom of Cys residue to form an S-NO bond (S-Nitrosothiol). S- nitrosylating agents involved are NO., NO^+, NO^- and metal-NO complexes (Fig. 6.3). NO. reacts with oxygen to form nitrogen oxides which further dissociates into NO^+. The electrophillic attack of this NO^+ on thiol group results in S- nitrosylation and thus the formation of S-nitrosothiol. Some evidences suggest the involvement of nucleophillic attack of NO^- on relatively electropositive sulfur of Cys residue. The reaction between thiyl radical (RS.) and NO. also causes nitrosylation (Wang et al. 2006; Lindermayr and Durner 2009). Apart from these reactions transnitrosylation has also been reported in which there is a direct exchange of NO^+ from an S-nitrosylated protein to reactive thiolate group of target protein. Low molecular weight S-nitrosothiol such as nitrosoglutathione (GSNO) are involved in transnitrosylation and serve as an endogenous reservoir of NO in cells (Wang et al. 2006). GSNO is formed by S- nitrosylation of glutathione (GSH). Recent studies have led to the identification of a GSH dependent GSNOR, GSNO reductase which converts GSNO into ammonia and GSH disulphide (GSSH). The contribution of GSNO in regulation of S-nitrosothiol content is also favoured by the evidence that came from the study of *Arabidopsis thaliana* mutants raised with impaired expression of GSNOR. These mutants showed increased level of S-nitrosothiols (Feechan et al. 2005; Rusterucci et al. 2007).

S-nitrosylation has recently emerged as an important post translational modification of proteins and is becoming an intensive field of research in plants. Like other post translational modifications, S-nitrosylation is also reversible. The S-nitrosothiol is very labile and redox sensitive. The S-NO bond formed in S- nitrosylation is reversed by the action of internal reducing agents such as GSH, ascorbate and reduced metal ions. The cumulative effect of nitrosylation and denitrosylation results in regulation of protein function. Specific enzymes catalyzing denitrosylation have been suggested (Astier et al. 2011).

The counting of protein substrates getting nitrosylated is increasing regularly. The substrates reported till date include protein kinases, phosphatases, ion channels, metabolic and regulatory enzymes, cytoskeletal and structural proteins, transcription factors, oxidoreductases, defence related proteins and respiratory proteins.

S-*Nitrosylation and CBF Pathway*

In response to cold stress, change in expression of genes occurs in plants and an array of transcription factors is involved. Nitric oxide (NO) is a small gaseous signaling molecule known to be involved in plant development and in plant's responses to abiotic and biotic stresses (Besson-Bard et al. 2008; Neill et al. 2008). Nitric oxide is accumulated in plants during cold stress and it can influence the transcriptional activity of genes to confer freezing tolerance (Zhao et al. 2009). The NO is produced in plants upon short exposure to cold and the expression of CBF genes also changes rapidly (15 min) in response to cold. Recent studies showed that NO participate in CBF dependent cold induced gene expression (Cantrel et al. 2011). The expression of some selected transcription factors was analyzed in *nil nia2* mutant plants with impaired nitrate reductase (NR) activity in arabidopsis. The expression of AtCBF1 and AtCBF3 was impaired in these transgenic lines whereas the expression of AtCBF2 remained unaffected. The decreased expression of CBF1 and CBF3 is correlated with impaired expression of COR15a, LT130 and LT178 genes, the downstream target genes in CBF dependent cold induced pathway. At the same time, the expression of ZAT12, a transcription factor not belonging to CBF family, was not affected (Cantrel et al. 2011).

There are reports of one more transcription factor getting modified with NO, AtMYB2 (an R2R3 type MYB transcription factor) from *Arabidopsis thaliana* (Serpa et al. 2007). Most plant MYB genes encode proteins of the R2R3-MYB class with more than 125 members in Arabidopsis (Stracke et al. 2001). R2R3-MYB transcription factors have a modular structure, with an N terminal DNA-binding domain (the MYB domain) and an activation or repression domain usually located at the C terminus. R2R3 DNA binding domains are formed by R2 and R3 adjacent MYB repeats and have a very conserved Cys at position 53 in DNA recognition helix of R2. R2R3 DNA binding domains and MYB proteins from animals and fungi have a conserved Cys at position 53. The reduction of this Cys is essential for cMyb DNA- binding and any mutation, alkylation or oxidation of this Cys inhibits the DNA-binding property (Guehmann et al. 1992; Brendeford et al. 1998). In maize, the study of P1 regulator of flavonoid biosynthesis showed the importance of an intra-molecular disulphide bond between Cys 53 and Cys 49 present in R2R3 MYB proteins under non reducing conditions in DNA binding (Heine et al. 2004). AtMYB2 gets nitrosylated at this Cys 53 residue and its DNA binding ability is impaired after this post translational modification (Serpa et al. 2007).

CBFs contain myc recognition regions in their promoters and ICE binds to these sequences and promotes the transcription of CBF. But myc recognition region is not the only conserved region found in promoter of CBF. There are also myb recognition domains. So it was proposed and confirmed that CBF genes are also regulated by MYB transcription factors (Shinwari et al. 1998). MYB 15 has been shown to bind to the myb recognition sequences in the promoters of CBF1, 2 and 3 in Arabidopsis. MYB 15 is a negative regulator of CBF dependent cold stress signaling pathway (Agarwal et al. 2006). Transgenic lines raised overexpressing MYB 15 showed reduced transcripts of CBF1, 2 and 3 and decreased freezing tolerance. Consistently, *myb15* mutant plants with loss of function of MYB15 showed increased levels of all three CBFs and thus increased freezing tolerance (Agarwal et al. 2006). MYB15 is also a member of R2R3 MYB family like AtMYB2. There are no direct reports of MYB15 getting nitrosylated yet. But if like AtMYB2, it also gets nitrosylated in response to cold stress, its DNA binding ability is also impaired then it will no longer bind to CBF and would not negatively regulate its expression. This means enhanced transcription of CBF and its target genes to confer freezing tolerance to plants.

PTMs and ICE-CBF Mediated Signaling Pathway

ICE is an upstream transcription factor in transcriptional cascade leading to activation of CBF and COR genes and ultimately resulting in cold tolerance. ICE is constitutively present in all tissues of Arabidopsis. It has been suggested that during cold stress ICE is activated due to some post-translational modification and binds to an MYC- recognition sequence in the promoter region lying upstream in CBF genes. CBF belongs to AP2/EREBP family of DNA binding proteins. This family of transcription factors recognize and bind to cold and dehydration responsive DNA regulatory element known as CRT/DRE cis element in the promoter of many cold responsive genes. A large number of low temperature induced genes have been identified and characterized in plants (Tsuda et al. 2000; Zhang et al. 2009). These include Late Embryogenesis Abundant (LEA), Dehyrins (DHN), Responsive to Abscisic acid (RAB), Low Temperature Responsive (LT) and Cold Responsive (COR) genes. As LEA family encoding highly hydrophilic proteins constitute majority of these genes, they are called COR/LEA or simply COR genes. The COR gene expression and freezing tolerance shows positive correlation. Among these gene products, many are structural proteins that are directly involved in protecting plants from stress. Significantly, multiple biochemical changes that are associated with cold acclimation like accumulation of simple sugars and the amino acid proline also take place due to this regulon. Moreover CBFs also regulate the expression of genes involved in phosphoinositide metabolism, osmolyte biosynthesis, ROS detoxification, membrane transport, hormone metabolism and signaling and confer cellular protection (Fowler and Thomashow 2002; Maruyama et al. 2004; Yamaguchi-Shinozaki and Shinozaki 2006).

The post translational modifications ubiquitination and sumoylation are known to control ICE1 dependent cold signaling. The RING finger protein high expression of osmotically responsive gene (HOS1), an E3 ligase physically interacts with ICE1 and mediates the ubiquitination of ICE1 both in vitro and *in vivo*. It has been found that cold induces the proteasome mediated degradation of ICE1 and this degradation requires HOS1. Overexpression of HOS1 decreases cold tolerance by repressing the expression of CBFs and their downstream genes, consistent to which in loss of function hos 1 mutant plants show increased cold responsive gene expression

(Dong et al. 2006). Recent studies suggest that serine 403 is a key residue for attenuation of cold stress response by HOS1-mediated degradation of ICE1 (Miura et al. 2011). The substitution of serine 403 by alanine enhanced the transactivational activity of ICE1 in Arabidopsis protoplasts and also the expression of cold induced genes, such as CBF3/DREB1A, COR 47 and KIN1 in plants resulting in cold tolerance. The serine 403 is not the main target residue for ubiquitylation, but it inhibits the polyubiquitylation of ICE1 *in vivo* as ICE1 (S403A) protein showed reduced polyubiquitylation mediated by E3 ligase HOS1 *in vivo* but could be polyubiquitylation in vitro. Thus, ICE1 (S403A) may facilitate protein stability which led to enhanced expression of CBF3/DREB1A and its regulon genes.

Sumoylation/desumoylation of proteins has been shown to have a pivotal role in plant responses to abiotic and biotic stress responses and in ABA and salicylic acid signaling (Miura et al. 2007). During sumoylation, SUMO (small ubiquitin-related modifier) proteins conjugates to protein substrates by SUMO E3 ligases and in desumoylation, SUMO proteins get removed from their target proteins by SUMO proteases. Sumoylation might protect target proteins from proteasomal degradation because sumoylation prevents ubiquitination (Ulrich 2005). An Arabidopsis SUMO E3 ligase, SIZ1 is required for the accumulation of SUMO conjugates during cold stress. *siz1,* null mutant shows hypersensitivity to chilling and freezing stresses due to the reduced cold induced expression of CBF/DREBs particularly of CBF3/DERB1A and other genes of regulon such as COR15A and COR47. The *siz1* mutants show unaltered expression of ICE1. Instead SIZ1 sumoylates ICE1 at K393 residue, the principal site for SUMO conjugation, and blocks its HOS1 mediated polyubiquitinization, stabilizing and enhancing the activity of the transcription factor. Furthermore, sumoylated ICE1 represses MYB15, which is a negative regulator of CBF3/DREB1A and confers freezing tolerance (Miura et al. 2007). The mechanisms by which a SUMO conjugation of ICE1 is induced by cold stimulus or by which this sumoylated ICE1 effects transactivation of specific genes to coordinate cold adaptation are yet to be established.

ICE is known to bind with MYC recognition sequences in promoters of CBF genes. Apart from MYC recognition sequences, many putative MYB binding sequences are also present in the promoters of CBF genes (Shinwari et al. 1998) proposing the role of MYB-like transcription factors in controlling CBF gene expression. Recently, MYB15, a member of R2R3-MYB family of transcription factors has been shown to bind to the promoters of CBF1, 2 and 3 genes (Agarwal et al. 2006). Transgenic plants overexpressing MYB15 showed reduced levels of CBF1, CBF2 and CBF3 transcripts in cold and thus decreased chilling tolerance. At the same time its knockout mutants showed increased levels of all these three CBFs and exhibited increased freezing tolerance. Thus MYB15 confirms its involvement in CBF regulon. Moreover, ICE1 is known to physically interact with AtMYB15. Although the functional consequence of this interaction lies in dark at present but strongly emphasize on the role of MYB15 in cold regulation cascade. All these regulators cause physiological and biochemical changes leading to cold tolerance (Fig. 6.4).

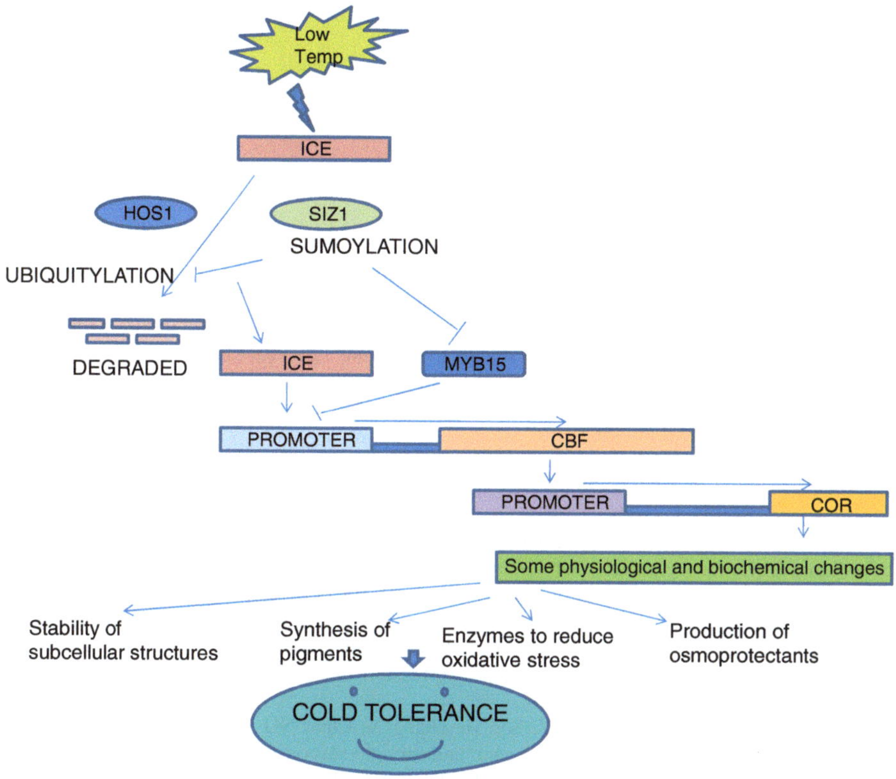

Fig. 6.4 Overview of cold stress induced CBF-ICE signaling pathway

Conclusions

Understanding the molecular mechanism involved in plant responses to cold stress is very important as it can help in manipulating plants to improve their cold tolerance. ICE-CBF pathway is most investigated pathway that operates in plants during cold acclimation and involves a transcriptional cascade. Like other transcriptional cascades, this is also incomplete without the knowledge of PTMs. Ubiquitination and sumoylation have confirmed role in this pathway by modifying ICE1 post translationally. There are hints of other two common PTMs, phosphorylation and nitrosylation to be involved. These need to be investigated and confirmed to get the complete information about the complex regulatory network involved in CBF dependent signaling pathway.

Acknowledgements Partial financial support from DST/SAP/DST PURSE grants of government of India to RD and research fellowship to PK is acknowledged.

References

Agarwal M, Hao Y, Kapoor A, Dong CH, Fujii H, Zheng X, Zhu JK (2006) A R2R3 type MYB transcription factor is involved in the cold regulation of CBF genes and in acquired freezing tolerance. J Biol Chem 281:37636–37645

Astier J, Rasul S, Koen E, Manzoor H, Besson-Bard A, Lamotte O, Jeandroz S, Durner J, Lindermayr C, Wendehenne D (2011) S-nitrosylation: An emerging post-translational protein modification in plants. Plant Sci 181:527–533

Auld KL, Brown CR, Casolari JM, Komili S, Silver PA (2006) Genomic association of the proteasome demonstrates overlapping gene regulatory activity with transcription factor substrates. Mol Cell 2:861–871

Bentem SF, Roitinger E, Anrather D, Csazar E, Hirt H (2006) Phosphoproteomics as a tool to unravel plant regulatory mechanisms. Physiol Plantarum 126:110–119

Bernier-Villamor V, Sampson DA, Matunis MJ, Lima CD (2002) Structural basis for E2-mediated SUMO conjugation revealed by a complex between ubiquitin-conjugating enzyme Ubc9 and RanGAP1. Cell 108:345–356

Besson-Bard A, Pugin A, Wendehenne D (2008) New insights into nitric oxide signaling in plants. Annu Rev Plant Biol 59:21–39

Bhaumik SR, Malik S (2008) Diverse regulatory mechanisms of eukaryotic transcriptional activation by the proteasome complex. Crit Rev Biochem Mol Bio 43:419–433

Brendeford EM, Andersson KB, Gabrielsen OS (1998) Nitric oxide (NO) disrupts specific DNA binding of the transcription factor c-Myb in vitro. FEBS Lett 425:52–56

Cantrel C, Vazquez T, Puyaubert J, Reze N, Lesch M, Kaiser WM, Dutilleul C, Guillas I, Zachowski A, Baudouin E (2011) Nitric oxide participates in cold-responsive phosphosphingolipid formation and gene expression in *Arabidopsis thaliana*. New Phytol 189:415–427

Catala R, Ouyang J, Abreu IA, Hu Y, Seo H, Zhang X, Chua NH (2007) The Arabidopsis E3 SUMO ligase SIZ1 regulates plant growth and drought responses. Plant Cell 19:2952–2966

Dong CH, Agarwal M, Zhang Y, Xie Q, Zhu JK (2006) The negative regulator of plant cold responses, HOS1, is a RING E3 ligase that mediates the ubiquitination and degradation of ICE1. Proc Natl Acad Sci USA 103:8281–8286

Downes B, Vierstra RD (2005) Post-translational regulation in plants employing a diverse set of polypeptide tags. Biochem Soc Trans 33:393–399

Elsasser S, Finley D (2005) Delivery of ubiquitinated substrates to protein unfolding machines. Nat Cell Biol 7:742–749

Feechan A, Kwon E, Yun BW, Wang Y, Pallas JA, Loake GJ (2005) A central role for S-nitrosothiols in plant disease resistance. Proc Natl Acad Sci USA 102:8054–8059

Fleming JA, Lightcap ES, Sadis S, Thoroddsen V, Bulawa CE, Blackman RK (2002) Complementary whole-genome technologies reveal the cellular response to proteasome inhibition by PS-341. Proc Natl Acad Sci USA 99:1461–1466

Fowler S, Thomashow MF (2002) Arabidopsis transcriptome profiling indicates that multiple regulatory pathways are activated during cold acclimation in addition to the CBF cold response pathway. Plant Cell 14:1675–1690

Gao M, Karim M (2005) Regulating the regulators: control of protein ubiquitination and ubiquitin-like modifications by extracellular stimuli. Mol Cell 19:581–593

Geiss-Fridelander R, Melchior F (2007) Concepts in sumoylation: a decade on. Nat Rev Mol Cell Biol 8:947–956

Gill G (2005) Something about SUMO inhibits transcription. Curr Opin Genet Dev 15:536–541

Glickman MH, Ciechanover A (2002) The Ubiquitin-proteasome proteolytic pathway: destruction for the sake of construction. Physiol Rev 82:373–428

Guehmann S, Vorbrueggen G, Kalkbrenner F, Moelling K (1992) Reduction of a conserved Cys is essential for Myb DNA-binding. Nucleic Acids Res 20:2279–2286

Hay RT (2005) SUMO: a history of modification. Mol Cell 18:1–12

Heine GF, Hernandez JM, Grotewold E (2004) Two cysteines in plant R2R3 MYB domains participate in REDOX-dependent DNA binding. J Biol Chem 279:37878–37885
Hershko A, Ciechanover A (1998) The ubiquitin system. Annu Rev Biochem 67:425–480
Hershko A, Eytan E, Ciechanover A, Haas AL (1982) Immunochemical analysis of the turnover of ubiquitin-protein conjugates in intact cells. Relationship to the breakdown of abnormal proteins. J Biol Chem 257:13964–13970
Hess DT, Matsumoto A, Kim SO, Marshall HE, Stamler JS (2005) Protein S-nitrosylation: purview and parameters. Nat Rev Mol Cell Biol 6(2):150–166
Hochstrasser M (2009) Origin and function of ubiquitin-like proteins. Nature 458:422–429
Jonak C, Kiegerl S, Ligterink W, Barker PJ, Huskisson NS, Hirt H (1996) Stress signaling in plants: a mitogen activated protein kinase pathway is activated by cold and drought. Proc Natl Acad Sci USA 93:11274–11279
Kerscher O, Felberbaum R, Hochstrasser M (2006) Modification of proteins by ubiquitin and ubiquitin-like proteins. Annu Rev Cell Dev 22:159–180
Kim SO, Merchant K, Nudelman R, Beyer WF Jr, Keng T, DeAngelo J, Hausladen A, Stamler JS (2002) OxyR: a molecular code for redox related signaling. Cell 109:383–396
Knight MR, Campbell AK, Smith SM, Trewavas AJ (1991) Imaging calcium dynamics in living plants using semi-synthetic recombinant aequorins. Nature 352:524–526
Konstantinova IM, Tsimokha AS, Mittenberg AG (2008) Role of proteasomes in cellular regulation. Int Rev Mol Biol 267:59–124
Levitt J (1980) Responses of plants to environmental stresses I. Chilling, freezing, and high temperature stresses. Academic, New York
Lindermayr C, Durner J (2009) S-nitrosylation in plants: pattern and function. J Proteomics 73:1–9
Maruyama K, Sakuma Y, Kasuga M, Ito Y, Seki M, Goda H, Shimada Y, Yoshida S, Shinozaki K, Yamaguchi-Shinozaki K (2004) Identification of cold-inducible downstream genes of the Arabidopsis DREB1A/CBF3 transcriptional factor using two microarray systems. Plant J 38:982–993
Mazzucotelli E, Mastrangelo AM, Crosatti C, Guerra D, Stanca AM, Cattiveli L (2008) Abiotic stress response in plants: When post-transcriptional and post-translational regulations control transcription. Plant Sci 174:420–431
Melchior F, Schergaut M, Pichler A (2003) SUMO: ligases, isopeptidases and nuclear pores. Trends Biochem Sci 28:612–618
Miura K, Jin JB, Hasegawa PM (2007) Sumoylation, a post-translational regulatory process in plants. Curr Opin Plant Biol 10:495–502
Miura K, Ohta M, Nakazawa M, Ono M, Hasegawa PM (2011) ICE1 Ser403 is necessary for protein stabilization and regulation of cold signaling and tolerance. Plant J 67:269–279
Mizoguchi T, Irie K, Hirayama T, Hayashida N, Yamaguchi-Shinozaki K, Matsumoto K, Shinozaki K (1996) A gene encoding a mitogen activated protein kinase kinase kinase is induced simultaneously with genes for a mitogen-activated protein kinase and an S6 ribosomal protein kinase by touch, cold, and water stress in *Arabidopsis thaliana*. Proc Natl Acad Sci USA 93:765–769
Monroy AF, Dhindsa RS (1995) Low temperature signal transduction: induction of cold acclimation-specific genes of alfalfa by calcium at 25°C. Plant Cell 7:321–331
Monroy AF, Sangwan V, Dhindsa RS (1998) Low temperature signal transduction during cold acclimation: protein phosphatase 2A as an early target for cold inactivation. Plant J 13:653–660
Myung J, Kim BK, Crews CM (2001) The ubiquitin-proteasome pathway and proteasome inhibitors. Med Res Rev 21:245–273
Neill S, Bright J, Desikan R, Hancock J, Harrison J, Wilson I (2008) Nitric oxide evolution and perception. J Exp Bot 59:25–35
Nishida I, Murata N (1996) Chilling sensitivity in plants and cyanobacteria: the crucial contribution of membrane lipids. Annu Rev Plant Physiol Plant Mol Biol 47:541–568
Pickart CM, Rose IA (1985) Functional heterogeneity of ubiquitin carrier proteins. J Biol Chem 260:1573–1581

Rusterucci C, Espunya MC, Diaz M, Chabannes M, Martinez MC (2007) S-nitrosoglutathione reductase affords protection against pathogens in Arabidopsis, both locally and systemically. Plant Physiol 143:1282–1292

Serpa V, Vernal J, Lamattina L, Grotewold E, Cassia R, Terenzi H (2007) Inhibition of AtMYB2 DNA-binding by nitric oxide involves cysteine S-nitrosylation. Biochem Biophys Res Commun 361:1048–1053

Shinwari ZK, Nakashima K, Miura S, Kasuga M, Seki M, Yamaguchi- Shinozaki K, Shinozaki K (1998) Biochem Biophys Res Commun 250:161–170

Spoel SH, Tada Y, Loake GJ (2010) Post-translational protein modification as a tool for transcription reprogramming. New Phytol 186:333–339

Stracke R, Werber M, Weisshaar B (2001) The R2R3-MYB gene family in *Arabidopsis thaliana*. Curr Opin Plant Biol 4:447–456

Thomashow MF (1999) Plant cold acclimation: freezing tolerant genes and regulatory mechanisms. Annu Rev Plant Physiol Plant Mol Biol 50:571–599

Thrower JS, Hoffman L, Rechsteiner M, Pickart CM (2000) Recognition of the polyubiquitin proteolytic signal. EMBO J 19:94–102

Trewavas AJ, Gilroy S (1991) Signal transduction in plant cells. Trends Genet 7:356–361

Tsuda K, Tsvetanov S, Takumi S, Mori N, Atanassov A, Nakamura C (2000) New members of a cold-responsive group-3 Lea / Rab-related Cor gene family from common wheat (Triticum aestivum L.). Genes Genet Syst 75:179–188

Ulrich HD (2005) Mutual interactions between the SUMO and ubiquitin systems: a plea of no contest. Trends Cell Biol 15:525–532

Voges D, Zwickl P, Baumeister W (1999) The 26S proteasome: a molecular machine designed for controlled proteolysis. Annu Rev Biochem 68:1015–1067

Wang Y, Yun BW, Kwon E, Hong JK, Yoon J, Loake GJ (2006) S-nitrosylation: an emerging redox-based post-translational modification in plants. J Exp Bot 57:1777–1784

Weissman AM (2001) Themes and variations on ubiquitylation. Nat Rev Mol Cell Biol 2:169–178

Yamaguchi-Shinozaki K, Shinozaki K (2006) Transcriptional regulatory networks in cellular responses and tolerance to dehydration and cold stresses. Annu Rev Plant Biol 57:781–803

Zhang CZ, Fei SZ, Warnke S, Li LJ, Hannapel D (2009) Identification of genes associated with cold acclimation in perennial ryegrass. J Plant Physiol 166:1436–1445

Zhao MG, Chen L, Zhang LL, Zhang WH (2009) Nitric reductase dependent nitric oxide production is involved in cold acclimation and freezing tolerance in Arabidopsis. Plant Physiol 151:755–767

Chapter 7
Regulation and Function of Protein S-Nitrosylation in Plant Stress

Gitto Thomas Kuruthukulangarakoola and Christian Lindermayr

Introduction

Physiological processes in plants are regulated by a complex network of signaling processes and the role of biological molecules that can mediate signals across this network is very vital. Nitric oxide (NO) is one such signaling molecule known to coordinate many physiological processes in almost all the organisms studied. Following the discovery of NO as a signaling molecule in animals (Ignarro et al. 1987), it was first identified in plants as a mediator of defense responses during disease resistance (Delledonne et al. 1998; Durner et al. 1998). Since then, studies have revealed the ubiquitous signaling nature of NO in regulating plethora of physiological processes in plants like germination (Bethke et al. 2006; Belenghi et al. 2007), stomatal closure (Neill et al. 2002a; Garcia-Mata et al. 2003; Sokolovski et al. 2005), flowering (He et al. 2004), senescence (Corpas et al. 2004; Guo and Crawford 2005), wounding responses (Huang et al. 2004), and abiotic stresses (Grun et al. 2006; Corpas et al. 2011). Ubiquitous behavior of NO in signaling processes puzzled the researchers to find an answer on how this sensitive and highly diffusible gaseous free radical can be regulated spatially and temporally. It is now known that plants scrutinize this regulation by controlling the NO bioactivity at different levels ranging from NO production to site-specific reactivity and finally, the NO turnover. Cellular redox status, a primary means of coordinating many signaling pathways (Spoel and Loake 2011), is also a crucial regulator of NO bioactivity.

Various upstream signaling pathways like extracellular adenosine triphosphate, phosphatidic acid, cyclic nucleotide phosphate, calcium and mitogen-activated protein kinases coordinates plant stress responses to induce NO production in plants

G.T. Kuruthukulangarakoola • C. Lindermayr (✉)
Institute of Biochemical Plant Pathology, Helmholtz Zentrum München,
Ingolstädter Landstrasse 1, 85764 Neuherberg, Germany
e-mail: lindermayr@helmholtz-muenchen.de

M. Sarwat et al. (eds.), *Stress Signaling in Plants: Genomics and Proteomics Perspective, Volume 1*, DOI 10.1007/978-1-4614-6372-6_7,

(Sueldo et al. 2010; Gaupels et al. 2011a; Ma and Berkowitz 2011). However, efforts to identify the mechanism through which these upstream signaling events regulate NO production are hampered due to the fact that an exact enzymatic source of NO production is yet to be revealed in plants. Nevertheless, NO accumulation in plants has witnessed to induce downstream signaling events (Besson-Bard et al. 2008; Aboul-Soud et al. 2009). More importantly, effectiveness of NO-signaling strongly relies on its spatial regulation that is conferred by the specificity of NO to react with selective targets and on its temporal regulation that is achieved by the reversibility of NO-induced target modification.

Mechanisms to regulate biological processes in eukaryotes are multilayered and interconnected. They range from transcriptional, post-transcriptional, and translational to post-translational regulation. Post-translational modifications (PTMs) change the properties of proteins by addition of a modifying chemical group (biomolecules or other small agents) to an amino acid residue. More than 200 different types of PTMs are known that can regulate protein properties like affecting the catalytic activity, changing the ligand binding affinity, altering protein structure and/or protein–protein interactions (Mann and Jensen 2003; Kho et al. 2004). NO can mediate several PTMs, such as tyrosine nitration or metal nitrosylation or protein S-nitrosylation. The nitration of free tyrosine or protein tyrosine residues generates 3-nitrotyrosine. This reaction has been utilized as a footprint for the *in vivo* formation of peroxynitrite and other reactive nitrogen species. Metal nitrosyls are formed by the reaction of NO with transition metals. The predominant mode of action of NO seems to be protein S-nitrosylation, a PTM that involves covalent attachment of NO moiety to the thiol side chain of redox sensitive cysteine residue. This covalent attachment of NO to the thiol group is reversible and is determined by the redox status of its micro-environment. Fluctuations in the cellular redox status are a typical phenomenon associated with stress-related response. Also redox-sensitive cysteine residues are often key regulators of protein function (Spoel and Loake 2011) and are present in all major classes of proteins. Thus, ubiquitous signaling behavior of NO and its ability to sense the changes in cellular redox status and reversibly modify functionally important redox-sensitive cysteine residue (Stamler et al. 2001) make protein S-nitrosylation an illustrative example for redox-regulated PTM.

Protein S-nitrosylation is the most studied NO-mediated signaling mechanism in plants. Several proteins prone to S-nitrosylation have been identified in plants (Lindermayr et al. 2005; Romero-Puertas et al. 2008; Abat and Deswal 2009; Palmieri et al. 2010; Maldonado-Alconada et al. 2011; Lin et al. 2012). These proteins are involved in different physiological and stress-related processes highlighting the ubiquitous regulatory role of S-nitrosylation. In the model plant *Arabidopsis thaliana* non-symbiotic hemoglobin (Perazzolli et al. 2004), S-adenosylmethionine synthetase (SAMS1) (Lindermayr et al. 2006), metacaspase 9 (Mc9) (Belenghi et al. 2007), glyceraldehyde 3-phosphate dehydrogenase (GAPDH) (Holtgrefe et al. 2008; Wawer et al. 2010), salicylic acid-binding protein 3 (SABP3) (Wang et al. 2009), non-expressor of pathogenesis-related gene1 (NPR1) (Lindermayr et al. 2010), glycine decarboxylase complex (GDC) (Palmieri et al. 2010) and NADPH oxidases (Yun et al.

Table 7.1 List of proteins regulated by S-nitrosylation

Name of the protein	Function of the protein	Effect of S-nitrosylation
Nonsymbiotic hemoglobin1 (**Hb1**)	NO dioxygenase activity	NO detoxification during hypoxia (Perazzolli et al. 2004)
S-adenosylmethionine synthetase (**SAMS1**)	Catalyzes the synthesis of the ethylene precursor S-adenosylmethionine	Inhibits the catalytic activity of SAMS1 (Lindermayr et al. 2006)
Metacaspase 9 (**Mc9**)	Proteolytic caspases activity	Inhibition of the proteolytic activity (Belenghi et al. 2007)
Peroxiredoxin II E (**PrxII E**)	Reduces H_2O_2 and alkyl hydroperoxides to H_2O and the corresponding alcohol. Also functions in detoxifying peroxynitrite ($ONOO^-$)	Inhibits hydroperoxide-reducing peroxidase activity and $ONOO^-$ detoxification activity (Romero-Puertas et al. 2007)
R2R3-MYB transcription factors	Upon DNA binding, R2R3-MYB induce many physiological and stress related pathways	Inhibits the DNA binding of R2R3-MYB transcription factor (Serpa et al. 2007)
Glyceraldehyde 3-phosphate dehydrogenase (**GAPDH**)	Oxidative catalysis during glycolysis	Inhibit glycolytic acitivty. Initiates nuclear localization and induce programmed cell death (putative function) (Holtgrefe et al. 2008)
Salicylic acid-binding protein 3 (**SABP3**)	SA binding to SABP3 activates its carbonic anhydrase (CA) activity and modulates plant defense response	Inhibits SA binding and CA activity (Wang et al. 2009)
Non-expressor of pathogenesis-related gene1 (**NPR1**)	Conversion of polymeric NPR1 (oxidized form) to monomeric NPR1 isomers (reduced form) helps the nuclear localization of monomers and binding to TGA1 transcription factor that leads to DNA binding and PR1 gene expression	Helps in achieving redox equilibrium between oxidized and reduced forms of NPR1 (Tada et al. 2008; Lindermayr et al. 2010)
Glycine decarboxylase complex (**GDC**)	Involved in the mitochondrial photorespiratory C2 cycle of C3 plants	Inhibited the photorespiratory function of GDC and induced ROI accumulation and cell death (Palmieri et al. 2010)
NADPH-dependent oxidases (**NADPH-oxidase**)	Synthesis of pathogen-induced ROI production and mediates fully developed hypersensitive response	Inhibits pathogen-induced ROI production (Yun et al. 2011)

2011) are the proteins reported to be regulated by S-nitrosylation. Proteins that are known to be S-nitrosylated in plants are listed in Table 7.1 along with their function and inhibitory effect by S-nitrosylation.

Regulation of Protein S-Nitrosylation

Multiple pathways of regulation and technical limitations to find these pathways have considerably slowed down the progress of understanding the regulatory mechanisms that govern S-nitrosylation. Despite these obstacles, considerable progress has been made over the last decade in unveiling its basics that has formed the foundation for a promising field of cellular signaling ahead. Regulation of protein S-nitrosylation has a hand-in-hand association with NO-bioactivity. Essential steps during S-nitrosylation regulation are i) cellular S-nitrosothiol formation ii) transnitrosylation and iii) denitrosylation. Cellular S-nitrosothiol formation is closely associated with NO production. In animals, the enzyme nitric oxide synthase (NOS) is the main source of NO production (Bredt and Snyder 1990; Jaffrey et al. 2001). Gupta et al. has recently reviewed about various sources of NO production in plants (Gupta et al. 2011). Unique chemical nature of NO that makes it physiologically stable, but with high target specific reversible reactivity is the prime basis of S-nitrosylation signaling event. Cellular redox status utilizes these unique features of NO and co-ordinates the spatio-temporal regulation through the controlled S-nitrosylation/denitrosylation mechanisms.

Mechanism of Protein S-Nitrosothiol Formation

Stress related responses in plants are often associated with increase in the steady-state levels of cellular NO in plants that induce protein S-nitrosylation. However, an exact *in vivo* reaction mechanism describing the formation of S-nitrosothiols from cellular NO is not known yet. The intrinsic biochemistry of NO suggests multiple reaction pathways for S-nitrosylation mechanisms with evidences supported by various *in vitro* studies. Most of these studies have used thiol-containing molecules like cysteine (CySH) and glutathione (GSH) as model compounds as the reaction targets of NO, which upon S-nitrosylation yield low molecular weight (LMW) S-nitrosothiols such as S-nitrosocysteine (CySNO) and S-nitrosoglutathione (GSNO) (Gow et al. 1997; Keszler et al. 2010). They form the integral part of total cellular nitrosothiol (RSNO) pool along with S-nitrosylated peptides and proteins.

Unique chemistry of NO allows it to exist in three redox-related forms, all with different biochemical properties; the reduced nitroxyl anion (NO^-), the NO radical ($^{\cdot}NO$) and the oxidized nitrosonium cation (NO^+) each with different oxidation state for the nitrogen atom, +1, +2 and +3, respectively (Arnelle and Stamler 1995). NO^- can further exist in two chemical forms; high energy singlet form and low energy triplet form, with zero or two unpaired electrons respectively (Lipton et,al. 1998). In mammals, neuronal nitric oxide synthase produces singlet NO^- that reacts with thiols to form S-nitrosothiols (Schmidt et al. 1996). However, this does not exclude the involvement of NO^- as a source for other S-nitrosylation pathways. Low energy triplet NO^- may react with dioxygen to form peroxynitrite (Lipton et al. 1998) that

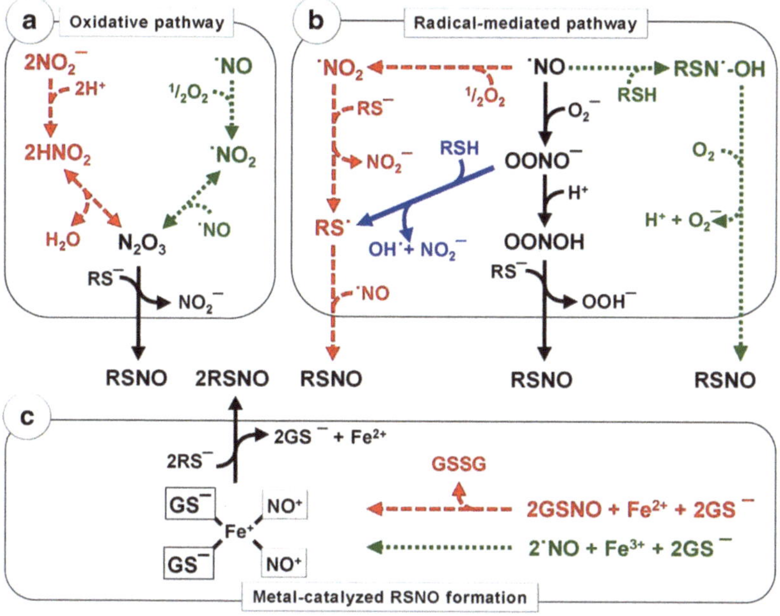

Fig. 7.1 Pathways leading to S-nitrosothiol (RSNO) formation. (**a**) N_2O_3 can be formed from protonated nitrite at very low pH (*dashed arrows*) and by the auto-oxidation of ˙NO in an O_2 rich environment (*dotted arrows*). N_2O_3 provides NO^+ equivalence to nucleophilic thiols to form RSNO (undashed arrows). (**b**) RS˙ radicals produced either by peroxynitrite radical (*undashed bold arrows*) or by the auto-oxidation products of ˙NO can directly react with ˙NO radical to form RSNO (*dashed arrows*). In the presence of thiolate anions (RS^-) protonation of peroxynitrite can also result in the formation of RSNO (*undashed arrows*). Furthermore, ˙NO can form an intermediate radical with thiols which then oxidizes to form RSNO (*dotted arrows*). (**c**) Chelatable iron pool can mediate the formation of dinitrosyliron complexes (*dashed and dotted arrows*) that yields NO^+ equivalence to form RSNO (*undashed arrows*)

in-turn may influence S-nitrosylation (Balazy et al. 1998; van der Vliet et al. 1998). Even though, the free radical ˙NO has reported to interact with cysteine thiols to form S-nitrosothiols in the presence of a suitable electron acceptor (Gow et al. 1997) this interaction did not happen with glutathione and is therefore doubtful to happen in physiological conditions (Keszler et al. 2010). Thiyl radicals (RS˙) that is a byproduct of stress-related redox chemical pathways also (see Fig. 7.1b) can react with ˙NO to form S-nitrosothiols (Jourd'heuil et al. 2003). The existence of free NO^+ is favored only at very high pH values and is therefore biologically non-viable. But oxidation products of ˙NO that are functionally equivalent to NO^+ exist under physiological conditions and can donate NO^+ to more nucleophilic thiols resulting in S-nitrosylation (Hughes 1999). In general, none of the three redox-related forms of NO is known to mediate S-nitrosothiol formation independently in its free form. Alternatively, various reaction mechanisms that lead to the formation of S-nitrosothiols from NO have been proposed. Their possible physiological relevance in the context of plant stress responses are discussed in the following section.

Oxidative S-Nitrosylation by Higher Oxides of NO: Formation of N_2O_3

Dinitrogen trioxide (N_2O_3) is generally considered as a nitrosylating agent that can directly mediate S-nitrosylation (Wink et al. 1994). N_2O_3 often donates NO^+ to the reduced (nucleophilic) thiolate anion (RS^-) to yield S-nitrosylated product (RSNO) (Fig. 7.1a). In biological systems N_2O_3 can be formed in two ways (Fig. 7.1a). In a pH-dependent pathway N_2O_3 is formed by the reversible dehydration of nitrous acid (HNO_2) (Guikema et al. 2005) (Fig. 7.1a). Since the pK_a of HNO_2 is approximately 3.4, at higher pH values HNO_2 dissociates into nitrite (NO_2^-). Hence N_2O_3 formation from HNO_2 occurs only at low acidic pH. The apoplast of plants is acidic nature (Yu et al. 2000) and might be mediating pH-dependent synthesis of N_2O_3. Furthermore, changes of the pH value in various plant compartments are associated with signaling in plants and regulate both physiological processes (Gibbon and Kropf 1994; Feijo et al. 1999) and stress related defense responses (Mathieu et al. 1996; Roos et al. 2006; Kader et al. 2007). In an acidified apoplast NO can be produced from nitrite (Bethke et al. 2004). Moreover, acidification of cytoplasm in tobacco suspension cultures induced defense related genes and interestingly NO responsive genes (Lapous et al. 1998). Therefore, combining all these factors argue for the possibility of a pH-dependent formation of S-nitrosothiol in plants.

Another mechanism to produce N_2O_3 is the direct oxidation of the radical $^{\cdot}NO$ by O_2 (Fig. 7.1a) (Wink et al. 1994; Goldstein and Czapski 1996). This aerobic formation of N_2O_3 depends on the concentration of available $^{\cdot}NO$ and O_2 because of the second order dependence of the reaction with respect to $^{\cdot}NO$ concentration and first order dependence of the reaction with respect to O_2 concentration (Goldstein and Czapski 1996). Even though an enzymatic source for the production of $^{\cdot}NO$ is not known until now, $^{\cdot}NO$ burst is a typical stress-associated phenomenon in plants (Desikan et al. 2002; Zeidler et al. 2004). It is possible that under these conditions oxidation of $^{\cdot}NO$ to $^{\cdot}NO_2$ occurs to counteract exceeding levels of cellular $^{\cdot}NO$. N_2O_3 is formed by the reversible reaction of $^{\cdot}NO_2$ with another molecule of $^{\cdot}NO$ (Fig. 7.1a). Due to the hydrophobic nature of $^{\cdot}NO$ and O_2 the reaction rate increases 300-fold in a hydrophobic environment (Liu et al. 1998). Thus, cellular hydrophobic milieu like lipid membranes and protein interiors can accelerate N_2O_3 formation. Consequently, increased S-nitrosylation of proteins has been reported under these conditions (Rafikova et al. 2002). However, some recent studies have contradicted this finding (Zhang et al. 2009; Keszler et al. 2010). They support the assumption that hydrophobic environment protonate thiol (RSH) and thereby hindering it from reducing to thiolate anion (RS^-) for accepting NO^+ from N_2O_3.

Radical Mediated S-Nitrosylation

Radicals play an important role in mediating cellular signaling processes during stress responses. Although there are many radicals proposed to mediate RSNO formation, their influence in the *in vivo* formation of RSNO is not known. $^{\cdot}NO$ can react directly with cysteine thiols to form free radical intermediate RSN˙OH (Fig. 7.1b) that in the presence of an electron acceptor like O_2 or NAD^+ get oxidized to S-nitrosocysteine (Gow et al. 1997). However, kinetic stimulation studies involving glutathione rather than single cysteine thiols have suggested this pathway either as a negligible pathway for RSNO formation or a pathway that might play only a role at low steady-state levels of NO (Keszler et al. 2010). Other molecules that mediate the formation of RSNO via radical interactions are $^{\cdot}NO/O_2$, peroxynitrite ($OONO^-$) radical and ˙NO/superoxide anion ($O_2^{\cdot-}$). In the $^{\cdot}NO/O_2$ pathway, autooxidation of ˙NO results in the formation of $^{\cdot}NO_2$, which can oxidize thiols (RSH) to thiyl radicals (RS˙) (Jourd'heuil et al. 2003). These thiyl radicals can directly react with ˙NO to form RSNO (Fig. 7.1b) (Jourd'heuil et al. 2003; Schrammel et al. 2003; Madej et al. 2008; Keszler et al. 2010). Peroxynitrite-dependent RSNO formation can occur in two different ways, either by a direct electrophilic attack of $OONO^-$ on the thiolate anion (van der Vliet et al. 1998) or through an intermediate thiyl radical formation(Goldstein and Czapski 1996; Keszler et al. 2010) (Fig. 7.1b). Peroxynitrite itself can be produced by the reaction of ˙NO and $O_2^{\cdot-}$. Interestingly, co-production of $^{\cdot}NO/O_2^{\cdot-}$ can mediate S-nitrosylation in a peroxynitrite-independent manner (Schrammel et al. 2003). In plants, many stress related responses are associated with rapid production of reactive oxygen and nitrogen species (Neill et al. 2002b; Modolo et al. 2005; Torres and Dangl 2005). Since, both reactive oxygen and nitrogen species have signaling function as well as toxic effects effective regulatory mechanisms are necessary. Plants possess antioxidants like glutathione and ascorbate that is shown to detoxify and regulate the free radical cellular levels (Dahm et al. 2006; Foyer and Noctor 2011). While $O_2^{\cdot-}$ influences radical mediated S-nitrosylation formation, $O_2^{\cdot-}$ production is under the influence of S-nitrosylation during stress (Yun et al. 2011) that highlights the mutual regulatory roles of these two signaling mechanisms in a balanced cellular homeostasis of free radicals. Thus, it is possible that radicals and their regulatory systems together might be playing a crucial role in RSNO formation in plants during stress conditions.

S-Nitrosylation Catalyzed by Metals

Transition metals, especially iron, are important elements for the proper regulation of physiological functions in plants. Both iron and NO are redox related species and can take part in reversible electron transfer processes depending on the redox environment. While, ferric iron (Fe^{3+}) can accept electrons from radical ˙NO resulting in the formation of ferrous (Fe^{2+}) and NO^+ ions, Fe^{2+} can donate electron to radical ˙NO to form Fe^{3+} and NO^- (Graziano and Lamattina 2005). Due to high affinity of iron for NO they form coordinate complexes named iron-nitrosyl complexes. Changes in the iron pool have shown to influence signaling processes mediated by S-nitrosylation in mammalian cell lines (Kim et al. 2000). Fe^{2+}, NO and low molecular weight thiols can form *in vivo* metal containing S-nitrosothiols called as dinitrosyl iron complexes

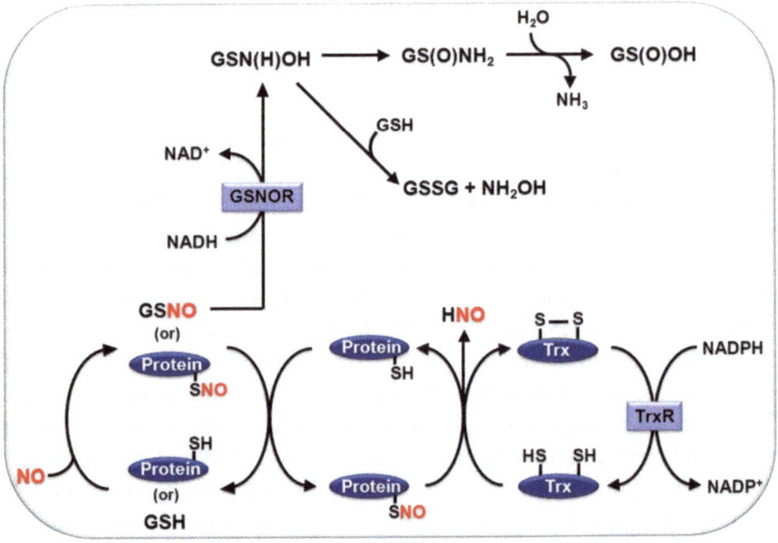

Fig. 7.2 Schematic illustration of different mechanisms used to regulate protein S-nitrosylation. RSNO comprises of low (e.g. GSNO) and high (S-nitrosylated proteins) molecular weight molecules. GSNO and S-nitrosylated proteins can mediate S-nitrosylation of specific free thiol groups of other proteins through transnitrosylation. Trx/TrxR mediates denitrosylation by reducing S-nitrosylated thiol group of the proteins. Regulation of protein-S-nitrosylation by GSNOR is indirect. It metabolizes the transnitrosylation mediator GSNO in an NADH-dependent pathway to an intermediate S-(N-hydroxyamino)glutathione (GSNHOH). This intermediate is converted to glutathione sulphinamide ($GSONH_2$), which is then spontaneously hydrolyzed to glutathione sulphinic acid (GSOH) and ammonia (NH_3). However, in the presence of GSH, GSNHOH is converted oxidized glutathione (GSSG) and hydroxylamine (NH_2OH) (Jensen et al. 1998; Liu et al. 2001; Hedberg et al. 2003; Staab et al. 2008)

(DNICs) (Fig. 7.1c) (Mulsch et al. 1993). DNICs are considered as endogenous NO carriers like LMW nitrosothiols. They have shown to transfer NO to the metal-centers of metalloproteins (Ueno et al. 2002) and/or can donate NO^+ equivalents to thiol groups to form RSNO (Bosworth et al. 2009) (Fig. 7.1c). Increased NO levels in plants elevate the levels of nitrosyl-iron complexes (Simontacchi et al. 2012). Moreover, under oxidative conditions, in NO-binding hemeproteins NO can be transferred from the heme group to intramolecular cysteine thiol residues (Luchsinger et al. 2003).

Transnitrosylation

Apart from the direct modification of thiol group by NO equivalents, both low molecular weight S-nitrosocysteine thiols and S-nitrosylated proteins can directly transfer the nitrosyl moiety (NO group) to non-S-nitrosylated cysteine thiols through a process termed transnitrosylation (Fig. 7.2) (Zhang and Means 1996; Pawloski et al. 2001; Dahm et al. 2006; Mitchell et al. 2007; Kornberg et al. 2010;

Nakamura et al. 2010). Transnitrosylation is largely responsible for *in vivo* S-nitrosocysteine thiol activity and NO signaling. Not all, but specific thiol containing cysteine residues of proteins are the targets of S-nitrosylation. Low molecular weight S-nitrosocysteine thiols like GSNO mediate transnitrosylation of specific cysteine thiols on multiple proteins. In contrast, transnitrosylation mediated by S-nitrosylated proteins (protein-protein transnitrosylation) occurs only with their binding partners, thus showing additional selectivity along with specificity (Kornberg et al. 2010). In the fly *Drosophila melanogaster*, glyceralde-hyde-3-phosphate dehydrogenase (GAPDH) that is physiologically S-nitrosylated at a specific cysteine residue can transfer its NO moiety to the protein sirtuin, as soon as they interact. Abolishing their interaction eliminates the transnitrosylation ability (Kornberg et al. 2010).

Among the low molecular weight S-nitrosocysteine thiols, GSNO is the major physiological NO-donor and is known for its ability to mediate transnitrosylation (Dahm et al. 2006). Treatment of proteins and cell cultures with GSNO significantly increased the S-nitrosylation levels of proteins in plants (Lindermayr et al. 2005; Palmieri et al. 2010). Cellular GSH pool is the key in regulating the levels of S-nitrosylated thiols, which under the oxidized conditions favors an increase in the levels of S-nitrosylated thiols (Dahm et al. 2006). Plant defense responses are often associated with rapid changes in the cellular redox environment that induce oxidation of GSH pool (Vanacker et al. 2000). Stress-related responses are accompanied by an apparent increase in the levels of S-nitrosylated thiols (Feechan et al. 2005; Lee et al. 2008; Chaki et al. 2011a, b), which also constitute higher GSNO levels. Increased GSNO levels can mediate S-nitrosylation of specific cysteine residues of proteins.

Several attempts have been made to reveal the factors that influence the S-nitrosylation specificity of cysteine residue. These studies did not reveal a linear sequence motif on the proteins that could mediate S-nitrosylation of specific cysteine residues. Rather they showed various factors that can enhances the chance for the cysteine residue to be S-nitrosylated like:

(a) Presence of a acid–base motif flanking the cysteine residue either in the primary sequence or in the tertiary structure (Stamler et al. 1997; Greco et al. 2006; Doulias et al. 2010).
(b) Cysteine residue in the hydrophobic vicinity due to enhanced ˙NO autooxidation to form S-nitrosylating species N_2O_3 (Nedospasov et al. 2000; Greco et al. 2006)
(c) Cysteine residues skewed towards accessible surface areas in the α-helices with charged amino acids within a 6°A distance (Doulias et al. 2010)
(d) Low pK_a values of cysteine residue (Foster et al. 2009) and
(e) Cysteine thiol in the proximity of the amino acid interacting with transnitro-sylating protein (Kornberg et al. 2010).

But, none of these factors were conclusive to define a general mechanism of S-nitrosylation specificity for cysteine residue (Doulias et al. 2010; Marino and Gladyshev 2010). Each factor has shown to be significant in one or the other situation of protein S-nitrosylation (Greco et al. 2006; Foster et al. 2009; Doulias et al. 2010; Kornberg et al. 2010). This suggest that each factor might be playing different role in mediating protein S-nitrosylation under different circumstances of nitrosothiol formation like N_2O_3-mediated S-nitrosylation, metal-catalyzed S-nitrosylation, transnitrosylation by GSNO and protein-protein transnitrosylation (Foster et al. 2009).

Protein Denitrosylation: A Regulator of S-Nitrosylation Signaling

Removing NO moiety from the S-nitrosylated cysteine residue of the proteins, known as denitrosylation, is very important for proper regulation of protein S-nitrosylation. While S-nitrosylation of proteins is generally considered to initiate stress-induced signaling pathways, denitrosylation is responsible for maintaining cellular S-nitrosylated levels of this protein during the response and finally to switch-off the same pathway to reconstitute the normal situation after stress response. However, denitrosylation can also function as a switch to induce pathways during stress response. Mitochondrial caspase-3 zymogens apoptotic activity is inhibited by S-nitrosylation in resting cells. Stress-induced denitrosylation of caspase-3 activates the protein (Mannick et al. 1999; Kim and Tannenbaum 2004; Reynaert et al. 2004; Erwin et al. 2005). Thus, denitrosylation can have dual roles in regulating signaling pathways. Even though several enzymes have been proposed to mediate denitrosylation, S-nitrosoglutathione reductase (GSNOR) and thioredoxin/thioredoxin reductase are the two systems that are characterized to have significant role in mediating this process in animals (Fig. 7.2) (Benhar et al. 2009; Lopez-Sanchez et al. 2010). A similar role of their counterparts in plants, especially of that of GSNOR is emerging and is of considerable interest.

GSNOR is Crucial in Regulating S-Nitrosothiol Levels

Search for an enzyme that can mediate metabolism of physiological NO molecule GSNO has led to the identification of GSNOR that is conserved in almost all living systems including plants (Liu et al. 2001; Sakamoto et al. 2002; Diaz et al. 2003). GSNOR was classified to class III alcohol dehydrogenase (ADH) and was originally found to function as glutathione dependent-formaldehyde dehydrogenase (FALDH) in plants. FALDH has been a well characterized enzyme in several plant species (Uotila and Koivusalo 1979; Martinez et al. 1996) before its GSNOR activity was discovered (Sakamoto et al. 2002; Achkor et al. 2003; Diaz et al. 2003). GSNOR metabolizes GSNO with NADH as an electron donor (Fig. 7.2) (Wilson et al. 2008). While NOS-like activity, nitrate reductase activity and other non-enzymatic sources for NO are associated with NO accumulation in plants (Gupta et al. 2011), GSNOR is associated with the removal of NO through

GSNO metabolism. This is evident in the *gsnor* knock-out and overexpression lines of *Arabidopsis* plant that showed increased and reduced nitrosothiol levels respectively (Feechan et al. 2005). Since oxidized GSH (GSSG) is a product of GSNO metabolism (Fig. 7.2), it is possible that the redox status of the glutathione pool has a great influence on protein S-nitrosylation. Under oxidizing conditions, GSNOR mediated metabolism might be less favored. Thus GSH and GSNOR indirectly mediate protein denitrosylation through GSNO metabolism. GSNOR, however, cannot metabolize S-nitrosylated moiety of proteins or peptides (Liu et al. 2001). There is an equilibrium that exists between low molecular weight S-nitrosothiols like GSNO and S-nitrosylated proteins and peptides (Seth and Stamler 2011). This equilibrium allows regulation of GSNO metabolism by GSNOR to indirectly regulate S-nitrosylated proteins (Fig. 7.2). In *gsnor* knock-out mutant plants, an increase in low molecular weight nitrosothiols resulted in a corresponding increase in the levels of high molecular weight S-nitrosothiols that is assumed to include proteins which is a clear indication of indirect effect of GSNOR regulation of protein S-nitrosylation (Liu et al. 2001; Liu et al. 2004; Yun et al. 2011).

GSNOR is receiving increasing attention for its role in plant stress responses. Physiological role of GSNOR is evident from the *Arabidopsis gsnor* knock-out mutant plants that showed delayed and stunned growth phenotype and altered flower development (Lee et al. 2008; Holzmeister et al. 2011). *Gsnor* mutant plants showed a reduced cell death phenotype after treatment with paraquat, a herbicide that is known to induce cell death phenotype in wild type plants via generation of reactive oxygen intermediates (ROI) (Chen et al. 2009). Interestingly, both wild type and *gsnor* mutant plants showed same levels of ROI accumulation after paraquat treatment. (Chen et al. 2009). Lack of sensitivity of the *gsnor* knock-out plants to increased ROI can be due to altered cellular ROI/NO homeostasis, which is very important for plant defense responses (Delledonne et al. 2001).

Arabidopsis gsnor knock-out mutants, challenged with avirulent *Pseudomonas syringae* pv. *tomato (Pst)* DC3000, showed low levels of salicylic acid accumulation that resulted in a compromised disease resistance (Feechan et al. 2005; Yun et al. 2011). However, these plants with high cellular RSNO levels showed an increased cell death induced by hypersensitive response (CDHR) through a pathway independent of SA and ROI production (Yun et al. 2011). On the other hand, even though SA-induced defense is compromised, increased CDHR rate prevented avirulent oomycete pathogens to complete its life cycle (Yun et al. 2011). These evidences highlight two different roles of GSNOR during defense response; positive regulator of SA-induced defense and negative regulator of CDHR-induced defense responses. Conversely, GSNOR transcript levels and GSNOR activity in *Arabidopsis* and tobacco respectively, were shown to be up regulated when treated with SA (Diaz et al. 2003). These studies indicate the possibility of a mutual regulation between GSNOR and SA during plant defense.

Interestingly, in another study on *gsnor* knock-out plants, there was no difference in the level of disease resistance against *Pseudomonas syringae* pv. *tomato (Pst)* DC3000 with respect to the wild type plants (Holzmeister et al. 2011). However, here the knock-out plants used were from different background ecotype of *Arabidopsis thaliana* plants and also procedures to inoculate them were also different. These contrary results have raised the questions on how GSNOR regulates disease resistance in various ecotypes. On the contrary, plants with reduced *gsnor* expression levels (antisense technology) have affirmed the negative regulatory role of GSNOR during disease resistance against oomycetes (Rusterucci et al. 2007). Further studies are required to show how this enzyme is regulated at transcript and protein levels during attempted pathogen invasions.

Transcripts of GSNOR, however, were down regulated transiently and systemically during wound-induced responses in *Arabidopsis* plants (Diaz et al. 2003). In tobacco plants, wound-induced down-regulation of GSNOR is mediated by jasmonic acid (JA) signaling pathway (Diaz et al. 2003). In Arabidopsis, GSNO accumulation is required to activate the JA-dependent wound responses, whereas the alternative JA-independent wound-signaling pathway did not involve GSNO. Furthermore, it was shown that GSNO acts synergistically with salicylic acid in systemic acquired resistance activation (Espunya et al. 2012). Plant stress responses induced by wounding are often associated with nitrosative stress and tyrosine-nitration (Chaki et al. 2011b). Stress experiments in sunflower plants have demonstrated that wound-induced nitrosative stress is mediated by down-regulation of GSNOR expression levels resulting in decreased activity and in considerable increase in cellular RSNO levels (Chaki et al. 2011b). In pea plants wounding enhanced RSNO levels, but surprisingly GSNOR activity too is also increased (Corpas et al. 2008). The same phenomenon was observed during cold stress (Corpas et al. 2008). The reason for this unexpected co-relation between RSNO levels and GSNOR activity is not clear. Furthermore, GSNOR is regulated in pea plants during cadmium stress, both on activity and transcript level (Barroso et al. 2006). However, a pathway that regulates GSNOR under cadmium stress is not known. Cadmium treatment also induced SA, JA and ethylene levels in pea plants (Rodriguez-Serrano et al. 2006) accompanied by a decrease in the glutathione content (Barroso et al. 2006).

Gene silencing studies in tobacco plants have demonstrated the significant role of GSNOR in plant-herbivore interaction (Wunsche et al. 2011). Silencing *GSNOR* compromised plant defense against herbivore with a decrease in the accumulation of JA and ethylene (Wunsche et al. 2011). However, this silencing did not affect transcriptional regulation of all the secondary metabolites that are regulated by JA signaling (Wunsche et al. 2011) implying the specificity of GSNOR in mediating defense response against the herbivore *Manduca sexta*. GSNOR is also required for thermo tolerance. It has been observed that *Arabidopsis* knock-out mutants of GSNOR were highly sensitive to hot temperatures (Lee et al. 2008). This heat sensitivity was associated with increased NO species in these knock-out plants. *NO-overproducing* mutants and wild-type plants treated with NO donors were also sensitive to high temperatures (Lee et al. 2008). Consequently, thermo tolerance was restored in *gsnor* mutants when treated with chemicals that scavenge NO. Furthermore, expression of heat-shock-proteins that are essential for thermo tolerance was not affected in *gsnor* mutant plants (Lee et al. 2008). Interestingly, neither expression nor activity of GSNOR was altered in wild-type plants due to heat stress

(Lee et al. 2008). This study suggests that though GSNOR do not regulate heat stress response in plants, its activity to regulate cellular RSNO levels is essential for thermo tolerance.

Denitrosylation Mediated by Trx/TrxR System

The thioredoxin/thioredoxin reductase (Trx/TrxR) system, present in almost all organisms, consists of oxidized and reduced forms of Trx, TrxR and NADPH/ $NADP^+$ (Fig. 7.2) (Lillig and Holmgren 2007). In animals, Trx/TrxR system was recently proved to mediate denitrosylation (Benhar et al. 2008; Benhar et al. 2010). Unlike GSNOR, Trx/TrxR system is proposed to mediate denitrosylation of S-nitrosylated proteins directly (Fig. 7.2). In a recent review, Trx from plants is mentioned to possess *in vitro* denitrosylation activity with reference to an unpublished data (Spoel and Loake 2011). Also, thioredoxin (TRX-5h) is a positive regulator of SA-induced defense response in plants (Tada et al. 2008), probably by denitrosylation.

Physiological Functions of Protein S-Nitrosylation in Plants During Stress Response

To get insight into the physiological function of protein S-nitrosylation the target proteins for this type of modification have to be identified. In plants, potential candidates for S-nitrosylation have been identified from GSNO-treated cell culture extracts, NO-treated plants, infected plants and plants undergoing HR (Fig. 7.3) (Lindermayr et al. 2005; Romero-Puertas et al. 2008; Maldonado-Alconada et al. 2011; Yun et al. 2011). Until now, considerable progress has been made to demonstrate the physiological role of S-nitrosylation for distinct proteins.

Regulation of Pathogen-Induced ROI Production by S-Nitrosylation

Defense related CDHR can be mediated by pathogen-induced accumulation of ROI, which was inhibited by high RSNO levels (Fig. 7.3) (Yun et al. 2011). In *Arabidopsis*, AtRBOHD (NADPH-oxidase) activity is required for the pathogen-induced ROI production (Torres et al. 2002). Interestingly, during hypersensitive response AtRBOHD activity is inhibited by S-nitrosylation of its cysteine residue (Cys890) (Yun et al. 2011). Cys890 is an evolutionary conserved amino acid residue in humans and *Drosophila* and is positioned closely behind the binding

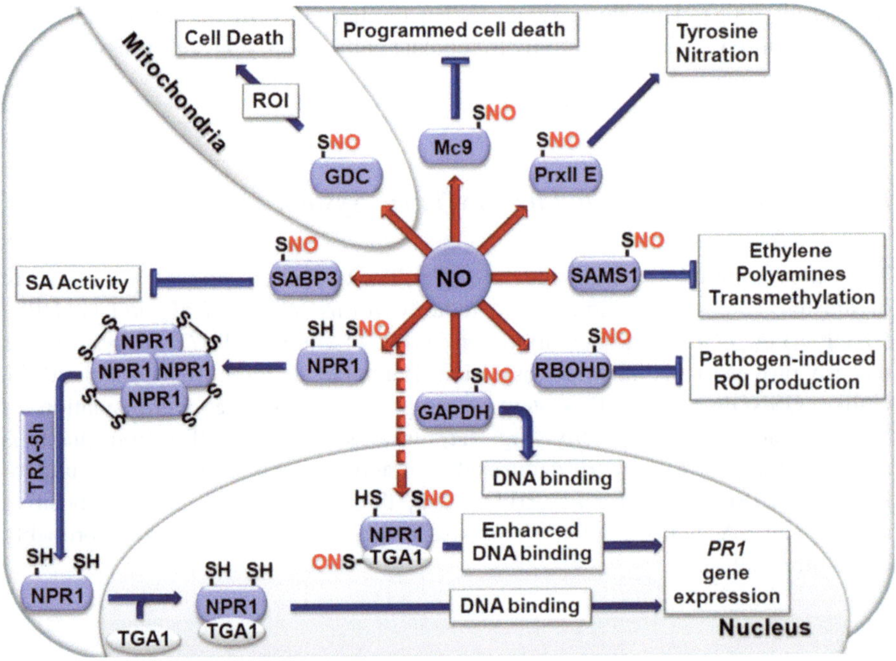

Fig. 7.3 Function of protein S-nitrosylation in plant stress response. Stress-induced accumulation of nitric oxide species can inhibit, activate or alter the function of proteins through S-nitrosylation. The activity of SABP3 (important mediator of SA signaling), Mc9 (cysteine protease activity), PrxII E (detoxifing peroxynitrite—regulate tyrosine nitration), SAMS1 (enzyme involved in ethylene and polyamine synthesis and transmethylation reactions) and RBOHD (synthesis of pathogen-induced ROI) is inhibited by S-nitrosylation. Furthermore, inhibition of GDC induces mitochondrial ROI production and cell death. S-Nitrosylation of mammalian GAPDH mediates its nuclear localization and induces cell death. Plant GAPDH can also be S-nitrosylated, but its role in cell death is not yet known. Monomer to oligomer transition of NPRI is proposed to be mediated by S-nitrosylation and reversible transition by thioredoxin and induce PR1 gene expression. Moreover, NO-treatment enhances the DNA binding activity of the NPR1/TGA1 complex

site of flavin adenine dinucleotide (FAD). FAD mediates the electron transfer from NADPH through heme to O_2 to produce O_2^- (Sumimoto et al. 2004). S-nitrosylation of Cys890 prevented FAD binding and ROI production of AtRBOHD (Yun et al. 2011). Mutation of Cys890 however abolished S-nitrosylation ability without disturbing FAD binding ability to its active site that in turn enhanced the pathogen-induced ROI production and corresponding CDHR in *Arabidopsis* plants (Yun et al. 2011). These data demonstrate that plants regulate ROI production by AtRBOHD through S-nitrosylation which might help restricting cell-death to the site of infection. It is, however, not clear whether pathogens utilize this negative regulatory role on ROI production as a strategy to overcome the plant defense. In sunflower, biotrophic oomycete *Plasmopara halstedii* induce accumulation of RSNO in susceptible but not in resistant cultivars (Chaki et al. 2009). Moreover,

CDHR is the key reaction in restricting the pathogen spreading in resistant cultivar (Radwan et al. 2005). Thus, it is possible that pathogen-induced RSNO accumulation in the susceptible cultivar might inhibit pathogen-induced CDHR through S-nitrosylation and inhibition of NADPH-oxidase. Analysis to see the difference in the activity of NADPH-oxidase in the two sunflower cultivars upon pathogen challenge and their role in regulating CDHR are required to verify these arguments.

S-Nitrosylation is a Regulator of SA-Dependent Signaling in Plant Defense Response

SA is an important signaling molecule in plant defense response. SA regulates the function of NPR1 (non-expressor of PR1), a co-activator of transcription of the pathogenesis-related gene 1 (*PR1*). Endogenous NPR1 is located in the cytoplasm in an oligomeric status. Upon SA-dependent activation NPR1 dissociates into its monomers, which are translocated into the nucleus (Mou et al. 2003; Pieterse and Van Loon 2004). Nitric oxide plays a crucial role in regulating oligomer/monomer transition. S-nitrosylation of NPR1 facilitates its oligomerization, which keeps it in the cytosol and is essential for NPR1 homeostasis upon SA induction (Tada et al. 2008). The monomerization of NPR1 is catalyzed by thioredoxin TRX-5h, which reduce NPR1 and allow the translocation into the nucleus. But surprisingly, in *Arabidopsis* mesophyll protoplasts nuclear localization of NPR1 is promoted by GSNO (Lindermayr et al. 2010). However, the S-nitrosylation–mediated oligomerization of NPR1 is not seen as an inhibitory effect of NPR1 signaling but rather as a step prior to monomer accumulation. From this point of view, the observed NO-mediated nuclear translocation of NPR1 is not contradictory to the results described by Tada et al. (2008). During defense responses, GSNOR plays a crucial role in regulating the cellular RSNO and SA levels which are essential for mediating oligomer to monomer transition of NPR1. Inside the nucleus NPRI interacts with the transcription factor TGA1 (TGACG motif binding factor) and activates PR1 gene expression (Despres et al. 2003). Both NPR1 and TGA1 were S-nitrosylated when treated with GSNO resulting in enhanced DNA binding of the NPR1/TGA1 complex (Lindermayr et al. 2010). But S-nitrosylation can also serve as a negative regulator of SA-signaling. Binding of SA to SA-binding protein 3 (SAPB3) can activate its carbonic anhydrase activity and thereby positively regulate the plant defense response. S-nitrosylation of SABP3 reduced its SA binding ability resulting in reduction of the CA activity of the enzyme (Wang et al. 2009). S-nitrosylation of SABP3 might be used either by the plant as a negative feedback loop to modulate SA signaling or by the pathogen as a strategy to suppress the plant defense response.

S-Nitrosylation Might be Crucial in Mitochondrial-CDHR

A role of S-nitrosylation in mitochondria mediated CDHR has also been demonstrated. Inhibition of the activity of glycine decarboxylase complex (GDC), a key enzyme involved in the mitochondrial photorespiratory C2 cycle of C3 plants, resulted in ROI accumulation and cell death (Palmieri et al. 2010). Interestingly, GDC is S-nitrosylated/S-glutathionylated at specific cysteine residues when incubated with physiological concentrations of GSNO resulting in inhibition of its activity (Palmieri et al. 2010). Moreover, this inhibition is part of the stress-related response of Arabidopsis to the bacterial elicitor hairpin. In sum, these data reinforce the model of cross talk between NO/ROS and mitochondria in the activation of stress-related responses in plants (Palmieri et al. 2010).

S-Nitrosylation Positively Regulates Tyrosine Nitration

Interestingly, protein S-nitrosylation regulates nitration of tyrosine residues. Stress related processes are often associated with the accumulation of $^{\cdot}NO$ and $O_2^{\cdot-}$ radicals. Diffusion-limited reaction of $^{\cdot}NO$ and $O_2^{\cdot-}$ radicals results in the formation of peroxynitrite ($OONO^-$), an effective tyrosine nitrating compound. Defense related responses in plants are accompanied by $OONO^-$ accumulation (Saito et al. 2006; Gaupels et al. 2011b). In plants however, detoxification of $OONO^-$ is carried out by Peroxyredoxin II E (PrxII E). During HR responses PrxII E gets S-nitrosylated and its activity is inhibited (Romero-Puertas et al. 2008). This allows the accumulation of peroxynitrite which can mediate tyrosine nitration. Consequently, higher tyrosine nitrate levels can be found in plants undergoing biotic stress (Saito et al. 2006). In sunflower-mildew interaction, susceptible cultivars with increased levels of RSNO showed increased tyrosine nitrate levels whereas resistant cultivars did not (Chaki et al. 2009). Moreover, enhanced RSNO levels are accompanied by accumulation of nitrated tyrosine residues in sunflower after mechanical wounding (Chaki et al. 2011b). This correlation between RSNO levels and tyrosine nitration is again seen in sunflower plants stressed with high temperature (Chaki et al. 2011a). All these evidences point out the regulatory role of S-nitrosylation over other NO-related mechanisms.

Stress Induced Function of GAPDH Is Regulated by S-Nitrosylation

S-Nitrosylation of a protein can also alter its function. Glyceraldehyde-3-phosphate dehydrogenase (GAPDH), for instance, is an enzyme of the glycolytic pathway and is present in almost all species. In plants, however, GAPDH plays also an important role in regulating cellular metabolic process during stress conditions to remove toxic ROI accumulation (Sweetlove et al. 2002; Graham et al. 2007; Rius et al. 2008). Interestingly, GADPH has been identified as a candidate prone to S-nitrosylation and treatment of the enzyme with GSNO inhibited its enzymatic activity (Lindermayr et al. 2005). Surprisingly, plant GADPH was observed to move into the nucleus and it was shown that GAPDH can bind to a partial gene sequence of the NADP-dependent malate dehydrogenase (Holtgrefe et al. 2008).

In rat cells S-nitrosylated GAPDH interacts with the E3-ubiqitin-ligase Siah1, translocates into the nucleus and mediates cell death (Sen et al. 2008). If there is a similar function in plants have to be proven. In *Nicotiana tabacum*, NtOSAK is a protein kinase that is induced partially by NO during salt stress. GAPDH, a binding partner of NtOSAK has found to be S-nitrosylated and the level of S-nitrosylation was proportional to the increase in the kinase activity of NtOSAK. However, S-nitrosylation of GAPDH did not have any impact on the activity of NtOSAK and hence might not be the mediator of NO signaling in regulating NtOSAK (Wawer et al. 2010). Additionally, in *Drosophila melanogaster,* S-nitrosylated GADPH can transfer its NO moiety to other proteins through transnitrosylation (Kornberg et al. 2010).

Regulation of Ethylene Biosynthesis by S-Nitrosylation

Ethylene is a natural regulator of growth, development and stress-related processes in plants. Ethylene emission from *Arabidopsis* cell culture was significantly reduced when treated with NO donors indicating an opposite effect of NO on ethylene production (Lindermayr et al. 2006). Such a negative correlation has been also observed in plant foliage and fruits when they are switching from growth stage (low ethylene, high NO) to ripening stage (high ethylene, low NO) (Leshem and Haramaty 1996; Leshem et al. 1998). S-adenosylmethionine synthetase (SAMS) is an enzyme that catalyzes the biosynthesis of S-adenosylmethionine (SAM), a precursor of ethylene. Among three known isoforms of SAMS, one isoform SAMS1 can be regulated by S-nitrosylation (Lindermayr et al. 2006). This regulatory check-point of SAMS1 through S-nitrosylation might link NO signaling with ethylene signaling. But, NO can also inhibit ethylene production in a pathway independent of S-nitrosylation (Leshem et al. 1998; Lindermayr et al. 2005; 2006).

Inhibition of Metacaspase Activity by S-Nitrosylation

Plant metacaspases are cysteine-dependent proteases, which contain a specific cysteine residue that can serve as a nucleophile for the substrate to mediate peptide bond hydrolysis. They are related to animal caspases, a family of proteins involved in the execution of programmed cell death. In plants, fungi, and protozoa metacaspases are homologs of caspases that belong to the D cysteine protease superfamily. *Arabidopsis* has nine metacaspases groups that are classified into two types based on their difference in the N-terminal region (Coll et al. 2010). For type II metacaspase 9 (MC9) has been demonstrated that S-nitrosylation of their active site cysteine residue results in suppression of its autoprocessing and proteolytic activity (Belenghi et al. 2007). A similar NO-dependent regulation has been described for animal caspase3 (Mitchell and Marletta 2005). While S-nitrosylated caspase3 is inactive under normal physiological conditions, denitrosylation activates its proteolytic activity to trigger programmed cell death (Mannick et al. 1999). The regulatory role of S-nitrosylation of MC9 is still unknown. Probably inhibition of MC9 by NO is responsible for avoiding its inappropriate activation.

Conclusions

NO plays important role virtually in all physiological, patho-physiological and stress related responses in plants. Recent advances in NO research have identified S-nitrosylation as a key regulatory mechanism for its mediation. Here we discussed the regulatory function of S-nitrosylation for different stress-related signaling pathways. However, we are still at the beginning of understanding the function of S-nitrosylation in plants and also great effort has to be done to understand how S-nitrosylation is regulated. A combination of proteomics and bioinformatics approach will boost the identification of potential S-nitrosylation targets. Furthermore, biochemical and genetic studies will provide insight into the physiological function of S-nitrosylated proteins/enzymes. Moreover, identification of the exact sources of NO for S-nitrosylation as well as characterization of the denitrosylation process will help to understand the S-nitrosylation mechanism.

References

Abat JK, Deswal R (2009) Differential modulation of S-nitrosoproteome of *Brassica juncea* by low temperature: change in S-nitrosylation of Rubisco is responsible for the inactivation of its carboxylase activity. Proteomics 9:4368–4380

Aboul-Soud MA, Aboul-Enein AM, Loake GJ (2009) Nitric oxide triggers specific and dose-dependent cytosolic calcium transients in *Arabidopsis*. Plant Signal Behav 4:191–196

Achkor H, Diaz M, Fernandez MR, Biosca JA, Pares X, Martinez MC (2003) Enhanced formaldehyde detoxification by overexpression of glutathione-dependent formaldehyde dehydrogenase from *Arabidopsis*. Plant Physiol 132:2248–2255

Arnelle DR, Stamler JS (1995) NO^+, NO, and NO^- donation by S-nitrosothiols: implications for regulation of physiological functions by S-nitrosylation and acceleration of disulfide formation. Arch Biochem Biophys 318:279–285

Balazy M, Kaminski PM, Mao KY, Tan JZ, Wolin MS (1998) S-nitroglutathione, a product of the reaction between peroxynitrite and glutathione that generates nitric oxide. J Biol Chem 273:32009–32015

Barroso JB, Corpas FJ, Carreras A, Rodriguez-Serrano M, Esteban FJ, Fernandez-Ocana A, Chaki M, Romero-Puertas MC, Valderrama R, Sandalio LM, Del Rio LA (2006) Localization of S-nitrosoglutathione and expression of S-nitrosoglutathione reductase in pea plants under cadmium stress. J Exp Bot 57:1785–1793

Belenghi B, Romero-Puertas MC, Vercammen D, Brackenier A, Inze D, Delledonne M, Van Breusegem F (2007) Metacaspase activity of *Arabidopsis thaliana* is regulated by S-nitrosylation of a critical cysteine residue. J Biol Chem 282:1352–1358

Benhar M, Forrester MT, Hess DT, Stamler JS (2008) Regulated protein denitrosylation by cytosolic and mitochondrial thioredoxins. Science 320:1050–1054

Benhar M, Forrester MT, Stamler JS (2009) Protein denitrosylation: enzymatic mechanisms and cellular functions. Nat Rev Mol Cell Biol 10:721–732

Benhar M, Thompson JW, Moseley MA, Stamler JS (2010) Identification of S-nitrosylated targets of thioredoxin using a quantitative proteomic approach. Biochemistry 49:6963–6969

Besson-Bard A, Pugin A, Wendehenne D (2008) New insights into nitric oxide signaling in plants. Annu Rev Plant Biol 59:21–39

Bethke PC, Badger MR, Jones RL (2004) Apoplastic synthesis of nitric oxide by plant tissues. Plant Cell 16:332–341

Bethke PC, Libourel IGL, Jones RL (2006) Nitric oxide reduces seed dormancy in *Arabidopsis*. J Exp Bot 57:517–526

Bosworth CA, Toledo JC Jr, Zmijewski JW, Li Q, Lancaster JR Jr (2009) Dinitrosyliron complexes and the mechanism(s) of cellular protein nitrosothiol formation from nitric oxide. Proc Natl Acad Sci USA 106:4671–4676

Bredt DS, Snyder SH (1990) Isolation of nitric oxide synthetase, a calmodulin-requiring enzyme. Proc Natl Acad Sci USA 87:682–685

Chaki M, Fernandez-Ocana AM, Valderrama R, Carreras A, Esteban FJ, Luque F, Gomez-Rodriguez MV, Begara-Morales JC, Corpas FJ, Barroso JB (2009) Involvement of reactive nitrogen and oxygen species (RNS and ROS) in sunflower-mildew interaction. Plant Cell Physiol 50:265–279

Chaki M, Valderrama R, Fernandez-Ocana AM, Carreras A, Gomez-Rodriguez MV, Lopez-Jaramillo J, Begara-Morales JC, Sanchez-Calvo B, Luque F, Leterrier M, Corpas FJ, Barroso JB (2011a) High temperature triggers the metabolism of S-nitrosothiols in sunflower mediating a process of nitrosative stress which provokes the inhibition of ferredoxin-NADP reductase by tyrosine nitration. Plant Cell Environ 34:1803–1818

Chaki M, Valderrama R, Fernandez-Ocana AM, Carreras A, Gomez-Rodriguez MV, Pedrajas JR, Begara-Morales JC, Sanchez-Calvo B, Luque F, Leterrier M, Corpas FJ, Barroso JB (2011b) Mechanical wounding induces a nitrosative stress by down-regulation of GSNO reductase and an increase in S-nitrosothiols in sunflower (*Helianthus annuus*) seedlings. J Exp Bot 62:1803–1813

Chen R, Sun S, Wang C, Li Y, Liang Y, An F, Li C, Dong H, Yang X, Zhang J, Zuo J (2009) The *Arabidopsis PARAQUAT RESISTANT2* gene encodes an S-nitrosoglutathione reductase that is a key regulator of cell death. Cell Res 19:1377–1387

Coll NS, Vercammen D, Smidler A, Clover C, Van Breusegem F, Dangl JL, Epple P (2010) *Arabidopsis* type I metacaspases control cell death. Science 330:1393–1397

Corpas FJ, Barroso JB, Carreras A, Quiros M, Leon AM, Romero-Puertas MC, Esteban FJ, Valderrama R, Palma JM, Sandalio LM, Gomez M, Del Rio LA (2004) Cellular and subcellular localization of endogenous nitric oxide in young and senescent pea plants. Plant Physiol 136:2722–2733

Corpas FJ, Chaki M, Fernandez-Ocana A, Valderrama R, Palma JM, Carreras A, Begara-Morales JC, Airaki M, Del Rio LA, Barroso JB (2008) Metabolism of reactive nitrogen species in pea plants under abiotic stress conditions. Plant Cell Physiol 49:1711–1722

Corpas FJ, Leterrier M, Valderrama R, Airaki M, Chaki M, Palma JM, Barroso JB (2011) Nitric oxide imbalance provokes a nitrosative response in plants under abiotic stress. Plant Sci Int J Exp Plant Biol 181:604–611

Dahm CC, Moore K, Murphy MP (2006) Persistent S-nitrosation of complex I and other mitochondrial membrane proteins by S-nitrosothiols but not nitric oxide or peroxynitrite: implications for the interaction of nitric oxide with mitochondria. J Biol Chem 281:10056–10065

Delledonne M, Xia Y, Dixon RA, Lamb C (1998) Nitric oxide functions as a signal in plant disease resistance. Nature 394:585–588

Delledonne M, Zeier J, Marocco A, Lamb C (2001) Signal interactions between nitric oxide and reactive oxygen intermediates in the plant hypersensitive disease resistance response. Proc Natl Acad Sci USA 98:13454–13459

Desikan R, Griffiths R, Hancock J, Neill S (2002) A new role for an old enzyme: nitrate reductase-mediated nitric oxide generation is required for abscisic acid-induced stomatal closure in *Arabidopsis thaliana*. Proc Natl Acad Sci USA 99:16314–16318

Despres C, Chubak C, Rochon A, Clark R, Bethune T, Desveaux D, Fobert PR (2003) The *Arabidopsis* NPR1 disease resistance protein is a novel cofactor that confers redox regulation of DNA binding activity to the basic domain/leucine zipper transcription factor TGA1. Plant Cell 15:2181–2191

Diaz M, Achkor H, Titarenko E, Martinez MC (2003) The gene encoding glutathione-dependent formaldehyde dehydrogenase/GSNO reductase is responsive to wounding, jasmonic acid and salicylic acid. FEBS Lett 543:136–139

Doulias PT, Greene JL, Greco TM, Tenopoulou M, Seeholzer SH, Dunbrack RL, Ischiropoulos H (2010) Structural profiling of endogenous S-nitrosocysteine residues reveals unique features that accommodate diverse mechanisms for protein S-nitrosylation. Proc Natl Acad Sci USA 107:16958–16963

Durner J, Wendehenne D, Klessig DF (1998) Defense gene induction in tobacco by nitric oxide, cyclic GMP, and cyclic ADP-ribose. Proc Natl Acad Sci USA 95:10328–10333

Erwin PA, Lin AJ, Golan DE, Michel T (2005) Receptor-regulated dynamic S-nitrosylation of endothelial nitric-oxide synthase in vascular endothelial cells. J Biol Chem 280:19888–19894

Espunya MC, De Michele R, Gomez-Cadenas A, Martinez MC (2012) S-Nitrosoglutathione is a component of wound- and salicylic acid-induced systemic responses in *Arabidopsis thaliana*. J Exp Bot 63:3219–3227

Feechan A, Kwon E, Yun BW, Wang Y, Pallas JA, Loake GJ (2005) A central role for S-nitrosothiols in plant disease resistance. Proc Natl Acad Sci USA 102:8054–8059

Feijo JA, Sainhas J, Hackett GR, Kunkel JG, Hepler PK (1999) Growing pollen tubes possess a constitutive alkaline band in the clear zone and a growth-dependent acidic tip. J Cell Biol 144:483–496

Foster MW, Forrester MT, Stamler JS (2009) A protein microarray-based analysis of S-nitrosylation. Proc Natl Acad Sci USA 106:18948–18953

Foyer CH, Noctor G (2011) Ascorbate and glutathione: the heart of the redox hub. Plant Physiol 155:2–18

Garcia-Mata C, Gay R, Sokolovski S, Hills A, Lamattina L, Blatt MR (2003) Nitric oxide regulates K+ and Cl- channels in guard cells through a subset of abscisic acid-evoked signaling pathways. Proc Natl Acad Sci USA 100:11116–11121

Gaupels F, Kuruthukulangarakoola GT, Durner J (2011a) Upstream and downstream signals of nitric oxide in pathogen defence. Curr Opin Plant Biol 14:707–714

Gaupels F, Spiazzi-Vandelle E, Yang D, Delledonne M (2011b) Detection of peroxynitrite accumulation in *Arabidopsis thaliana* during the hypersensitive defense response. Nitric Oxide 25:222–228

Gibbon BC, Kropf DL (1994) Cytosolic pH gradients associated with tip growth. Science 263:1419–1421

Goldstein S, Czapski G (1996) Mechanism of the nitrosation of thiols and amines by oxygenated ˙NO solutions: the nature of the nitrosating intermediates. J Am Chem Soc 118:3419–3425

Gow AJ, Buerk DG, Ischiropoulos H (1997) A novel reaction mechanism for the formation of S-nitrosothiol in vivo. J Biol Chem 272:2841–2845

Graham JW, Williams TC, Morgan M, Fernie AR, Ratcliffe RG, Sweetlove LJ (2007) Glycolytic enzymes associate dynamically with mitochondria in response to respiratory demand and support substrate channeling. Plant Cell 19:3723–3738

Graziano M, Lamattina L (2005) Nitric oxide and iron in plants: an emerging and converging story. Trends Plant Sci 10:4–8

Greco TM, Hodara R, Parastatidis I, Heijnen HF, Dennehy MK, Liebler DC, Ischiropoulos H (2006) Identification of S-nitrosylation motifs by site-specific mapping of the S-nitrosocysteine proteome in human vascular smooth muscle cells. Proc Natl Acad Sci USA 103:7420–7425

Grun S, Lindermayr C, Sell S, Durner J (2006) Nitric oxide and gene regulation in plants. J Exp Bot 57:507–516

Guikema B, Lu Q, Jourd'heuil D (2005) Chemical considerations and biological selectivity of protein nitrosation: implications for NO-mediated signal transduction. Antioxid Redox Signal 7:593–606

Guo FQ, Crawford NM (2005) Arabidopsis nitric oxide synthase1 is targeted to mitochondria and protects against oxidative damage and dark-induced senescence. Plant Cell 17:3436–3450

Gupta KJ, Fernie AR, Kaiser WM, Van Dongen JT (2011) On the origins of nitric oxide. Trends Plant Sci 16:160–168

He YK, Tang RH, Hao Y, Stevens RD, Cook CW, Am SM, Jing LF, Yang ZG, Chen LG, Guo FQ, Fiorani F, Jackson RB, Crawford NM, Pei ZM (2004) Nitric oxide represses the *Arabidopsis* floral transition. Science 305:1968–1971

Hedberg JJ, Griffiths WJ, Nilsson SJF, Hoog JO (2003) Reduction of S-nitrosoglutathione by human alcohol dehydrogenase 3 is an irreversible reaction as analysed by electrospray mass spectrometry. Eur J Biochem 270:1249–1256

Holtgrefe S, Gohlke J, Starmann J, Druce S, Klocke S, Altmann B, Wojtera J, Lindermayr C, Scheibe R (2008) Regulation of plant cytosolic glyceraldehyde 3-phosphate dehydrogenase isoforms by thiol modifications. Physiol Plant 133:211–228

Holzmeister C, Frohlich A, Sarioglu H, Bauer N, Durner J, Lindermayr C (2011) Proteomic analysis of defense response of wildtype *Arabidopsis thaliana* and plants with impaired NO-homeostasis. Proteomics 11:1664–1683

Huang X, Stettmaier K, Michel C, Hutzler P, Mueller MJ, Durner J (2004) Nitric oxide is induced by wounding and influences jasmonic acid signaling in *Arabidopsis thaliana.* Planta 218:938–946

Hughes MN (1999) Relationships between nitric oxide, nitroxyl ion, nitrosonium cation and peroxynitrite. Biochim Biophys Acta Bioenergetics 1411:263–272

Ignarro LJ, Buga GM, Wood KS, Byrns RE, Chaudhuri G (1987) Endothelium-derived relaxing factor produced and released from artery and vein is nitric oxide. Proc Natl Acad Sci USA 84:9265–9269

Jaffrey SR, Erdjument-Bromage H, Ferris CD, Tempst P, Snyder SH (2001) Protein S-nitrosylation: a physiological signal for neuronal nitric oxide. Nat Cell Biol 3:193–197

Jensen DE, Belka GK, Du Bois GC (1998) S-Nitrosoglutathione is a substrate for rat alcohol dehydrogenase class III isoenzyme. Biochem J 331(Pt 2):659–668

Jourd'heuil D, Jourd'heuil FL, Feelisch M (2003) Oxidation and nitrosation of thiols at low micromolar exposure to nitric oxide - Evidence for a free radical mechanism. J Biol Chem 278:15720–15726

Kader MA, Lindberg S, Seidel T, Golldack D, Yemelyanov V (2007) Sodium sensing induces different changes in free cytosolic calcium concentration and pH in salt-tolerant and -sensitive rice (*Oryza sativa*) cultivars. Physiol Plant 130:99–111

Keszler A, Zhang YH, Hogg N (2010) Reaction between nitric oxide, glutathione, and oxygen in the presence and absence of protein: How are S-nitrosothiols formed? Free Radic Biol Med 48:55–64

Kho Y, Kim SC, Jiang C, Barma D, Kwon SW, Cheng J, Jaunbergs J, Weinbaum C, Tamanoi F, Falck J, Zhao Y (2004) A tagging-via-substrate technology for detection and proteomics of farnesylated proteins. Proc Natl Acad Sci USA 101:12479–12484

Kim JE, Tannenbaum SR (2004) S-nitrosation regulates the activation of endogenous procaspase-9 in HT-29 human colon carcinoma cells. J Biol Chem 279:9758–9764

Kim YM, Chung HT, Simmons RL, Billiar TR (2000) Cellular non-heme iron content is a determinant of nitric oxide-mediated apoptosis, necrosis, and caspase inhibition. J Biol Chem 275:10954–10961

Kornberg MD, Sen N, Hara MR, Juluri KR, Nguyen JV, Snowman AM, Law L, Hester LD, Snyder SH (2010) GAPDH mediates nitrosylation of nuclear proteins. Nat Cell Biol 12:1094–1100

Lapous D, Mathieu Y, Guern J, Lauriere C (1998) Increase of defense gene transcripts by cytoplasmic acidification in tobacco cell suspensions. Planta 205:452–458

Lee U, Wie C, Fernandez BO, Feelisch M, Vierling E (2008) Modulation of nitrosative stress by S-nitrosoglutathione reductase is critical for thermotolerance and plant growth in Arabidopsis. Plant Cell 20:786–802

Leshem Y, Haramaty E (1996) The characterization and contrasting effects of the nitric oxide free radical in vegetative stress and senescence of *Pisum sativum Linn.* foliage. J Plant Physiol 148:258–263

Leshem YY, Wills RBH, Ku VV (1998) Evidence for the function of the free reduced gas – nitric oxide (˙NO) as an endogenous maturation and senescence regulating factor in higher plants. Plant Physiol Biochem 36:825–826

Lillig CH, Holmgren A (2007) Thioredoxin and related molecules - from biology to health and disease. Antioxid Redox Signal 9:25–47

Lin A, Wang Y, Tang J, Xue P, Li C, Liu L, Hu B, Yang F, Loake GJ, Chu C (2012) Nitric oxide and protein S-nitrosylation are integral to hydrogen peroxide-induced leaf cell death in rice. Plant Physiol 158:451–464

Lindermayr C, Saalbach G, Durner J (2005) Proteomic identification of S-nitrosylated proteins in *Arabidopsis*. Plant Physiol 137:921–930

Lindermayr C, Saalbach G, Bahnweg G, Durner J (2006) Differential inhibition of *Arabidopsis* methionine adenosyltransferases by protein S-nitrosylation. J Biol Chem 281:4285–4291

Lindermayr C, Sell S, Muller B, Leister D, Durner J (2010) Redox regulation of the NPR1-TGA1 system of *Arabidopsis thaliana* by nitric oxide. Plant Cell 22:2894–2907

Lipton SA, Choi YB, Sucher NJ, Chen HS (1998) Neuroprotective versus neurodestructive effects of NO-related species. Biofactors 8:33–40

Liu X, Miller MJ, Joshi MS, Thomas DD, Lancaster JR Jr (1998) Accelerated reaction of nitric oxide with O2 within the hydrophobic interior of biological membranes. Proc Natl Acad Sci USA 95:2175–2179

Liu L, Hausladen A, Zeng M, Que L, Heitman J, Stamler JS (2001) A metabolic enzyme for S-nitrosothiol conserved from bacteria to humans. Nature 410:490–494

Liu LM, Yan Y, Zeng M, Zhang J, Hanes MA, Ahearn G, Mcmahon TJ, Dickfeld T, Marshall HE, Que LG, Stamler JS (2004) Essential roles of S-nitrosothiols in vascular homeostasis and endotoxic shock. Nitric Oxide 116:617–628

Lopez-Sanchez LM, Corrales FJ, Lopez-Pedrera C, Aranda E, Rodriguez-Ariza A (2010) Pharmacological impairment of S-nitrosoglutathione or thioredoxin reductases augments protein S-nitrosation in human hepatocarcinoma cells. Anticancer Res 30:415–421

Luchsinger BP, Rich EN, Gow AJ, Williams EM, Stamler JS, Singel DJ (2003) Routes to S-nitroso-hemoglobin formation with heme redox and preferential reactivity in the beta subunits. Proc Natl Acad Sci USA 100:461–466

Ma W, Berkowitz GA (2011) Ca^{2+} conduction by plant cyclic nucleotide gated channels and associated signaling components in pathogen defense signal transduction cascades. New Phytol 190:566–572

Madej E, Folkes LK, Wardman P, Czapski G, Goldstein S (2008) Thiyl radicals react with nitric oxide to form S-nitrosothiols with rate constants near the diffusion-controlled limit. Free Radic Biol Med 44:2013–2018

Maldonado-Alconada AM, Echevarria-Zomeno S, Lindermayr C, Redondo-Lopez I, Durner J, Jorrin-Novo JV (2011) Proteomic analysis of *Arabidopsis* protein S-nitrosylation in response to inoculation with *Pseudomonas syringae*. Acta Physiologiae Plantarum 33:1493–1514

Mann M, Jensen ON (2003) Proteomic analysis of post-translational modifications. Nat Biotechnol 21:255–261

Mannick JB, Hausladen A, Liu L, Hess DT, Zeng M, Miao QX, Kane LS, Gow AJ, Stamler JS (1999) Fas-induced caspase denitrosylation. Science 284:651–654

Marino SM, Gladyshev VN (2010) Structural analysis of cysteine S-nitrosylation: a modified acid-based motif and the emerging role of trans-nitrosylation. J Mol Biol 395:844–859

Martinez MC, Achkor H, Persson B, Fernandez MR, Shafqat J, Farres J, Jornvall H, Pares X (1996) Arabidopsis formaldehyde dehydrogenase. Molecular properties of plant class III alcohol dehydrogenase provide further insights into the origins, structure and function of plant class p and liver class I alcohol dehydrogenases. Eur J Biochem 241:849–857

Mathieu Y, Lapous D, Thomine S, Lauriere C, Guern J (1996) Cytoplasmic acidification as an early phosphorylation-dependent response of tobacco cells to elicitors. Planta 199:416–424

Mitchell DA, Marletta MA (2005) Thioredoxin catalyzes the S-nitrosation of the caspase-3 active site cysteine. Nat Chem Biol 1:154–158

Mitchell DA, Morton SU, Fernhoff NB, Marletta MA (2007) Thioredoxin is required for S-nitrosation of procaspase-3 and the inhibition of apoptosis in Jurkat cells. Proc Natl Acad Sci USA 104:11609–11614

Modolo LV, Augusto O, Almeida IMG, Magalhaes JR, Salgado I (2005) Nitrite as the major source of nitric oxide production by *Arabidopsis thaliana* in response to Pseudomonas syringae. FEBS Lett 579:3814–3820

Mou Z, Fan W, Dong X (2003) Inducers of plant systemic acquired resistance regulate NPR1 function through redox changes. Cell 113:935–944

Mulsch A, Mordvintcev PI, Vanin AF, Busse R (1993) Formation and release of dinitrosyl iron complexes by endothelial cells. Biochem Biophys Res Commun 196:1303–1308

Nakamura T, Wang L, Wong CCL, Scott FL, Eckelman BP, Han XM, Tzitzilonis C, Meng FJ, Gu ZZ, Holland EA, Clemente AT, Okamoto S, Salvesen GS, Riek R, Yates JR, Lipton SA (2010) Transnitrosylation of XIAP Regulates Caspase-Dependent Neuronal Cell Death. Mol Cell 39:184–195

Nedospasov A, Rafikov R, Beda N, Nudler E (2000) An autocatalytic mechanism of protein nitrosylation. Proc Natl Acad Sci USA 97:13543–13548

Neill SJ, Desikan R, Clarke A, Hancock JT (2002a) Nitric oxide is a novel component of abscisic acid signaling in stomatal guard cells. Plant Physiol 128:13–16

Neill SJ, Desikan R, Clarke A, Hurst RD, Hancock JT (2002b) Hydrogen peroxide and nitric oxide as signaling molecules in plants. J Exp Bot 53:1237–1247

Palmieri MC, Lindermayr C, Bauwe H, Steinhauser C, Durner J (2010) Regulation of plant glycine decarboxylase by S-nitrosylation and glutathionylation. Plant Physiol 152:1514–1528

Pawloski JR, Hess DT, Stamler JS (2001) Export by red blood cells of nitric oxide bioactivity. Nature 409:622–626

Perazzolli M, Dominici P, Romero-Puertas MC, Zago E, Zeier J, Sonoda M, Lamb C, Delledonne M (2004) *Arabidopsis* nonsymbiotic hemoglobin AHb1 modulates nitric oxide bioactivity. Plant Cell 16:2785–2794

Pieterse CM, Van Loon LC (2004) NPR1: the spider in the web of induced resistance signaling pathways. Curr Opin Plant Biol 7:456–464

Radwan O, Mouzeyar S, Venisse JS, Nicolas P, Bouzidi MF (2005) Resistance of sunflower to the biotrophic oomycete *Plasmopara halstedii* is associated with a delayed hypersensitive response within the hypocotyls. J Exp Bot 56:2683–2693

Rafikova O, Rafikov R, Nudler E (2002) Catalysis of S-nitrosothiols formation by serum albumin: the mechanism and implication in vascular control. Proc Natl Acad Sci USA 99:5913–5918

Reynaert NL, Ckless K, Korn SH, Vos N, Guala AS, Wouters EF, Van Der Vliet A, Janssen-Heininger YM (2004) Nitric oxide represses inhibitory kappaB kinase through S-nitrosylation. Proc Natl Acad Sci USA 101:8945–8950

Rius SP, Casati P, Iglesias AA, Gomez-Casati DF (2008) Characterization of *Arabidopsis* lines deficient in GAPC-1, a cytosolic NAD-dependent glyceraldehyde-3-phosphate dehydrogenase. Plant Physiol 148:1655–1667

Rodriguez-Serrano M, Romero-Puertas MC, Zabalza A, Corpas FJ, Gomez M, Del Rio LA, Sandalio LM (2006) Cadmium effect on oxidative metabolism of pea (*Pisum sativum L.*) roots. Imaging of reactive oxygen species and nitric oxide accumulation in vivo. Plant Cell Environ 29:1532–1544

Romero-Puertas MC, Laxa M, Matte A, Zaninotto F, Finkemeier I, Jones AM, Perazzolli M, Vandelle E, Dietz KJ, Delledonne M (2007) S-nitrosylation of peroxiredoxin II E promotes peroxynitrite-mediated tyrosine nitration. Plant Cell 19:4120–4130

Romero-Puertas MC, Campostrini N, Matte A, Righetti PG, Perazzolli M, Zolla L, Roepstorff P, Delledonne M (2008) Proteomic analysis of S-nitrosylated proteins in *Arabidopsis thaliana* undergoing hypersensitive response. Proteomics 8:1459–1469

Roos W, Viehweger K, Dordschbal B, Schumann B, Evers S, Steighardt J, Schwartze W (2006) Intracellular pH signals in the induction of secondary pathways–the case of Eschscholzia californica. J Plant Physiol 163:369–381

Rusterucci C, Espunya MC, Diaz M, Chabannes M, Martinez MC (2007) S-nitrosoglutathione reductase affords protection against pathogens in *Arabidopsis*, both locally and systemically. Plant Physiol 143:1282–1292

Saito S, Yamamoto-Katou A, Yoshioka H, Doke N, Kawakita K (2006) Peroxynitrite generation and tyrosine nitration in defense responses in tobacco BY-2 cells. Plant Cell Physiol 47:689–697

Sakamoto A, Ueda M, Morikawa H (2002) *Arabidopsis* glutathione-dependent formaldehyde dehydrogenase is an S- nitrosoglutathione reductase. FEBS Lett 515:20–24

Schmidt HHHW, Hofmann H, Schindler U, Shutenko ZS, Cunningham DD, Feelisch M (1996) No ˙NO from NO synthase. Proc Natl Acad Sci USA 93:14492–14497

Schrammel A, Gorren ACF, Schmidt K, Pfeiffer S, Mayer B (2003) S-nitrosation of glutathione by nitricoxide, peroxynitrite, and $^{\bullet}NO/O_2^{\bullet-}$. Free Radic Biol Med 34:1078–1088

Sen N, Hara MR, Kornberg MD, Cascio MB, Bae BI, Shahani N, Thomas B, Dawson TM, Dawson VL, Snyder SH, Sawa A (2008) Nitric oxide-induced nuclear GAPDH activates p300/CBP and mediates apoptosis. Nat Cell Biol 10:866–873

Serpa V, Vernal J, Lamattina L, Grotewold E, Cassia R, Terenzi H (2007) Inhibition of AtMYB2 DNA-binding by nitric oxide involves cysteine S-nitrosylation. Biochem Biophys Res Commun 361:1048–1053

Seth D, Stamler JS (2011) The SNO-proteome: causation and classifications. Curr Opin Chem Biol 15:129–136

Simontacchi M, Buet A, Lamattina L, Puntarulo S (2012) Exposure to nitric oxide increases the nitrosyl-iron complexes content in sorghum embryonic axes. Plant Sci Int J Exp Plant Biol 183:159–166

Sokolovski S, Hills A, Gay R, Garcia-Mata C, Lamattina L, Blatt MR (2005) Protein phosphorylation is a prerequisite for intracellular Ca^{2+} release and ion channel control by nitric oxide and abscisic acid in guard cells. Plant J 43:520–529

Spoel SH, Loake GJ (2011) Redox-based protein modifications: the missing link in plant immune signaling. Curr Opin Plant Biol 14:358–364

Staab CA, Alander J, Brandt M, Lengqvist J, Morgenstern R, Grafstrom RC, Hoog JO (2008) Reduction of S-nitrosoglutathione by alcohol dehydrogenase 3 is facilitated by substrate alcohols via direct cofactor recycling and leads to GSH-controlled formation of glutathione transferase inhibitors. Biochem J 413:493–504

Stamler JS, Toone EJ, Lipton SA, Sucher NJ (1997) (S)NO signals: translocation, regulation, and a consensus motif. Neuron 18:691–696

Stamler JS, Lamas S, Fang FC (2001) Nitrosylation. the prototypic redox-based signaling mechanism. Cell 106:675–683

Sueldo DJ, Foresi NP, Casalongue CA, Lamattina L, Laxalt AM (2010) Phosphatidic acid formation is required for extracellular ATP-mediated nitric oxide production in suspension-cultured tomato cells. New Phytol 185:909–916

Sumimoto H, Ueno N, Yamasaki T, Taura M, Takeya R (2004) Molecular mechanism underlying activation of superoxide-producing NADPH oxidases: Roles for their regulatory proteins. Jpn J Infect Dis 57:S24–S25

Sweetlove LJ, Heazlewood JL, Herald V, Holtzapffel R, Day DA, Leaver CJ, Millar AH (2002) The impact of oxidative stress on *Arabidopsis* mitochondria. Plant J 32:891–904

Tada Y, Spoel SH, Pajerowska-Mukhtar K, Mou Z, Song J, Wang C, Zuo J, Dong X (2008) Plant immunity requires conformational charges of NPR1 via S-nitrosylation and thioredoxins. Science 321:952–956

Torres MA, Dangl JL (2005) Functions of the respiratory burst oxidase in biotic interactions, abiotic stress and development. Curr Opin Plant Biol 8:397–403

Torres MA, Dangl JL, Jones JD (2002) *Arabidopsis* gp91phox homologues AtrbohD and AtrbohF are required for accumulation of reactive oxygen intermediates in the plant defense response. Proc Natl Acad Sci USA 99:517–522

Ueno T, Suzuki Y, Fujii S, Vanin AF, Yoshimura T (2002) In vivo nitric oxide transfer of a physiological NO carrier, dinitrosyl dithiolato iron complex, to target complex. Biochem Pharmacol 63:485–493

Uotila L, Koivusalo M (1979) Purification of formaldehyde and formate dehydrogenases from pea seeds by affinity chromatography and S-formylglutathione as the intermediate of formaldehyde metabolism. Arch Biochem Biophys 196:33–45

Van Der Vliet A, Hoen PA, Wong PS, Bast A, Cross CE (1998) Formation of S-nitrosothiols via direct nucleophilic nitrosation of thiols by peroxynitrite with elimination of hydrogen peroxide. J Biol Chem 273:30255–30262

Vanacker H, Carver TL, Foyer CH (2000) Early H_2O_2 accumulation in mesophyll cells leads to induction of glutathione during the hyper-sensitive response in the barley-powdery mildew interaction. Plant Physiol 123:1289–1300

Wang YQ, Feechan A, Yun BW, Shafiei R, Hofmann A, Taylor P, Xue P, Yang FQ, Xie ZS, Pallas JA, Chu CC, Loake GJ (2009) S-nitrosylation of AtSABP3 antagonizes the expression of plant immunity. J Biol Chem 284:2131–2137

Wawer I, Bucholc M, Astier J, Anielska-Mazur A, Dahan J, Kulik A, Wyslouch-Cieszynska A, Zareba-Koziol M, Krzywinska E, Dadlez M, Dobrowolska G, Wendehenne D (2010) Regulation of *Nicotiana tabacum* osmotic stress-activated protein kinase and its cellular partner GAPDH by nitric oxide in response to salinity. Biochem J 429:73–83

Wilson ID, Neill SJ, Hancock JT (2008) Nitric oxide synthesis and signaling in plants. Plant Cell Environ 31:622–631

Wink DA, Nims RW, Darbyshire JF, Christodoulou D, Hanbauer I, Cox GW, Laval F, Laval J, Cook JA, Krishna MC et al (1994) Reaction kinetics for nitrosation of cysteine and glutathione in aerobic nitric oxide solutions at neutral pH. Insights into the fate and physiological effects of intermediates generated in the NO/O_2 reaction. Chem Res Toxicol 7:519–525

Wunsche H, Baldwin IT, Wu JQ (2011) S-Nitrosoglutathione reductase (GSNOR) mediates the biosynthesis of jasmonic acid and ethylene induced by feeding of the insect herbivore Manduca sexta and is important for jasmonate-elicited responses in *Nicotiana attenuata*. J Exp Bot 62:4605–4616

Yu Q, Tang C, Kuo J (2000) A critical review on methods to measure apoplastic pH in plants. Plant Soil 219:29–40

Yun BW, Feechan A, Yin M, Saidi NB, Le Bihan T, Yu M, Moore JW, Kang JG, Kwon E, Spoel SH, Pallas JA, Loake GJ (2011) S-nitrosylation of NADPH oxidase regulates cell death in plant immunity. Nature 478:264–268

Zeidler D, Zahringer U, Gerber I, Dubery I, Hartung T, Bors W, Hutzler P, Durner J (2004) Innate immunity in *Arabidopsis thaliana*: lipopolysaccharides activate nitric oxide synthase (NOS) and induce defense genes. Proc Natl Acad Sci USA 101:15811–15816

Zhang H, Means GE (1996) S-nitrosation of serum albumin: spectrophotometric determination of its nitrosation by simple S-nitrosothiols. Anal Biochem 237:141–144

Zhang H, Andrekopoulos C, Xu Y, Joseph J, Hogg N, Feix J, Kalyanaraman B (2009) Decreased S-nitrosation of peptide thiols in the membrane interior. Free Radic Biol Med 47:962–968

Chapter 8
In-Silico Approaches for Studying the MAP Kinase Signaling Pathways Involved in Resistance Against Alternaria Blight in *Brassica*

Gohar Taj, Sugandha Sharma, Priyanka Giri, Dinesh Pandey, and Anil Kumar

Introduction

Brassica family members are among the top ten economic crops in the world. It is a genus of plants in the mustard family (*Brassicaceae*) and, thus, these are in the same taxonomic family as *Arabidopsis thaliana*. This genus is remarkable for containing more important agricultural and horticultural crops than any other genus. *Brassica* oilseed crops comprise of *B. juncea*, *B. rapa*, *B. napus*, *B. carinata* are believed to be originated from central Asian-Himalayas with migration to India, China, Middle east, Europe and Northeastern Africa, respectively. *Brassica juncea* (Indian mustard) is an important oilseed crop grown in many countries throughout the world while *B. oleracea* and *B. rapa* comprise many of the vegetables in our daily diet. Young tender leaves of mustard greens are used in salads or mixed with other salad greens. Older leaves with stems may be eaten fresh, canned or frozen, for potherbs, and to a limited extent in salads. Although widely and extensively grown as vegetables, it is being grown more for its seeds which yield an essential oil and condiment. The future prospects of *Brassica* oil crops could be engineered oils with enhanced health and proactive vegetables against cancer as "functional foods" or engineered oils for technical purpose as lubricants, plastics and detergents (Kruger et al. 2002).

Plant diseases have been known from the beginning of organized agriculture and have been frequently been associated with hunger and suffering (Houlb 2001). *Brassica* crops are heavily challenged by various fungal pathogen, bacteria, virus

G. Taj (✉) • S. Sharma • P. Giri • D. Pandey • A. Kumar
Department of Molecular Biology & Genetic Engineering,
College of Basic Sciences & Humanities, GB Pant University of Agriculture & Technology,
Pantnagar 263 145, U S Nagar, India
e-mail: gohartajkhan@rediffmail.com

M. Sarwat et al. (eds.), *Stress Signaling in Plants: Genomics and Proteomics Perspective, Volume 1*, DOI 10.1007/978-1-4614-6372-6_8,

Fig. 8.1 *Brassica* leaves at different stages of disease progression. (**a**) Healthy leaf, (**b**) necrotic leaf, (**c**) leaf with necrosis and chlorosis, (**d**) completely chlorotic leaf

and insects, and to some extent oomycetes. Major diseases for *Brassica* are black spot or Alternaria blight, Downy mildew, White rust, *Verticillium* wilt, blackleg, light leaf spot and stem rot (Rimmer and Buchwaldt 1995; Dixelius et al. 2005). Among these, Alternaria blight is a most devastating fungal disease which affects the majority of cruciferous crops and is one among the important diseases of rapeseed mustard caused by *Alternaria brassicae*. This disease has been reported from all the continents of the world causing severe yield losses with no proven source of transferable resistance in any of the hosts (Meena et al. 2010). Alternaria blight disease appears as a brown necrotic spots on cotyledonary leaves and brown streaks on hypocotyl in seedlings leading to post emergence losses. Alternaria invades siliquae and penetrates the seeds. These also affect the chemical composition of seed including protein, total carbohydrates and ash. The pathogen *Alternaria brassicae* produces a depsipeptide chlorotic toxin, Dextruxin B and a polyketide like necrotic toxin. Necrotic toxin is responsible for cell death whereas chlorotic toxin is responsible to suppress defence system (Taj et al. 2004). The disease appears on the cotyledonary leaves in the form of light brown lesions which soon turns black due to the production of spore mass (Valkonen and Koponen 1990). The infection on the leaves starts as minute brown to blackish spots which may vary in size from pinpoint-size dark circular spots to black; brown spots of some inches and later the formation of concentric rings in the lesion and a zone of yellow halo around the lesion are very prominent (Fig. 8.1) (Verma and Saharan 1994; Meena et al. 2004).

Because of the lack of the availability of the sources of resistance against *Alternaria brassicae* within the family *Brassicaceae*, *Alternaria* blight consider the most damaging and widespread fungal disease of *Brassica* (Ghose et al. 2008). A special pattern of development of several *Alternaria* species on leaf disc of host and non host was known. When pathogen attacks on leaf firstly the germination of spore take place then germ tube formation and formation of appressorium was observed. The plant responses to attempted penetration varied with specific pathogen-plant interactions.

Plant Pathogen Interaction

Plant pathogen interaction involves the close communication between the host and the pathogen which invade the plant. This interaction mainly comprises the three steps *i.e.* recognition, signal transduction and response. At the molecular level, defense systems often depend on specific recognition of pathogen invasion (Chisholm et al. 2006) by plants through two types of immune receptors *viz.*, membrane-resident pattern recognition receptors and pathogen associated molecular patterns (PAMP).

Signal Transduction in Response to Pathogen Attack

Interaction between plants and pathogen induce a series of plant defence responses (Hammond-Kosack and Parker 2003) including activation of protein kinases, production of signaling compounds such as salicylic acid, nitric oxide, ethylene and jasmonic acid which causes activation of many downstream responses (Baker et al. 1997). Signal transduction is the means by which cells respond to extracellular information by interacting with other proteins through the various processes like protein-protein interaction, protein-lipid interaction *etc*. Plants are equipped with integrated signaling network that is capable of mediating the responses to internal and external factors such as hormones and nutrients and the environmental cues and stresses that has an overall effect on plant growth and development (Craig et al. 2008). Signal-transduction cascades mediate the sensing and processing of stimuli. These pathways often involve the phosphorylation of the receptor protein which in turn phosphorylates other cellular proteins. These signals are passed through the several cellular proteins where eventually a transcription factor is modified such that it activates transcription. The levels of different transcriptional expression during plant responses to pathogen attack are regulated by various regulatory elements which are best characterized in terms of pathogen defence mechanism. Molecular level studies have shown that there are at least three component involved in Alternaria blight disease development *i.e.*, two toxin and a phytohormone. Toxin facilitates host cell death often by triggering genetically programmed apoptotic pathway or by directly causing cell damage resulting necrosis (Lawrence et al. 2008). There are evidences in literature which support the view that phytohormones act through action of cell signal transduction (Pandey et al. 2002; Taj et al. 2004). Therefore, it would be logical to assume that knowledge of signal transduction play a crucial role in disease resistance. It is being felt that molecular biology tools can be utilized to study the key molecules of signaling pathways to obtain mechanism of pathogenesis of Alternaria blight. In recent years, especially from our lab many researchers have been identifying various components of the signal transduction pathway activated by the pathogen, *Alternaria brassicae* and its interaction with the host *Brassica* at molecular and cellular level during pathogenesis.

Mitogen Activated Protein Kinases (MAPK): A Key Molecule of Signal Transduction Pathway Involved in Defence Against Pathogen Attack

In plants signal transduction are mediated by a special class of family of serine/theronine protein kinases known as MAPK which transduce the environmental and developmental signals into adaptive and programmed responses. Signaling through MAPK is a fundamental and conserved process in eukaryotes and transduce external signals through protein phosphorylation (Bethke et al. 2009). An important consideration for the analysis of signaling is the sub-cellular localization of MAPK. A translocation of MAPK from cytoplasmic compartment into the nucleus occurs after mitogen stimulation. Thus the targets of MAPK signaling are located within cellular compartments. Moreover, MAPK provides a physical link in the signal transduction pathway from the cytoplasm to the nucleus (Davis et al. 1994). MAPK pathway is one of the main phosphorylation pathways that plants use in biotic and abiotic stress resistance like wounding and pathogen infection, temperature, stress or drought (Fig. 8.2).

The transmission of extracellular signals into their intercellular targets is mediated by a network of interacting proteins that regulate a large number of cellular processes (Pedley and Martin 2005). MAPK signaling cascades are composed of three main signaling element *viz.,* MAPKKK, MAPKK and a MAPK, which phosphorylate, and therefore activate, each other in a specific way *i.e.* MAPKKK activates MAPKK through phosphorylation on serine/theronine residues which then phosphorylates MAPK at threonine and tyrosine residue. Fully activated MAPK activates downstream target proteins through phosphorylation on serine theronine residues. MAPK components are particularly abundant in plants; there are 80 putative MAPKKKs, 10 MAPKKs and at least 20 MAPKs in *Arabidopsis* (Colcombet and Hirt 2008).

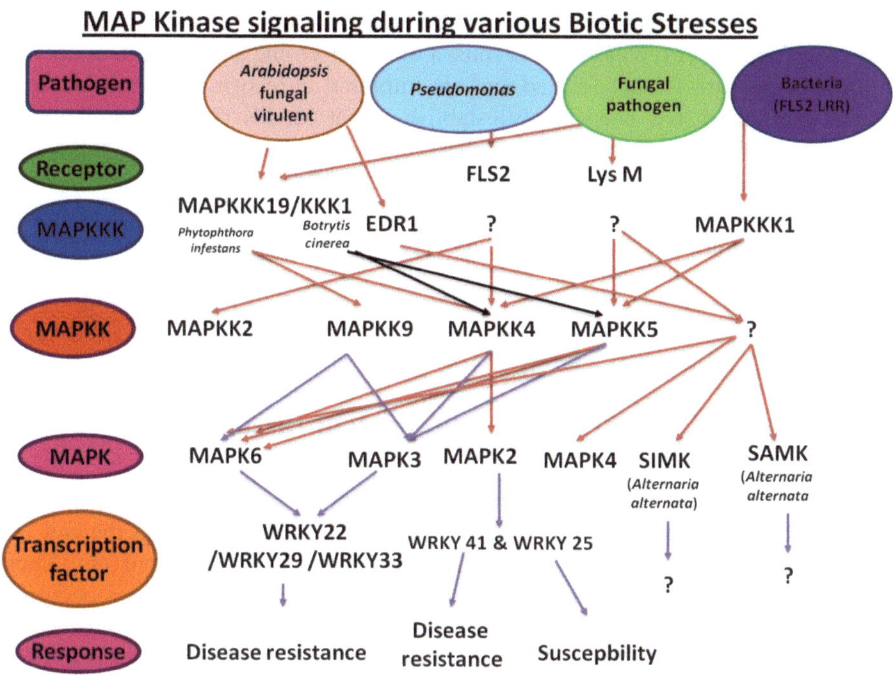

Fig. 8.2 MAPK cascades in the plant defence to bacterial and fungal pathogens

MAPK Cascade: A Combinatorial Interaction of Protein in Disease Resistance

Recent studies have shown that disease resistance against any pathogen in plant is governed by combinatorial interaction between proteins. Therefore, now days much effort has gone into finding the complete set of interacting proteins. Keeping the evolutionary conserved function of MAPK cascade it is possible that the *Alternaria* toxin also affects some of the key components of this highly conserved MAPK cascades. Based on the number of MAPKKKs, MAPKKs and MAPKs, there can be theoretically numerous combinations ($60\times10\times23=13800$) of pathways (Cvetkovska et al. 2005). Hence, there are tremendous chances of cross talks between these parallel and distinct pathways. An important consideration for the analysis of signaling is the subcellular localization of MAPK. The spatial organization of kinases and substrates determines what signals may be transmitted and received at various possible sites of action. The complement of cellular signaling proteins and cell state together determine the distribution of MAPK and other signaling molecules in a manner that can be regulated acutely and long term by extracellular signals. The specificity of different MAPK cascades functioning within the same cell is generated through interaction of two sites *i.e.* catalytic site and docking site. A typical MAPK consist of an active site and a common docking groove which are closely located and are implored in recognition and binding of target proteins. It has been firmly established that MAPK play a central role in pathogen defence in *Arabidopsis*, Tobacco, Tomato, Parsley, *Brassica* and Rice (Nadarajah and Sidek 2010). The activation of MAPK pathways is a specific response of the host cells to pathogen which may lead to successful symbiotic interaction, suggesting that MAPK may take part in the recognition of compatible partners.

Techniques for Studying Protein–Protein Interaction

Protein- protein interaction plays very important role in governing cell's homeostasis and maintaining the various biological and molecular function in the cell like signal transduction, defence responses, transportation, development etc. These interactions are of prime importance for almost every virtual processes occurring inside the living cell. There are basically two types of technique through which we can study the protein-protein interaction *viz.,* Wet-lab techniques and *in-silico* approach. Wet-lab technique includes yeast two hybrid system, affinity immunochip based system followed by mass spectrometry, phage display libraries, and protein microarray. There are various methods which comes under *in silico* interaction prediction like Machine Learning Approaches *viz*., Hidden Markov Model (HMM), Neural Network (NN), Support Vector Machine (SVM) etc.; Direct Sequence Analysis approach that takes into account surface information for predicting protein-protein interaction; Docking studies which attempts to use geometric and steric considerations to fit two proteins of known structure into a bound complex; through phylogenetic profiling- Proteins that interact are more likely to co-evolve, therefore it is possible to make inferences about interactions between pairs of proteins based on their phylogenetic distances or through phylogenetic profiling. In this chapter we will discuss some of these techniques to understand the predictive approach for studying the protein-protein interaction network and in turn better understanding of signaling cascades.

Wet-Lab Techniques for Studying Protein–Protein Interaction

Wet-lab techniques for studying protein -protein interactions mainly include protein-protein microarray and yeast two-hybrid approach, phage display libraries, affinity immunochip technique, mass spectrometry, phospho-proteomics approaches etc. Some of these conventional techniques are described below.

Yeast Two Hybrid Approach

Yeast two-hybrid (Y2H) system was pioneered by Stanley Fields and Song in 1998. In this technique Yeast is used as a model for finding eukaryotic protein interactions. Here a library is screened or a protein is characterized using bait (protein of interest) construct and the interactions are identified by the transcription of reporter gene such as HIS 3, MEL1 (a galactosidase), Lac Z reporter etc. Y2H basically measure direct physical interactions between protein pairs i.e. a binary approach. The main advantage of using Y2H is that, we can identify the novel interactions between proteins involved in the same biological function and which connect consequently the signal transduction pathways. Although the yeast two-hybrid system represents a powerful tool to identify protein- protein interactions, but in signal transduction the *in vivo* exploitation of molecular function is technically demanding. Lee et al. (2008) identified the protein-protein interactions between *Arabidopsis* MAPKs and MAPKKs using Y2H system to conduct a directed protein-protein interaction. They tested the novel interaction *in vitro* for enzyme-substrate functionality, using recombinant proteins.

Microarray Techniques

The Term Microarray signifies the arrangement of microspot in particular row or order on a solid support. It is based on complementary base pairing. As the binding of target sample to the probes on the spots increases, the intensity of hybridization also increases. The intensity of feature is compared to the intensity of same feature under different condition in order to assess the interaction. Microarray measures interaction between groups of proteins. Protein function based microarray investigates the biochemical properties as well as protein interactions and enzyme activity of protein printed on array. With the high density of arrays, genes that have altered expression in compatible and incompatible plant-pathogen interactions could be targeted for characterization by microarray analysis. Few examples that use microarray technique are:

Feilner et al. (2005); identified the Potential substrates for *Arabidopsis thaliana* Mitogen-activated Protein Kinases MAPK and MAPK6, which are activated by stress factor through high throughput identification technique. They generated protein microarrays including 1,690 *Arabidopsis* proteins which were obtained from the expression from an inflorescence meristem cDNA expression library. On the basis of threshold based quantification method they evaluated the microarray result and identified the 48 substrates of MAPK3 and 39 of MAPK6.

Popescu et al. (2009); acknowledged the MAPK target networks in *Arabidopsis thaliana* using functional protein microarrays. They determined which *Arabidopsis thaliana* MAPKKs preferentially activate 10 different MAPKs *in vivo* and used the activated MAPKs to probe high-density protein microarrays to determine their phosphorylation targets. Their analyses revealed known and novel signaling modules encompassing 570 MAPKs phosphorylation substrates; these substrates were enriched in transcription factors involved in the regulation of development, defense, and stress responses. This study predicted MAPKK–MAPK phosphorylation network and constitutes a valuable resource to understand the function and specificity of MAPK signaling systems.

Mass Spectrometry

In mass spectrometry technique the molecules in a test sample are converted to gaseous ions that are subsequently separated in a mass spectrometer according to their mass to charge ratio. It is used for the analysis of glycosylation sites, protein phosphorylation and identification of phosphopeptides, and disulphide linkages etc. Mass spectrometry is also used to identify the protein -protein interaction. The target proteins which interact are isolated through various isolation techniques like Immunoprecipitation, affinity chromatography, surface plasma resonance etc. then the direct interaction analysis is done by Matrix-assisted laser desorption ionization-Time of flight(MALDI-TOF). By the excitation of the sample molecules from the energy of a laser transferred light absorbing matrix, the MALDI produces gas phase protonated ions.

Kaur et al. (2011) performed the proteome analysis of the *Albugo candida-Brassica juncea* pathosystem and find out that the timing of the expression of defence related genes is a crucial determinant of pathogenesis. They performed the comparative proteomic study along with the Q-TF mass spectrometry in order to find out the variable resistance to *A. Candida* in *B. juncea* germplasm.

Taylor et al. (2009) performed dynamic analysis of MAPK signaling using a high-throughput microfluidic single-cell imaging platform. Microfluidics provides a powerful method for high-throughput imaging analysis with programmable control over the chemical environment, offering a new temporal dimension to live-cell imaging studies.

Phosphoproteomics Approach

Proteomics is the study of whole set of proteins which are specified by the genome, in other words it is a study of complete set of proteins present in a cell, tissue, organ, organism at a given interval of time. Phosphoproteomics is a branch of proteomics and it provides information about the protein or related pathways which might be activated because the change in phosphorylation status signifies the changes which occur in protein activity. Phosphorylation is a process which regulates the protein function, degradation and in turns regulates the protein- protein interaction. Phosphoproteomics approach combines various techniques like ion exchange chromatography, gel electrophoresis (SDS-PAGE), Mass Spectrometry etc.

Kav et al. (2007) used the proteomics approach to investigate Plant-Microbe Interaction. They performed the proteome level analysis to characterize the interactions between plants and fungi belonging to three categories *i.e.,* biotrophs (require live plant material, hemibiotrophs (initially act as biotrophs after which they switch to being a necrotroph), and necrotrophs (utilize dead plant material for survival).

Limitations of Wet-Lab Techniques

Each of the approaches has its own strengths and weakness in-terms of sensitivity and specificity, like wet-lab techniques are time consuming and require lot of efforts, for example, false positives results are the largest problem with Y2H technique. It can be caused due to non-specific binding of prey or the induction of transcription without interaction with the bait. As in the case of microarray technique, the data analysis and processing is little bit complicated and the main challenge with protein-microarray occur is that, when printing protein arrays involves keeping the proteins functional during array manufacture. Another restriction with protein microarray is that the proteins which are functional with multi-component complexes only, are difficult to analyse with this technique. Another wet-lab technique, Mass Spectrometry, also has some loopholes. It cannot distinguish stereoisomers and noncovalent complexes are often disrupted, and instrumentation is expensive too. Thus these limitations arises the need of *in-silico* techniques which gives more fast and accurate results and requires less time too.

Machine Learning Techniques

The term machine learning refers to a set of topics dealing with the creation and evaluation of algorithms that facilitate pattern recognition, classification, and prediction, based on models derived from existing data (Tarca et al. 2007). Machine learning is an interdisciplinary field with connections to artificial intelligence, information theory, statistics, pattern recognition, cognitive science and also with other various disciplines. Machine learning algorithms learn automatically from experience *i.e.* through training set and use different forms to represent knowledge. Protein-protein interaction (PPI) prediction as well as protein structure and function prediction is an important but tough and computationally complex problem in biological science. Shortcut prediction techniques which use sequence data to determine protein structure and function and in turns determination of PPI have been extensively considered in the literature. But unsatisfactory results motivated the consideration of techniques based on intelligent systems which have proved to be useful in various applications. Some of the examples of machine learning technique are described below.

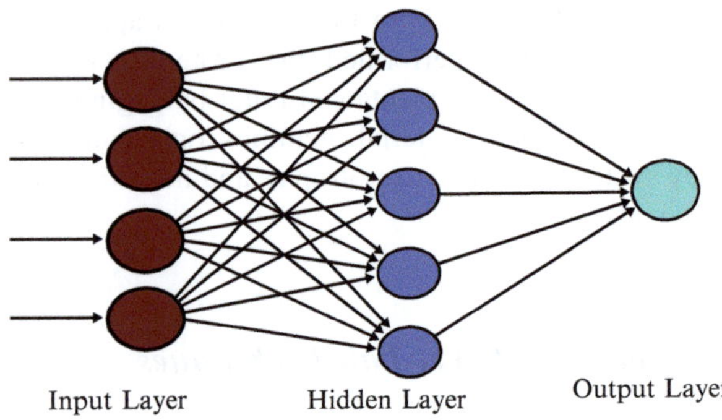

Fig. 8.3 A simple neural network with one hidden layer

Artificial Neural Network

Neurons are the set of interconnected units which are present in human brain. These neurons are connected to each other by *axons* and *dendrites* which send and receive signals respectively. Neurons send and receive information in the form of signals with various strengths. This system was the basis for building mathematical models for information processing, called artificial neural networks (ANN) or neural networks. Neural networks have shown very good performance in dealing with problems similar to protein structure and function prediction, and have been successfully used in predicting several structural and functional properties of proteins from local sequence data.

In an artificial neural network each connection between two neurons is assigned a fixed or computed value, based on the network architecture. Weights simulate the weakness and strength of signals in a human brain neural network. Neural networks used in protein structure and function prediction follow a simple feed forward architecture, consisting of one or two hidden layers. A feed forward network is composed of two or more layers of processing units. The first layer is called as input layer through which we give input to the network, the last is the output layer, and all the other layers between are termed hidden layers (Fig. 8.3). The hidden layers are used for data processing and computation. The most interesting feature of neural networks is their capability of learning which simply means that the output of the network is analyzed and used for improving the network by modifying the weights assigned to different nodes. Nodes are the different input points by which input has been given to the network. The first application of neural network in biotechnology was for secondary structure prediction of protein and was appeared in 1988. It was a natural extension into biology of artificial intelligence. In recent years the application of artificial intelligence in field of biology has grown extensively like transmembrane structure and function prediction, gene analyzing, splice site detection, transcription site detection, selection of SNP, secondary structure and class prediction *etc*.

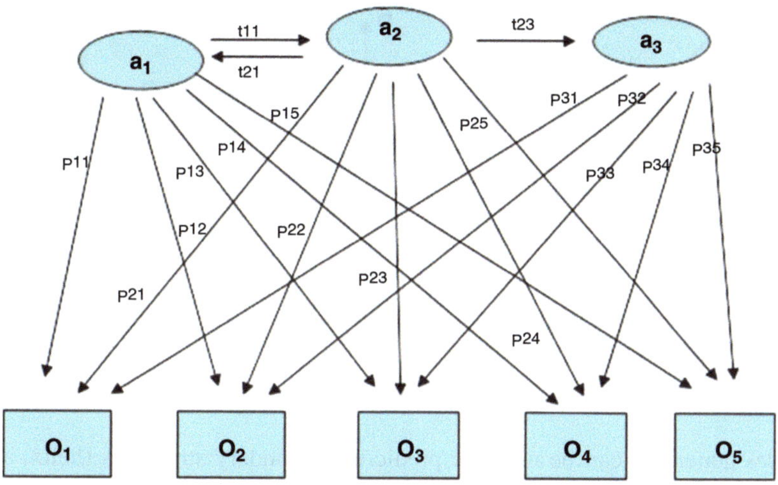

Fig. 8.4 Hidden Markov model based on probabilities

Hidden Markov Models

Hidden Markov models (HMM) are special cases of neural networks, stochastic grammars and Bayesian networks (Baldi and Brunak 2001). A Markov model is based on the probability mechanism *i.e.* a probabilistic model of symbol sequences in which the probability of the current event is governed by the previous event (Fig. 8.4).

Parameters of HMM (probabilistic) in the above example are:

a: states of HMM
t: state transition probabilities
o: possible observations
p: output probabilities

An HMM is a non-deterministic Markov model where knowledge of the emitted symbol does not determine the state-transition. This property of HMM emphasis that the more than one path through which the state-transition has occur should take into consideration in order to determine the probability of a given string. For the prediction of protein structure the Position-Specific Scoring matrices (PSSM) are formed by the conversion of multiple sequence alignment, by taking into account all matches, mismatches, and gaps in the alignment through the use of HMM. The PSSMs in turn can be used for searching distance homologues of the query sequence or for predicting protein structures (Eddy 1998; Karplus 1998). A set of 20–100 sequences is needed to train the HMMs (Mount 2001). HMM generally consider all insertions, deletions, and matches that appear in the related sequences and the associated transition probabilities in order to generate PSSMs. Hence the

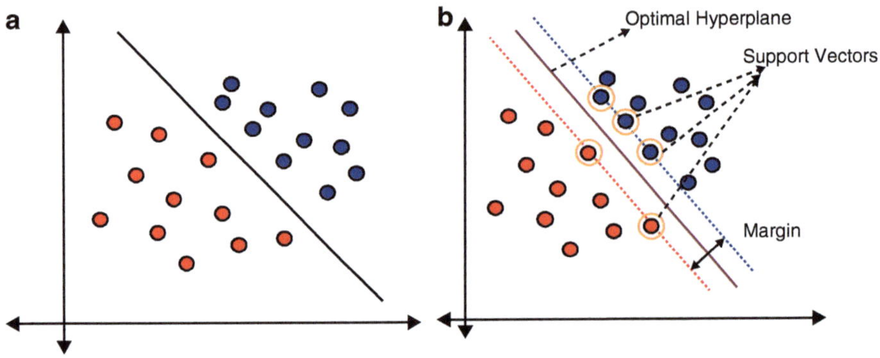

Fig. 8.5 (**a** and **b**) Schematic representation of support vector machines class boundary

matrices generated can be used for predicting secondary structures (helix, strand, and coil) or for modeling 3-D structures of proteins as well as revealing the unknown PPI networks. The most important limitations of HMMs are that they need to be trained on a larger set of sequences to correctly identify distant homologues. HMMs are unable to efficiently identify long-distance correlations between the amino acid residues of a sequence (Eddy 1998). Limitations of HMMs can be overcome by using them in conjunction with ANNs in hybrid architectures.

Support Vector Machine

Support Vector Machine (SVM) is useful technique for data classification which is widely being used in solving biological problems. SVM is rigorously based on statistical learning theory. Like ANN, it is also based on learning procedure by which it learns from the examples given in training data set. On the basis of the examples given in training data set the machine learns and partitioned the data of testing set into positive and negative data set.

SVM classify the two class linearly separable problems. In Fig. 8.5 two dimensional data points belongs to two different classes *i.e.,* red and blue circles. Samples along the dashed lines are called support vector. Only support vector information is sufficient to classify the unseen data. Among these decision boundaries, SVMs find the one that achieves maximum margin between the two classes. The margin is defined as the distance between a planar decision surface that separates two classes and the closest training samples to the decision surface. The working of SVM can be understood better by considering the example in (Fig. 8.6).

In the above example the SVM has been trained by the training set which includes the feature *i.e.* the characterization properties *viz.,* hydrophilicity and total charge along with the class label. It is an example of binary classifier which classified the testing set in two class *i.e.* negative and positive class.

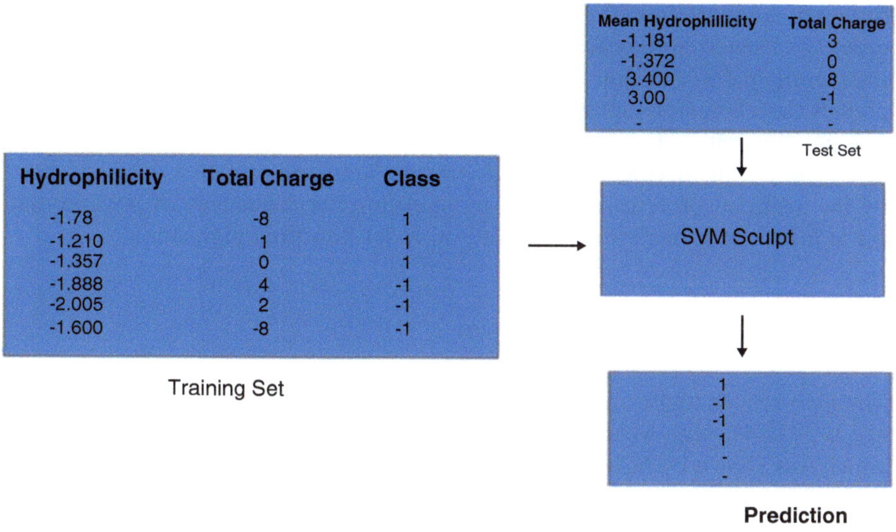

Fig. 8.6 An example of SVM

Protein–Protein Interaction Analysis Through Machine Learning Approach

By using machine learning technique we predicted the downstream interaction of transcription factors with MAPK3 in *Arabidopsis thaliana* (Taj et al. 2011). MAPK plays crucial role in cell-signaling mechanism of plant. MAPK cascade is composed of three signaling kinases, MAP3K, MAP2K and MAPK, which phosphorylate each other in a sequential manner. The response of plants and the interaction of these defending kinases to pathogen attack and other abiotic and biotic factors are currently poorly understood. Many experimental and high throughput methods have been developed to study these interaction but these methods are expensive and time consuming. Therefore, we identified the downstream interaction partners of MAPK3 in *Arabidopsis thaliana* using the information of protein sequences through Support Vector Machine (SVM) approach. The approach here used is supervised learning based on physiochemical properties of protein sequences through which we predicted whether the MAPK3 proteins interact with downstream transcription factor proteins *viz.,* Myb, bZIP, WRKY, Myb-related proteins, AP2/EREBP, and NAC with which its interaction is almost unknown.

Methodology

Here seven physicochemical feature groups of amino acids are selected to reflect these interaction modes and they are : *1)* Amino acid, dipeptide composition, *2)* Normalized Moreau-Broto autocorrelation, *3)* Moran autocorrelation,

4) Geary autocorrelation, *5)* Composition, transition, distribution, *6)* Sequence order, *7)* Pseudo amino acid composition (lamda=30). Based on examples of interacting pairs and non interacting pairs, we trained a binary classifier to predict the class (interacting or non-interacting) of a set of protein sequences. The positive data set is formed by the set of non redundant proteins that shows interaction with MAPK3 in *Arabidopsis thaliana* and negative data is formed by the artificial protein sequences by shuffling the sequences of positive data set at k-let count one which is based on K-let Shuffling algorithm.

Results

The Myb-related transcription factor family showed maximum interaction percentage *i.e.* 71.14% with MAPK3 while minimum interaction percentage was 21.15% which was shown by NAC transcription factor family. The interaction percentage shown by the gene loci of rest transcription factor family *i.e.* Myb, bZIP, AP2/EREBP, WRKY are 67.78%, 68.05%, 21.91% and 58.33% respectively (Taj et al. 2011). The results of our study clearly revealed the complexity of MAPK3 interaction with several variants of different transcription factors.

Sequence Analysis Approach

Direct Sequence Analysis approach takes into account surface information for predicting protein-protein interaction. This method comprises the various approaches like Phylogenetic profiling, Co-evolution, gene fusions, Gene neighborhood analysis *etc*.

Phylogenetic Profiling

This method finds pairs of protein families with similar patterns of presence or absence of a set of gene across large numbers of species. This method give an idea about those proteins which are likely to interact in the same biological condition but not necessarily imply physical interaction. Phylogenetic profiling can be done by the use of various on-line server and programs like PHYLIP, Musscle, CLUSTAL-W, T-cofee etc. There are various algorithms on which these programs are based like UPGMA, N-J, and Nearest Neighbour etc. In Phylogenetic profiling the first step is to identify the homologues pair of proteins of the target protein through various search tools, then building multiple sequence alignments with alignment tools such as Clustal. From these multiple sequence alignments, phylogenetic distance matrices are calculated for each protein in the hypothesized interacting pair. If the matrices are sufficiently similar in terms of Pearson correlation coefficient they are deemed likely to interact. There are various tools for phylogenetic analysis available both online as well as offline, some of them are:

Structure Based Approach

Structure based approach takes into account the structural information of the protein to predict the protein-protein interaction. Structure based approach mainly comprises the docking studies.

Docking

Docking is a computer simulation modeling the interaction between a ligand and a receptor active site. The technique of docking is to position the ligand in different orientations and conformations within the binding site to calculate optimal binding geometries and energies. Given two interacting molecules of known geometry the docking problem consists of finding their relative positions during the interaction. There are mainly two types of docking *i.e.* flexible docking and rigid body docking. Protein–protein docking is energetically governed by desolvation, Vander Waals interactions, electrostatics, and favorable specific interactions like hydrogen bonds. Protein-Protein docking can be used to study the signal transduction pathways, protein quaternary structure prediction, protein interaction prediction etc. Protein protein docking can be of two type *viz.,* bound docking and unbound docking (Fig. 8.7).

It is similar as protein ligand docking in respect that in protein protein docking too the conformations are searched and the scoring of energetic is performed. But the protein ligand docking is more trouble-free than protein-protein docking as in case of protein protein docking sites have large, flat surfaces, conservation may be, presence of hydrophobic cores, binding energies are usually dominated by geometry, hydrophobicity, and protein flexibility in terms of side chains and backbone is important. There are various docking programs through which protein protein docking can be performed *viz.,* MOE (Molecular Operating Environment), 3D-Dock, HEX, GRAMM, DOCK, AutoDock, FlexX, ZDOCK, Vakser etc.

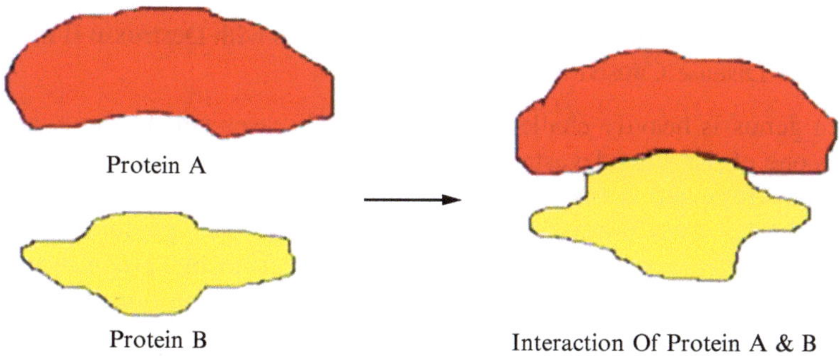

Fig. 8.7 Molecular docking of two hypothetical proteins

Use of Docking Approach to Study Protein–Protein Interaction

An Effort Towards the Identifications of Potential Interaction Partners of MAPK3 with the Transcription Factors

Through the Protein-protein docking the interactions of two proteins can be find which are similar in size. In protein-protein docking the interface between the two molecules tend to be flatter and smoother than those in protein-ligand interactions. Protein-protein interactions are usually more rigid; the interfaces of these interactions do not have the ability to alter their conformation in order to improve binding and ease movement. There are a range of evidence in literature in which docking is used to study the protein-protein interaction. We are trying to find the interaction of MAPK3 with various transcription factor *viz.,* NAC, bZip, WRKY, myb, Myb-related and AP2 transcription factors using the docking approach.

Methodology

Firstly the 3D structures of all the transcription factor along with the MAPK3 has been constructed using homology approach by using MOE (molecular Operating Environment) and the structures were validated by Ramachandran plot. Structural refinement through energy minimization model was performed using energy minimization tool of MOE, after that the interaction has been found through docking.

Results

The results clearly indicate that out of 71 members of WRKY transcription factor, 29 members of WRKY TF family are showing interaction with MAPK3 while 42 are showing non-interaction. The work is under progress in order to resolve the complexity of various signaling cascade mediated by MAP kinases (Fig. 8.8).

In-Silico Interaction Studies of Receptor Like Kinases with Destruxin B in Order to Combat the Disease Caused by Phyto-Pathogens

Brassica genus is heavily challenged by fungal diseases among these, alternaria blight is one of the most devastating fungal diseases which affects the majority of cruciferous crops and is one among the important diseases of rapeseed mustard caused by *Alternaria brassicae*. Destruxin B is the main phytotoxin which is produced by *Alternaria brassicae*. As signal is perceived by certain receptor, for that we have studied the interaction of Destruxin-B with receptor like kinases which is being expressed during fungal exposure *viz;* CHRK1, ERECTA, LRPK1, LYS, & WAK in plants.

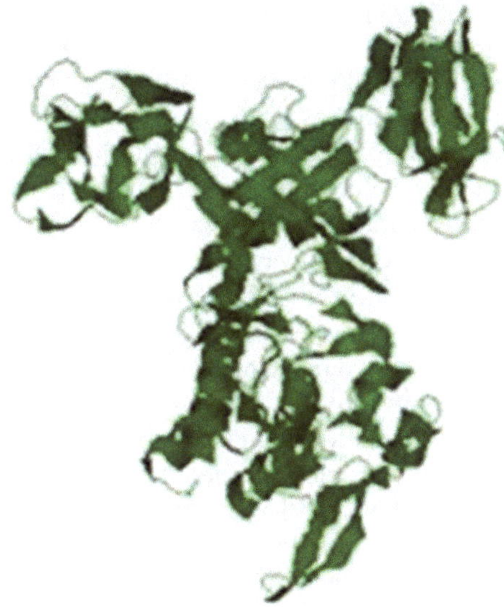

Fig. 8.8 Interaction of Lys receptor like kinase of *Arabidopsis* with destruxin B

Methodology

The structures of various receptor-like kinases have been formed by homology modelling using MOE 2011–2012. The active sites, secondary structures (alpha helix, beta sheets and coil) and the stabilization energy of the structures of these receptors has been found using MOE 2011–2012. And the motif, domain analysis has been done using various online tools. To find the interaction of these receptors with destruxin, the docking has been done and also in order to find out the complete cascade of disease pathogenesis the docking has been performed in between MAPK4, MAPKK, MAPKKK and with the Lys M receptor like kinase.

Result

The Lys M receptor kinase is showing strong interaction with destruxin B (Fig. 8.8). In order to find out the complete cascade of disease pathogenesis the docking has been performed in between MAPK4, MAPKK, MAPKKK and with the Lys M receptor like kinase.

Conclusion

The disease Alternaria blight critically hampers the productivity of rapeseed mustard which is caused by *Alternaria brassicae i.e.* a major pathogen of the disease, which is most prevalent and destructive. Elucidation of signal transduction pathways involved in plant pathogen interaction will help in identification of key molecules affected during defense or pathogenesis. MAPK cascade are known to determine pathogenicity and defense besides affecting a plethora of other developmental processes in plants. As the *Brassica* and *Arabidopsis* both belongs to the same family *Brassicaceae*, and there are evidences in literature which shows that some of the components of MAPK are shown to be in plant defense in *Arabidopsis thaliana* thus attempts can be made to elucidate the key components of signal transduction pathway which involved in disease resistance through variety of tools and technique. It is being felt that molecular biology tools *i.e.* high throughput technique *viz.,* yeast two hybrid system, affinity immunochip based system, mass spectrometry, phosphoproteomics approach, etc. are being used to identify the interacting proteins that are involved in signal transduction pathways. But these conventional techniques are expensive, time consuming and requires lot of labour, thus these problems turned the researchers to move towards the *in-silico* methodology like ANN, SVM, HMM, Phylogenetic profiling, Docking, etc. are being used to predict the interactions as well as interactors of signaling cascades. There are various approaches cited in literature through which the interaction of MAPK in signaling cascade has been successfully predicted and thus these studies helped the researchers to identify the molecular targets of pathogen which causes the disease.

Future Perspective

Signal transduction pathways involve thousands of the interactive proteins and other elements. With the help of genomics and proteomics approach as well as the advance *in-silico* approaches the identification of potential genes which are involved in signal transduction pathways as well as the analysis of sister pathways become much more efficient in terms of time as well as resources. It would be very interesting to know the orthologs of some these MAPK pathways which play an important role in determining pathogenesis of alternaria blight of *Brassica.* In the absence of the knowledge of genomic information and MAPK orthologs in *Brassica*, it has been realized that *Arabidopsis thaliana* can successfully serve as useful model plant to derive the knowledge on the global regulation of MAP kinase cascade in the event of pathogenesis of Alternaria blight. The genetic engineering of these pathways may produce the various fruitful result like it increase the transferability *i.e.* through the comparative genomics approach it can be elucidated that the signaling element involved in one species can also work in other species in the same manner or following the different strategies. As well as the knowledge of the interacting protein can increase the possibilities for achieving the high sensitivity and specificity in terms of disease resistance and defense against the pathogen.

Acknowledgments Authors are grateful to Sub-DIC, Bioinformatics unit at G.B. Pant University of Agriculture and Technology, Pantnagar, India for providing computational facility. This study was supported by Department of Biotechnology, Govt. of India under Programme Mode Support Project.

References

Baker B, Zambryski P, Staskawicz B, Dinesh-Kumar SP (1997) Signaling in plant microbe interactions. Science 276:726–733

Baldi P, Brunak S (2001) Bioinformatics: the machine learning approach. MIT, Cambridge

Bethke G, Unthan T, Uhrig JF, Poschi Y, Gust AA, Scheel D, Lee J (2009) Flg22 regulates the release of an ethylene response factor substrate from MAP kinase 6 in *Arabidopsis* via ethylene signaling. Proc Natl Acad Sci USA 106:8067–8072

Chisholm ST, Coaker G, Day B, Staskawicz BJ (2006) Host-microbe interactions: shaping the evolution of the plant immune response. Cell 124:803–814

Colcombet J, Hirt H (2008) Arabidopsis MAPKs: a complex signaling network involved in multiple biological processes. Biochem J 413:217–226

Craig EA, Stevens MV, Vaillancourt RR, Camenisch TD (2008) MAP3Ks as central regulators of cell fate during development. Dev Dyn 237:3102–3114

Cvetkovska M, Ramptitsch C, Bykova N, Xing T (2005) Genomic analysis of MAPK cascade in *Arabidopsis* defence responses. Plant Mol Biol Rep 23:331–343

Davis H, Fitt BDL, Evans RL (1994) A typical green leaf blotch lesions on barley leaves infected by *Rhynchosporium secalis*. New Phytol 127:139–145

Dixelius C, Happstadius I, Berg G (2005) Verticilium wilt on *Brassica* oilseed crops- a swedish perspective. J Swedish Seed Association 115:36–48

Eddy SR (1998) Profile hidden Markov models. Bioinformatics 14:755–763

Feilner T, Hultschig C, Lee J, Meyer S, Immink RGH, Koenig A, Possling A, Seitz H, Beveridge A, Scheel D, Cahill DJ, Lehrach H, Kreutzberger J, Kersten BH (2005) High throughput identification of potential Arabidopsis mitogen-activated protein kinases substrates. Mol Cell Proteomics 4:1558–1568

Ghose K, Dey S, Barton H, Loake GJ, Basu D (2008) Differential profiling of selected defence-related genes induced on challenge with *Alternaria brassiciciola* in resistant white mustard and their comparative expression pattern in susceptible India mustard. Mol Plant Pathol 9:763–775

Hammond-Kosack KE, Parker JE (2003) Deciphering plant pathogen communication: fresh perspectives for molecular resistance breeding. Curr Opin Biotechnol 14:177–193

Houlb EB (2001) The arms race is ancient history in *Arabidopsis*, the wildflower. Nat Rev Genet 2:516–527

Karplus K (1998) Hidden Markov models for detecting remote protein homologies. Bioinformatics 14:846–856

Kaur P, Jost R, Sivasithamparam K, Barbetti MJ (2011) Proteome analysis of *Albugo candida–Brassica juncea* pathosystem reveals that the timing of the expression of defense related genes is crucial determinant of pathogenesis. J Exp Bot 62:1285–1298

Kav NNV, Srivastava S, Yajima W, Sharma N (2007) Application of proteomics to investigate plant-microbe interactions. Curr Proteomics 4:28–43

Krüger J, Thomas CM, Golstein C, Dixon MS, Smoker M, Tang S, Mulder L, Jones JD (2002) A tomato cysteine protease required for Cf-2 dependent disease resistance and supression of autonecrosis. Science 296:744–747

Lawrence CB, Mitchell TK, Craven KD, Cho Y, Robert AC Jr, Kim KH (2008) At death's door: *Alternaria* pathogenicity mechanisms. Plant Pathol J 24:101–111

Lee SJ, Huh WK, Bhargava A, Eillis BE (2008) Comprehensive analysis of protein protein interactions between Arabidopsis MAPKs and MAPK kinases helps define potential MAPK signaling modules. Plant Signal Behav 3:1037–1041

Meena PD, Meena RL, Chattopadhyay C, Kumar A (2004) Identification of critical stage for disease development and biocontrol of Alternaria blight of Indian mustard (*Brassica juncea*). J Phytopathol 152:204–209

Meena PD, Awasthi RP, Chattopadhyay C, Kolte SJ, Kumar A (2010) Alternaria blight: a chronic disease in rapeseed-mustard. J oilseed *Brassica* 1(1):1–11

Mount DW (2001) Bioinformatics: sequences and genome analysis. Cold Spring Harbour, New York

Nadarajah K, Sidek HM (2010) The green MAPKs. Asian J Plant Sci 9:1–10

Pandey D, Kumar A, Garg GK (2002) Antagonism between *Alternaria brassicae* toxin and zeatin in cell culture of *Brassica juncea* (L.) Czern and Coss cv. Divya. Physiol Mol Biol Plants 7(2):181–184

Pedley KF, Martin GB (2005) Role of mitogen-activated protein kinases in plant immunity. Curr Opin Plant Biol 8:541–547

Popescu SC, Popescu GV, Snyder M, Dinesh SP (2009) Integrated analysis of co-expressed MAP kinase substrates in *Arabidopsis thaliana*. Plant Signal Behav 4:524–527

Rimmer SR, Buchwaldt L (1995) Diseases. In: Kimber DS, McGregor DI (eds) Brassica oilseeds: production and utilization. CAB International, Wallingford, pp 65–92

Taj G, Kumar A, Bansal KC, Garg GK (2004) Introgression of osmotin gene for creation of resistance against Alternaria blight by perturbation of cell cycle machinery. Int J Biotechnol 3:291–298

Taj G, Sharma S, Gaur VS, Kumar A (2011) Prediction of interaction of transcription factors with MAPK3 in *Arabidopsis thaliana* using protein sequence information. Int J Bioinform Res 3:167–177

Tarca AL, Carey VJ, Chen XW, Romero R, Draghici S (2007) Machine learning and its application to biology. PLOS Comp Biol 3:953–963

Taylor RJ, Falconnet D, Niemisto A, Rasey SA, Prinz S (2009) Dynamic analysis of MAPK signaling using a high-throughput microfluidic single-cell imaging platform. Proc Natl Acad Sci USA 106:3758–3763

Valkonen JPT, Koponen H (1990) The seed-borne fungi of chinese cabbage (*Brassica perkinensis*), their pathogenicity and control. Plant Pathol 39:510–516

Verma PR, Saharan GS (1994) *Alternaria brassicae* (Berk) Sacc *A. brassicicola*. Acta Phytopathol Entomol Hung 29:283–314

Chapter 9
Plant Cell Signaling in Metal Stress

Imran Haider Shamsi, Essa Ali, Lixi Jiang, Wenjing Liu, Chengliang Sun, Chongwei Jin, and Xianyong Lin

Introduction

Cell Signaling

Cell signaling is the component of a complex system of communication that directs basic cellular activities and synchronizes cell actions. The capability of cells to pick out and properly counter to their environment is the root of development, resistance to different environmental stresses, tissue repair and normal tissue homeostasis.

Environmental Stresses

Living creatures are often exposed to adverse environmental conditions (Fig. 9.1). Most of the animals can escape such conditions by moving or migrating, but sessile organisms such as higher plants only have limited possibilities of avoiding such conditions (Sreenivasulu et al. 2007; Madlung and Comai 2004). In addition to this,

I.H. Shamsi
MOE Key Laboratory of Environment Remediation and Ecosystem Health,
College of Environmental and Resource Sciences, Department of Agronomy,
Key Laboratory of Crop Germplasm Resource, College of Agriculture and Biotechnology,
Institute of Crop Science, Zhejiang University, 310058, Hangzhou,
People's Republic of China

E. Ali • L. Jiang
Institute of Crop Science, College of Agriculture and Biotechnology, Zhejiang University,
310058, Hangzhou, People's Republic of China

W. Liu • C. Sun • C. Jin • X. Lin (✉)
MOE Key Laboratory of Environment Remediation and Ecosystem Health, College
of Environmental and Resource Sciences, Zhejiang university, 310058, Hangzhou,
People's Republic of China
e-mail: xylin@zju.edu.cn

M. Sarwat et al. (eds.), *Stress Signaling in Plants: Genomics and Proteomics Perspective, Volume 1*, DOI 10.1007/978-1-4614-6372-6_9,

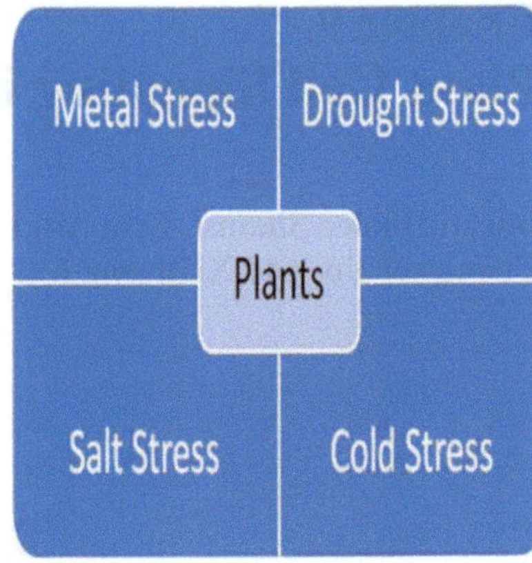

Fig. 9.1 Plants and different environmental stresses (by I.H. Shamsi et al.)

Agricultural soil worldwide is mostly contaminated with toxic materials that restrict the crop plants to reach their full genetic potential and cause significant loss by reducing the crop productivity (Yadav 2010). It has been projected that two-thirds of the yield potential of major crops are consistently lost due to the hostile ecological factors. Alternatively, the world population is estimated to be approximately 10 billion by the year of 2050, which will witness serious food deficiencies. So as to survive and adapt to such environmental stresses more preferably to the metal stresses, plants are outfitted to sense and counter the mechanical and chemical (metals) stimulation (Chehab et al. 2009; Monshausen and Gilroy 2009).

Metals Toxicity

Metal and metal compounds are innate constituents of all ecosystems, linked biological and non-biological system. Seventeen nutrient elements (Fig. 9.2) are recognized to be essential for higher plants, among which 14 are mineral elements that are acquired by plants from soil (Epstein and Bloom 2004). Rise in the levels of vital or non nutritional metal ions (that lead to cellular damage) triggers a broad range of cellular responses including alterations in gene expression and synthesis of metal-detoxifying peptides. In trace amounts, a number of these ions are required for metabolism, growth, and development. However, problems crop up when cells are confronted with an excess of these fundamental ions (Avery 2001; Schutzendubel and Polle 2002). Numerous studies have reported that toxic and carcinogenic effects induced when living organisms are exposed to certain metals especially arsenic (Mishra et al. 2008), lead (Flora et al. 2007), cadmium (Watanabe et al. 2003), chromium (Dayan and Paine 2001), mercury (Lee et al. 2001), Aluminum (Kochian et al. 2005), Zinc (Hartwig et al. 2002) and Copper (Fernandes and Henriques 1991). A growing amount of results provide evidence that toxic metals have the ability to interact with nuclear proteins and genetic material causing oxidative deterioration of natural macromolecules. Detailed studies in the past few decades showed that metals

Fig. 9.2 Essential elements required for the plant growth and development (by I.H. Shamsi et al.)

like iron, cadmium, aluminum, zinc, and copper hold the ability to turn out reactive free radicals, which start chain reaction resulting in oxidation of lipid termed as lipid peroxidation, protein oxidation and nucleic acid oxidation (Kalia and Flora 2005).

Accumulation and Role of Some Important Metals in Plant Stress

Accumulations of metals in plants can occur following their uptake from contaminated soil (Mills et al. 2005). Yew et al. (2010) analyzed the metal content of mitochondria isolated from Arabidopsis cell culture by inductively coupled plasma mass spectrometry (ICP-MS). Which showed Cu, Fe, Zn, and Mn to be the prime species of transition metals in Arabidopsis mitochondria, with trace levels of Co and molybdenum (Mo).

To balance the concentration of indispensable metal ions within physiological limits and to lessen the injurious effects of nonessential metals, plants launch a complex network of homeostatic mechanisms to control their absorption, accumulation, trafficking, and neutralization (Clemens 2001). Specialized transporter proteins mediate the movement of metals through membranes in the form of channels, carriers, or pumps (Williams et al. 2000). All living organisms have to keep sufficient supply of the metal ions required for essential cellular processes, but it is critical that these be prevented from accumulating to toxic stage. Tolerance in plants grown on metal-polluted soil can be fulfilled either by excluding the uptake mechanisms from the roots, or by metal efflux, compartmentation, and detoxification next to that uptake (Sichul et al. 2007).

Role of Zinc in Stress Signaling

The micronutrient zinc plays an important role in physiological and metabolic processes of plants (Ramesh et al. 2004). Zinc serves as a cofactor for more than 300 enzymes, including RNA polymerase, alcohol dehydrogenase, copper/zinc superoxide dismutase, and carbonic anhydrase (Guerinot and Eide 1999). However, uptake of high concentrations of Zn is found to be toxic to plant growth and development therefore, plants need to keep very tight control over the internal concentrations of zinc in a process called zinc homeostasis.

Zn homeostasis is a tightly regulated process because Zn can be both essential and deleterious to plants depending on its concentration. The effects of Zn on plants have been widely reported (Broadley et al. 2007; Wang et al. 2009). Zn deficiency reduces antioxidative enzyme activity and thereby results in reactive oxygen species (ROS) accumulation and oxidative damage (Sharma et al. 2004). Aside from Zn deficiency, excess Zn can also inhibit plant growth and development by disequilibrating the uptake and redistribution of mineral nutrients and by disturbing the antioxidant defense system and metabolic processes such as photosynthesis, transpiration, and antioxidative enzyme activity. The mechanisms of Zn toxicity are not fully understood; however, they may involve competition for catalytic sites or for transporter proteins (Gonzalez et al. 2005). Zn toxicity also inhibits the uptake of other nutrient elements, such as iron (Fe). Plants exposed to excess Zn become Fe deficient (Wintz et al. 2003). Deficiency of these elements can lead to ROS accumulation and oxidative stress (Bonnet et al. 2000).

Role of Aluminum (Al) in Stress Signaling

Aluminum (Al) toxicity in plants is one of the major constrains to crops growth and development (Ryan et al. 2001). It has been proposed that Al ions boost phospholipids and proteins peroxidation in cell membranes (Yamamoto et al. 1997). Release of organic acids is an important mechanism for plants to defend against Al toxicity. Many plants were found to secrete organic acids in response to Al stress, such as, citrate is released from the roots of maize (Pineros et al. 2002) and soybean (Yang et al. 2001), oxalate from buckwheat (Ma et al. 1997), and malate is released by Al-tolerant genotypes of wheat (Ryan et al. 1995). However, it remains unclear how organic acid secretion and its activation is regulated by Al (Ryan et al. 2001).

Over 20 genes induced by Al stress have been isolated from plants, including Arabidopsis (Richards et al. 1998), tobacco (Ezaki et al. 1997), and wheat (Delhaize et al. 1999). The biological role of Al-induced genes in plants is unclear. Because most of the Al-induced genes seem to be general stress genes that are induced by a different plant stresses, including low phosphate (Ezaki et al. 1995), wounding (Snowden et al. 1995), oxidative stress (Richards et al. 1998) and pathogen infection (Hamel et al. 1998). Some of the induced genes are well known as antioxidation enzymes (e.g. glutathione S-transferase, peroxidase, and superoxide dismutase). By comparison with other stress genes, these genes may play a role in protecting cells against Al stress, but experimental evidence on this point is lacking. Recently, 11 plant Al-induced genes were expressed in yeast (Saccharomyces cerevisiae) cells and showed that the tobacco (NtGDI1) gene and Arabidopsis gene (AtBCB), conferred Al resistance in yeast cells (Ezaki et al. 1999). Overexpression of another Al-induced gene encoding the wheat phosphatidylserine synthase enzyme also gave Al resistance in yeast. A range of alternative toxicity mechanisms have also been proposed (Kochian 1995). Chelation strategies are very useful, but combining them with additional Al tolerance mechanisms within the plant would be expected to provide more effective protection.

Role of Cadmium (Cd^{2+}) in Stress Signaling

Cadmium (Cd^{2+}) is categorized as a long biological half-life heavy metal. It is rapidly taken up by roots and enters the food chain, resulting in toxicity for living organisms (Sanita and Gabbrielli 1999), restrains seed germination, decreases plant growth and photosynthesis, and interferes with the distribution of nutrients (Rodriguez et al. 2006). Despite of the fact that Cd^{2+} is an ecological threat; the mechanisms by which it exerts its toxic effects in plants are not fully understood. In plant cells, Cadmium enters through Fe^{2+}, Ca^{2+}, and Zn^{2+} transporters/channels (Clemens 2006). Once Cd^{2+} enters to the cytosol, it stimulates the production of phytochelatins (PCs), a glutathione-derived class of peptides containing repeated units of Glu and Cys, which bind the metal ions and transport them into the vacuole (Sanita and Gabbrielli 1999). It is confirmed that high (millimolar) concentrations of Cd^{2+} induce reactive oxygen species (ROS) bursts in plants, which might have a role in signaling and/or degenerative steps leading to cell death (Garnier et al. 2006). Treatment with a lower, non-toxic Cd^{2+} concentration also caused increase in ROS production in pea (Pisum sativum) leaves and roots (Rodriguez et al. 2006) and Arabidopsis (Arabidopsis thaliana) cell cultures (Horemans et al. 2007).

Cadmium (Cd^{2+}) also affects the content of polyvalent cations through competition for binding sites of proteins or transporters (Gussarson et al. 1996). Thus, Cd produced a decrease of calcium (Ca) content in different plant species (Sandalio et al. 2001). Ca is involved in the regulation of plant cell metabolism and signal transduction (Rentel and Knight 2004) and modulates cellular processes by binding proteins such as calmodulin (CaM), which in turn regulates the activity of target proteins (Roberts and Harmon 1993). Cd can be detoxified by phytochelatins, whose synthesis is induced by Cd and other metals and is accompanied by a decrease in the concentration of glutathione (Zenk 1996).

Role of Copper (Cu) in Stress Signaling

Copper is an indispensable trace element for all higher plants, and plays an important role in metabolic processes in plants (Maksymiec 1997). Photosynthetically active tissues have a high demand for the copper (Cu) (Raven et al. 1999). Plant chloroplasts contain three major Cu proteins including Plastocyanin (PC), Polyphenol oxidase (PPO) and Copper/zinc superoxide dismutase (Cu/ZnSOD) (Pilon et al. 1999). It has been suggested that Cu enters the cytoplasm through the members of the COPT (Cu transporter) family (Burkhead et al. 2009; Garcia et al. 2011).

The mechanism of Cu toxicity to photosynthetic electron transport has been widely studied, and inhibition of the photosystem II (PSII) have been suggested (Yruela et al. 1996). Decrease in chlorophyll (Chl) concentration (Quartacci et al. 2000) has been observed to accompany Cu excess associated with structural changes in chloroplasts, such as reduction of thylakoid membranes (Elefteriou and Karataglis 1989). Excess Cu may interfere with the biosynthesis of the photosynthetic machinery and may alter the pigment and protein components of photosynthetic membranes (Maksymiec et al. 1994). Cu-induced lipid peroxidation has also been suggested to be the reason for the membrane degeneration (Gallego et al. 1996). Besides this Cu deficiency also include reduction in biomass, lack of photosynthetic activity, chlorosis, rolling of leaves and defects in plant morphology, desiccation, and delay in flowering (Marschner 1995).

Role of Iron (Fe) in Stress Signaling

In the oxidative environment of Earth, organisms must contend with the problem of transporting, storing, and assembling Iron into active cofactors and, at the same time, protect themselves against oxidative damage due to the interactions of iron with dioxygen and reactive oxygen species. The problem of balancing iron homeostasis and oxidative stress is most acute in photosynthetic organisms (Shcolnick and Keren 2006). On the one hand, the photosynthetic electron transfer chain utilizes radicals and reduced metal species as part of its normal catalysis, all prone to cause oxidative damage if not handled properly (Curie et al. 2001).

Iron (Fe) starvation frequently occurs in aquatic habitats and severely limits biomass production of photosynthetic organisms (Tortell et al. 1999). Among other metabolic processes, photosynthetic/respiratory electron transport chain with its high number of Fe cofactors is especially susceptible to the deleterious effects of Fe limitation (Michel et al. 2003). Under Fe deficiency, graminaceous plant species release phytosiderophores, which are hexadentate metal chelators with high affinity for complex formation with Fe (III). The whole Fe (III)-phytosiderophore complex is subsequently taken up by Fe deficiency-inducible transporters of the YS1/YSL protein family (Curie et al. 2001). As assayed by two-electrode voltage clamp, most of these metal-phytosiderophore chelates were also transported via the maize

(Zea mays) metal-phytosiderophore transporter ZmYS1 when expressed in Xenopus oocytes (Schaaf et al. 2004). Whether phytosiderophore release is also directly or indirectly up-regulated under metal stresses other than Fe is currently under debate.

Role of Nickel (Ni) in Stress Signaling

Although Ni is recognized as an essential mineral nutrient element for higher plants, but its physiological significance, especially to woody perennials, has received little attention. The existence of Ni deficiency in crops was only recently discovered, wherein mouse ear, a century-old malady of pecan trees, and replant disease were found to be Ni deficiencies (Wood et al. 2004). Severe Ni deficiency has subsequently been identified in containerized river birch (Ruter 2005) and has putatively been identified in several other crops (Wood et al. 2006). There are several enzyme systems (carbon monoxide dehydrogenase, NiFe-hydrogenase, acetyl-CoA decarbonylase methyl-coenzyme M reductase, Ni-dependent glyoxylase, superoxide dismutase, and methyleneurease) in lower plants that are activated by Ni (Mulrooney and Hausinger 2003); however, the activation of urease appears, to date, to be the only enzymatic function of Ni in higher plants (Gerendas et al. 1999). Urease contains two Ni ions at the active site (Ciurli 2001). Ni can also replace Zn or Fe, and other metal ions, in certain other metalloenzymes of lower plants (Mulrooney and Hausinger 2003). Circumstantial evidence indicates that ureide transporting species, such as pecan, possess a higher Ni requirement than amide-transporting species (Wood et al. 2006), thus raising the possibility that ureide transporters might possess enzymes, other than urease, that require Ni for activation or for enhanced activity.

Role of Phosphorus (P) in Stress Signaling

Plants are strongly affected by phosphorus deficiency because phosphorus is an essential constituent of nucleic acids and membrane phospholipids. Furthermore, phosphorus plays a key role in energy transfer, as a regulator of enzyme activity, and in signal transduction. Thus, of course, low phosphorus availability activates a series of morphological and physiological responses that maximize phosphorus acquisition (Raghothama 1999) and help to maintain internal phosphorus homeostasis (Ticconi and Abel 2004).

Plants absorb Pi from the soil as an inorganic orthophosphate ion, but their availability is strictly limited by reactions of inorganic and organic phosphates with soil constituents. A significant amount (20–80 %) of the nutrient may be found in organic forms (Richardson 1994). In response to persistent Pi deficiency plants have developed many adaptive mechanisms to increase the availability and uptake of Pi. Production and secretion of phosphatases to release Pi from organic forms is one of such adaptive mechanisms (Duff et al. 1989). An increase in the Pi uptake rate of roots and cell cultures following a period of Pi starvation has also been well documented in plants (Shimogawara and Usuda 1995). A combination of phosphatase activity and enhanced Pi uptake may help plants to acquire required amounts of Pi from the rhizosphere (Plaxton and Carswell 1999). The induction of acid phosphatases (ACPs) is a distinct and universal response of higher plants to Pi starvation (Duff et al. 1994). Due to this unique feature, phosphatase activity has been used as a potential marker of the Pi status of plants (Ascencio 1994).

Generation of ROS and RNS

Reactive oxygen species (ROS) and Reactive Nitrogen Species (RNS) are the products of metal catalyzed reactions (Cadenas 1989). Reactive oxygen species at elevated concentration are important mediators of damage to cellular components, including lipids, membranes, proteins and nucleic acids (termed oxidative stress) (Poli et al. 2004). A number of studies have focused on metal-induced toxicity and carcinogenicity, highlighting their role in the production of reactive oxygen and nitrogen species in biological systems, and their significance. Metal-mediated generation of free radicals may cause a range of modifications to DNA bases as well as can boost lipid peroxidation. Oxidative stress is a condition where production of oxygen radicals beyond a threshold for proper antioxidant detoxification, has been linked as a pathologic condition in several cellular disorders. Besides ROS, metal exposure can also affect the generation of RNS (Flora 2009).

Suppressions of ROS and RNS

The detrimental effects of ROS and RNS are balanced by the antioxidant action of non-enzymatic antioxidants and antioxidant enzymes (Halliwell 1996). Harmful free radical-mediated oxidations occur in aerobic organism as a result of normal oxygen metabolism. An antioxidant is a substance capable of preventing or slowing the oxidation of other molecules. In general, an antioxidant can protect against metal toxicity by trapping free radicals thus terminating the chain reaction, by chelating metal ion and preventing the reaction with reactive oxygen species or by chelating metal and maintaining it in a redox state leading to its incompetency to reduce molecular oxygen (Flora 2009).

Genes Responsible for Detoxification of Free Radicals

Several genes play their role in the neutralization of free radicals. They take part in metal detoxification by changing the pH of cytosolic part of cytoplasm (Clemens 2001). Metallothioneins (MTs) and phytochelatins (PCs) are considered as well known heavy metal-binding ligands of plant cells (Cobbett and Goldsborough 2002). Metallothioneins are Cys-rich polypeptides encoded by a family of genes. On the contrary, phytochelatins are a family of enzymatically synthesized Cys-rich peptides. The stimulation of PC synthesis depends on the posttranscriptional activation of already existing PC synthase and is independent of transcriptional control. Numerous heavy metal-inducible genes have been documented in plants (Hagen et al. 1988; Lescure et al. 1991; Berna and Bernier 1999), but astonishingly, very little is known about the gene expression profiling in response to heavy metals. One cis-acting element has been acknowledged in the promoters of tobacco (Nicotiana tabacum) parA and soybean (Glycine max) GH2/4, two genes regulated by cadmium and auxin (Ellis et al. 1993; Kusaba et al. 1996). This element is related to the as-1 element previously identified in the cauliflower mosaic virus (CaMV) 35S promoter (Katagiri et al. 1989). In the green algae Chlamydomonas reinhardtii, the copper response element (CuRE) with consensus (5#-GTAC-3#) was identified in the promoter of CPX1 and CYC6. A copper response regulator, CRR1, binds to these sites and mediates target gene expression under copper-deficient conditions (Quinn and Merchant 1995; Quinn et al. 2000). Recently, Quinn et al. (2003) demonstrated that both CuRE and CRR1 are required for a response to nickel, which recommend that nickel interferes with a component in the nutritional copper signal transduction pathway.

Mechanism of Metal Stress Signaling In Plants

Plants come across a wide range of environmental insults during its life spin and have developed mechanisms to increase their tolerance to such conditions. These mechanisms include physical, biochemical and molecular adaptations that begin after the onset of harsh conditions (Fig. 9.3). The first step in switching on such molecular responses is to perceive the stress as it occurs and to relay information about it through a signal transduction pathway. These pathways ultimately lead to physiological changes or to the expression of genes and resultant adjustment of molecular and cellular processes (Heather and Marc 2001).

The signal pathways are accountable for the sensing and transduction of the "metal signal" in the cell, ultimately driving the induction of transcription factors (TFs) following genes expression that facilitate plants to neutralize the metal stress. In order to respond to stress signals, plant cells must be able to perceive these signals and convert them into appropriate responses, which in turn confer on plants the ability to tolerate unfavorable conditions. Plant tolerance mechanisms require a

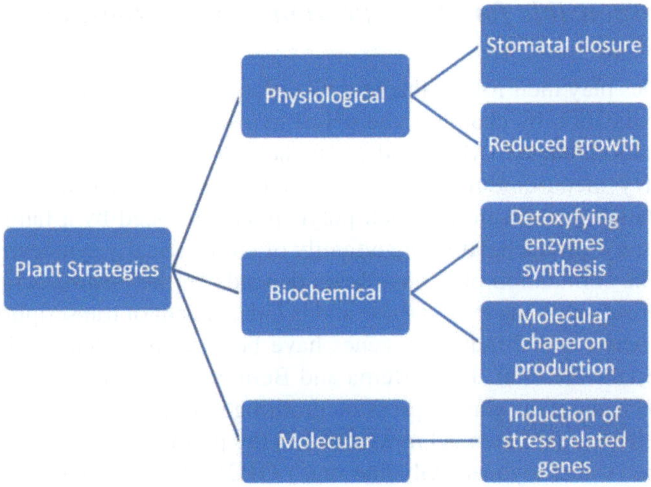

Fig. 9.3 Plant strategies to cope up with different environmental stresses (by I.H. Shamsi et al.)

coordination of complex physiological and biochemical processes, including changes in global gene expression, protein modification and primary and secondary metabolite compositions (Baraym et al. 2009). Waldemar (2007) showed a general modal for local and systemic stress signaling in plants (Fig. 9.4). In the last decade functional genomics approaches have partially unraveled the complex mechanisms that drive from stress perception and transduction, through a cascade of signaling molecules, to the expression modulation of genes responsible for plant stress response (Kitano 2002). In addition, the elucidation of the function of newly identified stress-responsive non-coding RNA will facilitate understanding of the complex response to stress (Jones and Dangl 2006).

Biological signaling processes may be mediated by complex networks in which network components and network sectors interact with each other in complex ways. Studies of complex networks benefit from approaches in which the roles of individual components are considered in the context of the network. To appreciate the regulation of a specific biological process, it is of extreme importance to illuminate what structural features of the signaling network regulating the process direct the performance of the signaling network as a whole (Baraym et al. 2009; Kitano 2002).

Signaling Pathways

There are a large number of intracellular signaling pathways responsible for transmitting information within the cell. They fall into two main categories. The majority respond to external stimuli arriving at the cell surface, usually in the form

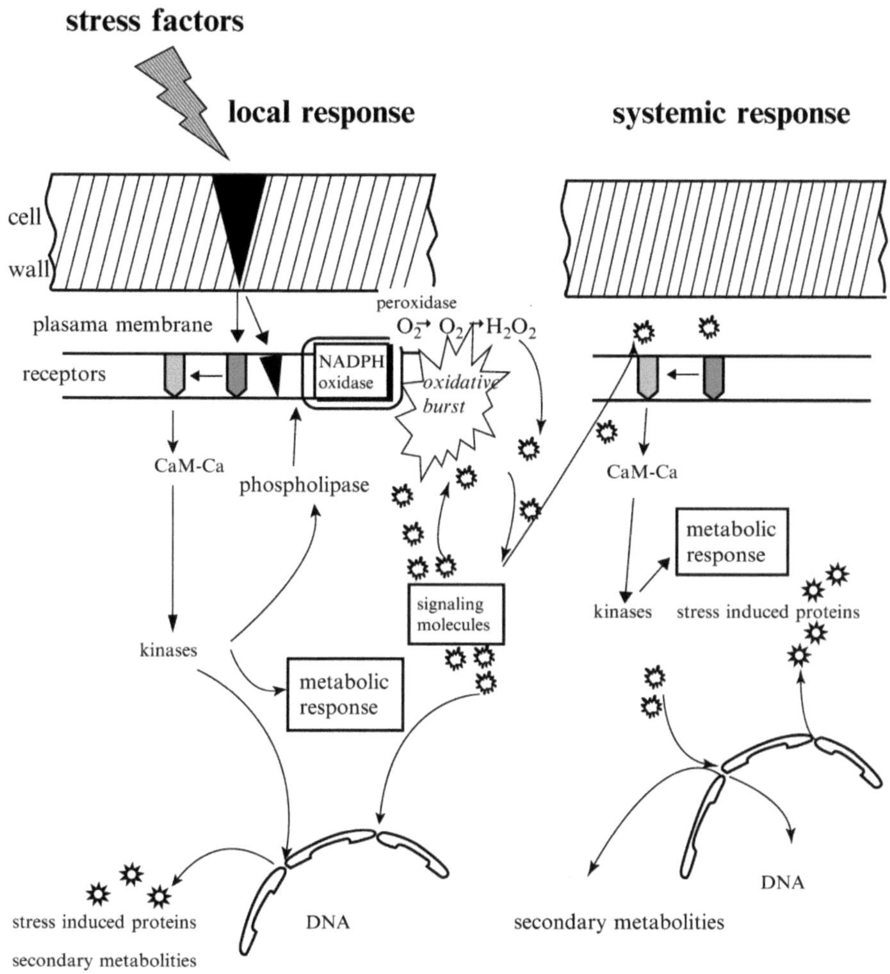

Fig. 9.4 General mode for local and systemic stress signaling in plants. From Maksymiec W. Signaling responses in plants to heavy metal stress. Acta physiologiae plantarum. 2007;29(3):177. Reprinted with permission from Springer-Verlag

of a chemical signal (hormone or metal ions), which is received by receptors at the cell periphery that function as molecular antennae embedded in the plasma membrane. These receptors then transfer information across the membrane using a variety of transducers and amplifiers that engage a diverse repertoire of intracellular signaling pathways. The other categories are the pathways that are activated by signals generated from within the cell. There are a number of metabolic messengers that act from within the cell to initiate a variety of signaling pathways. All of these signaling pathways generate an internal messenger that is responsible for relaying information to the sensors that then engage the effectors that activate cellular responses. The main signaling pathways are outlined below (Michael 2009).

Cyclic AMP Signaling Pathway

One of the first signaling systems to be characterized was the cyclic AMP signaling pathway, which led to the second messenger concept that applies to many other signaling systems. The idea is that the external stimulus arriving at the cell surface is the first messenger, which is then transformed at the cell surface by adenylyl cyclase (AC) into a second messenger, cyclic AMP, which is part of the signaling cascade that then activates downstream effectors.

Nitric Oxide (NO)/Cyclic GMP Signaling Pathway

Nitric oxide synthase (NOS) generates the gas NO that acts both through cyclic GMP and nitrosylation reactions. NO has a particularly important role in modulating the activity of other pathways such as Ca^{2+} signaling.

Redox Signaling

Many receptors act through NADPH oxidase (NOX) to form reactive oxygen species, such as the superoxide radical and hydrogen peroxide (H_2O_2), which act to regulate the activity of specific signaling proteins such as tyrosine phosphatases, transcription factors and ion channels.

Mitogen-Activated Protein Kinase (MAPK) Signaling

This is a classical example of a protein phosphorylation cascade that often begins with Ras and consists of a number of parallel pathways that function to control many cellular processes and particularly those related to cell proliferation, cell stress and apoptosis.

Phospholipase D (PLD) Signaling Pathway

This is a lipid-based signaling system that depends upon the hydrolysis of phosphatidylcholine by phospholipase D (PLD) to give phosphatidic acid (PA), which functions as a second messenger to control a variety of cellular processes.

Janus Kinase/Signal Transducer and Activator of Transcription Signaling Pathway

This is a fast track signal transduction pathway for transferring information from cell-surface receptors into the nucleus. The Janus kinases (JAKs) are tyrosine kinases that phosphorylate the signal transducers and activators of transcription (STATs), which carry the information into the nucleus.

AMP Signaling Pathway

This pathway is regulated by adenosinemonophosphate (AMP), which functions as a metabolic messenger to activate an important pathway for the control of cell proliferation.

Key Players Involved in Signaling Pathways (Fig. 9.5)

After sensing the toxic metal, the plant cell activates specific genes and/or the synthesis of proteins (Table 9.1), to counteract the stress stimuli, for example PvSR2 (Zhang et al. 2001), which is connected with plant tolerance; they can also induce genes characteristic of other stress factors. Such an effect has been known for the action of signaling molecules like jasmonates, salicylic acid, ethylene and abscisic acid (Farmer et al. 2003) induced by biotic and abiotic stress factors. Some of the presented papers indicate a similar effect of JA and heavy metals on gene expression. Xiang and Oliver (1998) presented a common response of Cd/Cu ions and JA in transcription of genes involved in glutathione metabolism. A similar effect was observed in the case of VSP2 (vegetative-storage protein) and MAPK (mitogen-activated protein kinase) transcripts induction (Kim et al. 2003) gradually clarifying the way of heavy metal action through the signaling pathway on the gene level.

A signal transduction cascade is therefore responsible for the differential gene regulation. The main players of this cascade are described below (Fig. 9.5).

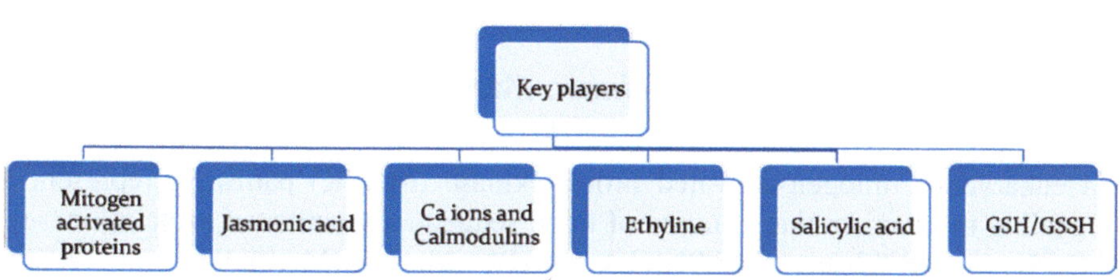

Fig. 9.5 Chemicals involved in metal stress signaling (by I.H. Shamsi et al.)

Table 9.1 Specific gene induction by heavy metals

Genes/protein	Description	Metals
• At-SLP2,3	Subtilisin-like serine proteases induced by jasmonate	Cd
• 70kDa	Protien equivalent to Hsp70 (heat shock protiens)	Cd
• KSAP	Cd stress associated 51 Kda protein, homology to fusarium - induced stress protein	Cd
• gsh, gsh2 and gr1	Glutathione metabolic genes	Cu, Cd
• PR-1	Pathogenesis-related protein induced by TMV, SA	Cu
• TA1-18	Protein homolog to PR2	Cu, Cd
• CBP20-PR	Protein of −20 kDa belong PR-4 protien induced by pathogen and wound	Cd, Zn
• VSP2 • GST1	Vegetative storage protein gene regulated by JA, wounding, sugars Glutathione S- transferase gene	Cu
• OsMSRMK2	Multiple responsive MAPK gene induced by elicitor, high salt, sucrose, drought, UV, JA, H_2O_2, ABA	Cu, Cd, Hg
• OsBWMK1 • OsAOC • OsAOS	MAPK Gene induced by blast infection, wounding, drought, sucrose JA, heating, H_2O_2, ABA Allene oxide cyclase gene Allene oxide synthase gene	Cu, Cd, Hg
• OsMAPK2	Oryza sativa MAPK genes developmentally regulated and stress responsive, 42,50,64-kDa kinase	Cu
• OsEDR1	MAPKKK gene induced by cut, JA, SA, ABA, ethylene, H_2O_2, elicitors, sugar, drought, ozone, developmentally regulated	Cu, Cd, Hg
• SIM SAMK MMK2,3	Kinases belonging to MAPKs, induced by wounding, drought, tough, cold, elicitors, ACC	Cu, Cd
• 16kDa	Protein related to PR-10 protien induced by H_2O_2, ethylene, SA	Pb, Cu
• GRP	Glycine-rich protein induced by heat, NaCl, cold wound stress	Hg
• HSP	Heat-shock protein induced by heat and cold, glucanase chitinase (protein belonging to PR)induced by wounding, NaCL and UV	Hg
• PvPRP2	Gene of kDa proline-rich protein responsive to wounding, drough, UV, salt, ABA and elevated temperature specially expressed in leaves	As, Cd, Hg, Zn Cd, Fe, Zn, Cu Cd, Fe, Zn, Cu, Ga, In, La
• Wali 3,5	Genes show homology wound- induced and encoding protienase inhibitor	
• Wali 4	Gene encoding of phenylalanine ammonialyase	

From Maksymiec W. Signaling responses in plants to heavy metal stress. Acta physiologiae plantarum. 2007;29(3):177. Reprinted with permission from Springer-Verlag

Mitogen-Activated Protein Kinase (MAPK)

In eukaryotes, mitogen-activated protein kinase (MAPK) pathways represent a signaling mechanism that consists of three sequentially activated protein kinases: MAPK kinase kinase (MAPKKK), MAPK kinase (MAPKK), and MAPK (Schaeffer and Weber 1999; Widmann et al. 1999). MAPKKKs are Ser/Thr protein kinases that phosphorylate MAPKKs. Once phosphorylated, and therefore activated, MAPKKs are responsible for the phosphorylation of MAPKs on Thr and Tyr residues. This phosphorylation renders these enzymes active. MAPKs are able to phosphorylate numerous substrates in different cellular compartments. In plants, the MAPK cascade is involved in response to a variety of environmental, hormonal and developmental stimuli (Jonak et al. 2002). It has recently been shown that stress due to excessive Cd and Cu activates different kinase enzymes belonging to the MAPK family (Jonak et al. 2004). The phosphorylation cascade is therefore thought to be involved in Cd signaling to the nucleus.

Calcium Ions and Calmodulin

Calcium ions and calmodulin are well-known second messengers of external stimuli, and the participation of this system in heavy metal signaling has also been hypothesized (Suzuki et al. 2001). Indeed, Ca concentration in cells greatly increases during Cd stress (Dalcorso et al. 2008), and it stimulates calmodulin-like proteins that interact with Ca ions. Changing their conformation in response to Ca binding, calmodulin proteins regulate a variety of mechanisms, including ion transport, gene regulation, metabolism and stress tolerance that coordinate, at least in part, the plant response to Cd (Yang and Poovaiah 2003). The Ca/calmodulin system is also involved in sensing other heavy metals, and in fact, transgenic plants expressing a tobacco calmodulin-like protein exhibit increased Ni tolerance and Pb accumulation. Regulation in hormone synthesis has also been observed during heavy metal stress (Arazi et al. 2000).

Jasmonic Acid

The major function of Jasmonic acid and its various metabolites are regulating plant responses to abiotic and biotic stress as well as plant growth and development. Regulated plant growth and development processes include growth inhibition, senescence, tendril coiling, flower development and leaf abscission. Treatments with Cd or Cu, for instance, enhance jasmonic acid content in Arabidopsis, Oryza and bean (Maksymiec 2007).

Ethylene

The biosynthesis of ethylene boosted under unfavorable conditions, which may leads to considerable yield losses. Ethylene is distinctive among plant growth regulators in a sense that it is in gaseous state, which give it volatile nature when plants are exposed to stress. Increases in the level of ethylene can mimic numerous symptoms of plant stress, or can induce acclimation processes with aid in plant tolerance and endurance to stress. Ethylene released by one plant is also perceived by another plant. Thus, in a way plants talk to each other about stresses through ethylene. Ethylene synthesis is shown to be increased upon treatment with Cd, Cu, Fe, Zn, and in the case of Cd and Cu, this increase is due to an upregulation of ACC synthase transcription and enhanced activity (Maksymiec 2007).

Salicylic Acid (SA)

Salicylic acid (SA) is a plant Phenolic compound, considered as a hormone-like endogenous regulator and its role is well documented in defense against different environmental stresses. Exposure to Cd has been shown to stimulate SA accumulation in roots. (Yalpani et al. 1994; Maksymiec 2007)

GSH/GSSG

One of the important mechanisms that are thought to be involved in metal sensing is the reduced glutathione-oxidized glutathione ratio (GSH/GSSG). Glutathione can control the differential expression of antioxidant enzymes, such as chalcone synthase, phenylalanine ammonia lyase, and superoxide dismutase or glutathione reductase, usually induced by heavy metal stress. During Cd stress, a reduction in GSH/GSSG ratio has been observed in different plant species, with the consequent activation of the response genes (Romero et al. 2007).

It is noteworthy that plant cells probably transduce heavy metal signaling in different ways for different heavy metals. The main differentiation is probably due to the fact that some metals do not have any known function and could induce deleterious effects even at low concentration. Conversely, other metal ions take part in the normal cell metabolism and are shown to be toxic only at high concentrations. A good example of this is the activation of the phosphorylation cascade of MAPK proteins induced by Cu and Cd. Cu stress rapidly activates SIMK, MMK2, MMK3 and SAMK kinases, and their activation is probably the consequence of oxidative stress generated by the metal ion. Conversely, activation of the above-cited MAPKs in response to Cd ions is rather slower than to Cu. This could be due to the fact that Cd stimulates an oxidative stress as a secondary effect, which is responsible for the MAPKs activation, delaying the phosphorylation cascade (Jonak et al. 2004).

Conclusion

Metal stress signals appear to be transduced through a variety of pathways that overlap and cross-talk. Activation of phosphorylation cascades, Ca-calmodulin system, ROS signaling and stress-related hormones eventually converge regulating transcription factors that are deputed to the activation of gene sets responsible for response to stress.

Future Prospects

Although our effort of metal stress signal transduction in plants covers only a portion of the relevant studies, it is evident that the subject is very complex and a lot of work to do. Understanding regarding the mechanisms that allow plants to cope with metal stress would help in creating new tools. Genetic approaches are important tools for analyzing complex processes. It is therefore of prime importance to further analyze the processes of metal detoxification and signaling pathways in plants, to discover useful targets for biotechnological applications to increase plant suitability in the metal polluted sites.

Acknowledgment This work was supported by National Natural Science Foundation of China (31272237, 30971859) and The International Plant Nutrition Institute.

References

Arazi T, Kaplan B, Sunkar R, Fromm H (2000) Cyclicnucleotide- and Ca2+/calmodulin-regulated channels in plants: targets for manipulating heavy-metal tolerance, and possible physiological roles. Biochem Soc Trans 28:471–5

Ascencio J (1994) Acid phosphatase as a diagnostic tool. Commun Soil Sci Plant Anal 25:1553–1564

Avery SV (2001) Metal toxicity in yeasts and the role of oxidative stress. Adv Appl Microbiol 49:111–142

Baraym Y, Harmon D, De BB (2009) Systems biology. Attractors and democratic dynamics. Science 323:1016–1017

Berna A, Bernier F (1999) Regulation by biotic and abiotic stress of a wheat germin gene encoding oxalate oxidase, a H2O2 producing enzyme. Plant Mol Biol 39:539–549

Bonnet M, Camares O, Veisseire P (2000) Effects of zinc and influence of Acremonium lolium on growth parameters, chlorophyll a fluorescence and antioxidant enzyme activities of ryegrass (Lolium perenne L. cv. Apollo). J Exp Bot 346:945–953

Broadley MR, White PJ, Hammond JP, Zelko I, Lux A (2007) Zinc in plants. New Phytol 173:677–702

Burkhead JL, Reynolds KA, Abdel GSE, Cohu CM, Pilon M (2009) Copper homeostasis. New Phytol 182:799–816

Cadenas E (1989) Biochemistry of oxygen toxicity. Ann Rev Biochem 58:79–110

Chehab EW, Eich E, Braam J (2009) Thigmomorphogenesis: a complex plant response to mechano-stimulation. J Exp Bot 60:43–56

Ciurli S (2001) Electronic structure of the nickel ions in the active site of urease. Chemistry (Easton) 2001:99–100

Clemens S (2001) Molecular mechanisms of plant metal tolerance and homeostasis. Planta 212:475–486

Clemens S (2006) Toxic metal accumulation, responses to exposure and mechanism of tolerance in plants. Biochimie 88:1707–1719

Cobbett CS, Goldsborough PB (2002) Phytochelatins and metallothioneins: roles in heavy metal detoxification and homeostasis. Annu Rev Plant Biol 53:159–182

Curie C, Panaviene Z, Loulergue C, Dellaporta SL, Briat JF, Walker EL (2001) Maize yellow stripe1 encodes a membrane protein directly involved in Fe(III) uptake. Nature 409:346–349

Dalcorso G, Farinati S, Maistri S, Furini A (2008) How plants cope with cadmium: staking all on metabolism and gene expression. J Integr Plant Biol 50:1268–1280

Dayan AD, Paine AJ (2001) Mechanisms of chromium toxicity, carcinogenicity and allergenicity: review of the literature from 1985 to 2000. Hum Exp Toxicol 20:439–51

Delhaize E, Hebb DM, Richards KD, Lin JM, Ryan PR, Gardner RC (1999) Cloning and expression of a wheat (Triticum aestivum L.) phosphatidylserine synthase cDNA: overexpression in plants alters the composition of phospholipids. J Biol Chem 274:7082–7088

Duff SMG, Moorhead GBG, Lefebvre DD, Plaxton WC (1989) Phosphate starvation inducible "Bypasses" of adenylate and phosphate dependent glycolytic enzymes in Brassica nigra suspension cells. Plant Physiol 90:275–1278

Duff SMG, Sarath G, Plaxton WC (1994) The role of acid phosphatase in plant phosphorus metabolism. Plant Physiol 90:791–800

Elefteriou EP, Karataglis S (1989) Ultrastructural and morphological char acteristics of cultivated wheat growing on copper-polluted fields. Bot Acta 102:134–140

Ellis JG, Tokuhisa JG, Llewellyn DJ, Bouchez D, Singh K, Dennis ES, Peacock WJ (1993) Does the ocs-element occur as a functional component of the promoters of plant genes? Plant J 4:433–443

Epstein E, Bloom AJ (2004) Mineral nutrition of plants: principles and perspectives, 2nd edn. Sinauer Associates, Sunderland

Ezaki B, Yamamoto Y, Matsumoto H (1995) Cloning and sequencing of the cDNAs induced by aluminium treatment and Pi starvation in tobacco cultured cells. Plant Physiol 93:11–18

Ezaki B, Koyanagi M, Gardner RC, Matsumoto H (1997) Nucleotide sequence of a cDNA for GDP dissociation inhibitor (GDI) which is induced by aluminum (Al) ion stress in tobacco cell culture (accession no. AF012823) (PGR 97–133). Plant Physiol 115:314

Ezaki B, Sivaguru M, Ezaki Y, Matsumoto H, Gardner RC (1999) Acquisition of aluminum tolerance in Saccharomyces cerevisiae by expression of the BCB or NtGDI1 gene derived from plants. FEMS Microbiol Lett 171:81–87

Farmer EE, Almeras E, Krishnamurthy V (2003) Jasmonates and related oxylipins in plant responses to pathogenesis and herbivory. Curr Opin Plant Biol 6:372–378

Fernandes JC, Henriques FS (1991) Biochemical, physiological, and structural effects of excess copper in plants. Bot Rev 57:246–273

Flora SJ (2009) Structural, chemical and biological aspects of antioxidants for strategies against metal and metalloid exposure. Oxid Med Cell Longev 2:191–206

Flora SJS, Saxena G, Mehta A (2007) Reversal of lead-induced neuronal apoptosis by chelation treatment in rats: role of ROS and intracellular Ca2+. J Pharmacol Exp Ther 322:108–16

Gallego SM, Benavides MP, Tomaro ML (1996) Effect of heavy metal ion excess on sunflower leaves: evidence for involvement of oxidative stress. Plant Science 121:151–159

Garcia MA, Andres CN, Perea A, Del VS, Penarrubia L, Puig S (2011) The intracellular Arabidopsis COPT5 transport protein is required for photosynthetic electron transport under severe copper deficiency. Plant J 65:848–860

Garnier L, Simon PF, Thuleau P, Agnel JP, Blein JP, Ranjeva R, Montillet JL (2006) Cadmium affects tobacco cells by a series of three waves of reactive oxygen species that contribute to cytotoxicity. Plant Cell Environ 29:1956–1969

Gerendás J, Polacco JC, Freyermuth SK, Sattelmacher B (1999) Significance of nickel for plant growth and metabolism. J Plant Nutr Soil Sci 162:241–256

Gonzalez GM, Azcon AC, Mooney M, Valderas A, MacDiarmid CW, Eide DJ, Ferrol N (2005) Characterization of a Glomus intraradices gene encoding a putative Zn transporter of the cation diffusion facilitator family. Fungal Genet Biol 42:130–140

Guerinot ML, Eide D (1999) Zeroing in on zinc uptake in yeast and plants. Curr Opin Plant Biol 2:244–249

Gussarson M, Asp H, Adalsteinsson S, Jensen P (1996) Enhancement of Cd effects on growth and nutrient composition of birch (Betula pendula) by buthionine sulphoxinine (BSO). J Exp Bot 47:211–215

Hagen G, Uhrhammer N, Guilfoyle TJ (1988) Regulation of expression of an auxin-induced soybean sequence by cadmium. J Biol Chem 263:6442–6446

Halliwell B (1996) Antioxidants in human health and disease. Ann Rev Nutr 16:33–50

Hamel F, Breton C, Houde M (1998) Isolation and characterization of wheat aluminum-regulated genes: possible involvement of aluminum as a pathogenesis response elicitor. Planta 205:531–538

Hartwig A, Asmuss M, Ehleben I, Herzer U, Kostelac D, Pelzer A, Schwerdtle T, Burkle A (2002) Interference by toxic metal ions with DNA repair processes and cell cycle control: molecular mechanisms. Environ Health Perspect 110:797–799

Heather K, Marc RK (2001) Abiotic stress signaling pathways: specificity and cross-talk. Trends Plant Sci 6:1360–1385

Horemans N, Raeymaekers T, Van BK, Nowocin A, Blust R, Broos K, Cuypers A, Vangronsveld J, Guisez Y (2007) Dehydroascorbate uptake is impaired in the early response of Arabidopsis plant cell cultures to cadmium. J Exp Bot 58:4307–4317

Jonak C, Okresz L, Bogre L, Hirt H (2002) Complexity, cross talk and integration of plant MAP kinase signaling. Curr Opin Plant Biol 5:415–24

Jonak C, Nakagami H, Hirt H (2004) Heavy metal stress. Activation of distinct mitogen-activated protein kinase pathways by copper and cadmium. Plant Physiol 136:3276–83

Jones JD, Dangl JL (2006) The plant immune system. Nat 444:323–329

Kalia K, Flora SJS (2005) Strategies for safe and effective treatment for chronic arsenic and lead poisoning. J Occup Health 47:1–21

Katagiri F, Lam E, Chua NH (1989) Two tobacco DNA-binding proteins with homology to the nuclear factor CREB. Nat 340:727–730

Kim JA, Agrawal GK, Rakwal R, Han KS, Kim KN, Yun CH, Heu S, Park SY, Lee YH, Jwa NS (2003) Molecular cloning and mRNA expression analysis of a novel rice (Oryza sativa L.) MAPK kinase kinase, OsEDR1, an ortholog of Arabidopsis AtEDR1, reveal its role in defense/stress signaling pathways and development. Bioch Biophys Res Commun 300:868–876

Kitano H (2002) Systems biology: a brief overview. Science 295:1662–1664

Kochian LV (1995) Cellular mechanisms of aluminum toxicity and resistance in plants. Annu Rev Plant Physiol Plant Mol Biol 46:237–260

Kochian LV, Pineros MA, Hoekenga OA (2005) The physiology, genetics and molecular biology of plant aluminum resistance and toxicity. Plant Soil 274:175–195

Kusaba M, Takahashi Y, Nagata T (1996) A multiple-stimuli-responsive as- 1-related element of parA gene confers responsiveness to cadmium but not to copper. Plant Physiol 111:1161–1167

Lee YW, Ha MS, Kim YK (2001) Role of reactive oxygen species and glutathione in inorganic mercury-induced injury in human glioma cells. Neurochem Res 26:1187–93

Lescure AM, Proudhon D, Pesey H, Ragland M, Theil EC, Briat JF (1991) Ferritin gene transcription is regulated by iron in soybean cell cultures. Proc Natl Acad Sci USA 88:8222–8226

Ma JF, Zheng SJ, Hiradate S, Matsumoto H (1997) Detoxifying aluminium with buckwheat. Nature 390:569–570

Madlung A, Comai L (2004) The effect of stress on genome regulation and structure. Ann Bot 94:481–95

Maksymiec W (1997) Effect of copper on cellular processes in higher plants. Photosynthetica 34:321–342

Maksymiec W (2007) Signaling responses in plants to heavy metal stress. Acta Physiol Plant 29:177–187

Maksymiec W, Russa R, Urbanic S, Baszynski T (1994) Effect of excess Cu on the photosynthetic apparatus of runner bean leaves treated at two different growth stages. Physiol Plant 91:715–721

Marschner H (1995) Mineral nutrition of higher plants. Academic, London

Michael JB (2009) Cell signaling pathways. Cell signaling biology. Portland Press, London

Michel KP, Berry S, Hifney A, Pistorius EK (2003) Adaptation to iron deficiency: a comparison between the cyanobacterium Synechococcus elongatus PCC 7942 wild-type and a DpsA-free mutant. Photosynth Res 75:71–84

Mills RF, Francini A, Ferreira PS, Baccarini PJ, Aylett M, Krijger GC, Williams LE (2005) The plant P1B-type ATPase AtHMA4 transports Zn and Cd and plays a role in detoxification of transition metals supplied at elevated levels. FEBS Lett 579:783–791

Mishra D, Mehta A, Flora SJS (2008) Reversal of hepatic apoptosis with combined administration of DMSA and its analogues in guinea pigs: role of glutathione and linked enzymes. Chem Res Toxicol 21:400–407

Monshausen GB, Gilroy S (2009) Feeling green: Mechanosensing in plants. Trends Cell Biol 19:228–235

Mulrooney SB, Hausinger RP (2003) Nickel uptake and utilization by microorganisms. FEMS Microbiol Rev 27:239–261

Pilon SEA, Hwang S, Mel LC, Zhu Y, Tai JC, Bravo RC, Chen Y, Leustek T, Terry N (1999) Overexpression of ATP sulfurylase in Indian mustard leads to increased selenate uptake, reduction, and tolerance. Plant Physiol 119:123–132

Pineros M, Magalhaes J, Carvalho-Alves VM, Kochian LV (2002) The physiology and biophysics of an aluminum tolerance mechanism based on root citrate exudation in maize. Plant Physiol 29:1194–1206

Plaxton WC, Carswell MC (1999) Metabolic aspects of the phosphate starvation response in plants. In HR Lerner, ed, Plant Responses to Environmental Stresses: From Phytohormones to Genome Reorganization. Marcel Dekker, New York, pp 349–372

Poli G, Leonarduzzi F, Biasi E (2004) Chiarpotto Oxidative stress and cell signaling. Curr Med Chem 11:1163–1182

Quartacci MF, Pinzino C, Sgherri CLM, Dalla VF, Navari IF (2000) Growth in excess copper induces changes in the lipid composition and fluidity of PSII-enriched membranes in wheat. Plant Physiol 108:87–93

Quinn JM, Merchant S (1995) Two copper-responsive elements associated with the Chlamydomonas Cyc6 gene function as targets for transcriptional activators. Plant Cell 7:623–638

Quinn JM, Barraco P, Eriksson M, Merchant S (2000) Coordinate copper and oxygen-responsive Cyc6 and Cpx1 expression in Chlamydomonas is mediated by the same element. J Biol Chem 275:6080–6089

Quinn JM, Kropat J, Merchant S (2003) Copper response element and Crr1-dependent Ni21-responsive promoter for induced, reversible gene expression in Chlamydomonas reinhardtii. Eukaryot Cell 2:995–1002

Raghothama KG (1999) Phosphate acquisition. Annu Rev Plant Physiol Mol Biol 50:665–693

Ramesh SA, Choimes S, Schachtman DP (2004) Over-expression of an Arabidopsis zinc transporter in hordeum vulgare increases short-term zinc uptake after zinc deprivation and seed zinc content. Plant Mol Biol 54:373–385

Raven JA, Ewans CWE, Korb RE (1999) The role of trace metals in photosynthetic electron transport in O2-evolving organisms. Photosynth Res 60:111–149

Rentel MC, Knight MR (2004) Oxidative stress-induced calcium signaling in Arabidopsis. Plant Physiol 135:1471–1479

Richards KD, Schott EJ, Sharma YK, Davis KR, Gardner RC (1998) Aluminum induces oxidative stress genes in Arabidopsis thaliana. Plant Physiol 116:409–418

Richardson AE (1994) Soil microorganisms and phosphorus availability. In: Pankhurst CE, Doube BM, Gupta VVSR, Grace PR (eds) Soil biota management in sustainable farming systems. CSIRO, Australia, pp 50–62

Roberts DM, Harmon AC (1993) Calcium-modulated proteins: targets of intracellular calcium signals in higher-plants. Annu Rev Plant Physiol Plant Mol Biol 43:375–414

Rodriguez SM, Romero PMC, Zabalza A, Corpas FJ, Gomez M, Delrio LA, Sandalio LM (2006) Cadmium effect on oxidative metabolism of pea (Pisum sativum L.) roots: imaging of reactive oxygen species and nitric oxide accumulation in vivo. Plant Cell Environment 29:1532–1544

Romero PMC, Corpas FJ, Rodriguez SM, Gomez M, Del RLA, Sandalio LM (2007) Differential expression and regulation of antioxidative enzymes by cadmium in pea plants. J Plant Physiol 164:1346–1357

Ruter JM (2005) Effect of nickel applications for the control of mouse ear disorder on river birch. J Environ Hort 23:17–20

Ryan PR, Delhaize E, Randall PJ (1995) Characterisation of Al-stimulated efflux of malate from the apices of Al wheat roots. Planta 196:103–110

Ryan PR, Delhaize E, Jones DL (2001) Function and mechanism of organic anion exudation from plant roots. Annu Rev Plant Physiol 52:527–560

Sandalio LM, Dalurzo HC, Gomez M, Romero MC, Delrio LA (2001) Cadmium-induced changes in the growth and oxidative metabolism of pea plants. J Exp Bot 52:2115–2126

Sanità di Toppi L, Gabbrielli R (1999) Response to cadmium in higher plants. Env Exp Bot 41: 105–130

Schaaf G, Ludewig U, Erenoglu BE, Mori S, Kitahara T, Von WN (2004) ZmYS1 functions as a proton-coupled symporter for phytosiderophore and nicotianamine-chelated metals. J Biol Chem 279:9091–9096

Schaeffer HJ, Weber MJ (1999) Mitogen-activated protein kinases: specific messages from ubiquitous messengers. Mol Cell Biol 19:2435–2444

Schutzendubel A, Polle A (2002) Plant responses to abiotic stresses: heavy metal-induced oxidative stress and protection by mycorrhization. J Exp Bot 53:1351–1365

Sharma PN, Kumar P, Tewari RK (2004) Early signs of oxidative stress in wheat plants subjected to zinc deficiency. J Plant Nutr 27:449–461

Shcolnick S, Keren N (2006) Metal homeostasis in cyanobacteria and chloroplasts: balancing benefits and risks to the photosynthetic appa ratus. Plant Physiol 141:805–810

Shimogawara K, Usuda H (1995) Uptake of inorganic phosphate by suspension-cultured tobacco cells: kinetics and regulation by Pi starvation. Plant Cell Physiol 36:341–351

Sichul L, Yu YK, Youngsook L, Gynheung A (2007) Rice P1B-type heavy-metal ATPase, OsHMA9, is a metal efflux protein1. Plant Physio 145:831–842

Snowden KC, Richards KD, Gardner RC (1995) Aluminum induced genes: induction by toxic metals, low calcium, and wounding and pattern of expression in root tips. Plant Physiol 107:341–348

Sreenivasulu N, Sopory SK, Kavikishor PB (2007) Deciphering the regulatory mechanisms of abiotic stress tolerance in plants by genomic approaches. Gene 388:1–13

Suzuki N, Koizumi N, Sano H (2002) Screening of cadmium responsive genes in Arabidopsis thaliana. Plant Cell Environ 24:1177–1188

Ticconi CA, Abel S (2004) Short on phosphate: plant surveillance and countermeasures. Trends Plant Sci 9:548–555

Tortell PD, Maldonado MT, Granger J, Price NM (1999) Marine bacteria and biogeochemical cycling of iron in the oceans. FEMS Microbiol Ecol 29:1–11

Waldemar M (2007) Signaling responses in plants to heavy metal stress. Acta physio plant 29:177–187

Wang C, Zhang SH, Wang PF, Hou J, Zhang WJ, Li W, Lin ZP (2009) The effect of excess Zn on mineral nutrition and antioxidative response in rapeseed seedlings. Chemosphere 75:1468–1476

Watanabe M, Henmi K, Ogawa K, Suzuki T (2003) Cadmium dependent generation of reactive oxygen species and mitochondrial DNA breaks in photosynthetic and non- photosynthetic strains of Euglena gracilis. Comp Biochem Physiol Toxicol Pharmacol 134:227–234

Widmann C, Gibson S, Jarpe MB, Johnson GL (1999) Mitogen-activated protein kinase: conservation of a three-kinase module from yeast to human. Physiol Rev 79:143–180

Williams LE, Pittman JK, Hall JL (2000) Emerging mechanisms for heavy metal transport in plants. Biochim Biophys Acta 1465:104–126

Wintz H, Fox T, Wu YY, Feng V, Chen WQ, Chang HS, Zhu T, Vulpe C (2003) Expression profiles of Arabidopsis thaliana in mineral deficiencies reveal novel transporters involved in metal homeostasis. J Biol Chem 278:47644–47653

Wood, BW, Reilly CC, Nyczepir AP (2006). Field deficiency of nickel in trees: Symptoms and causes Proc. VInt. Symp. Mineral Nutr. Decid. Fruit Crops, 16–21 Jan. 2005, Talca, Chile. Acta Horticulturae.

Wood BW, Reilly CC, Nyczepir AP (2004) Mouse-ear of pecan: I. Symptomology and occurrence. Hort Science 38:87–94

Xiang C, Oliver DJ (1998) Glutathione metabolic genes coordinately respond to heavy metals and jasmonic acid in Arabidopsis. Plant Cell 10:1539–1550

Yadav SK (2010) Heavy metals toxicity in plants: an overview on the role of glutathione and phytochelatins in heavy metal stress tolerance of plants. South Afric J Bot 76:167–179

Yalpani N, Enyedi AJ, Leon J, Raskin I (1994) Ultraviolet light and ozone stimulate accumulation of salicylic acid and pathogenesis related protein and virus resistance in tobacco. Planta 193:373–376

Yamamoto Y, Hachiya A, Matsumoto H (1997) Oxidative damage to membranes by a combination of aluminum and iron in suspension-cultured tobacco cells. Plant Cell Physiol 38:1333–1339

Yang T, Poovaiah BW (2003) Calcium/calmodulin-mediated signal network in plants. Trends Plant Sci 8:505–512

Yang ZM, Nian H, Sivaguru M, Tanakamaru S, Matsumoto H (2001) Characterization of aluminium-induced citrate secretion in aluminium-tolerant soybean (Glycine max) plants. Physiol Plant 113:64–71

Yew FT, Nicholas O, Nicolas L, Harvey M (2010) Divalent metal ions in plant mitochondria and their role in interactions with proteins and oxidative stress-induced damage to respiratory function. Plant Physiol 152:747–761

Yruela I, Gatzen G, Picorel R, Holzwarth AR (1996) Cu (II)-inhibitory effect on photosystem II from higher plants: a picosecond time-resolved fluorescence study. Biochemistry 35:9469–9474

Zenk MH (1996) Heavy metal detoxification in higher plants: a review. Gene 179:21–30

Zhang Y, Chai TY, Dong J, Zhao W, An CC, Chen ZL, Burkard G (2001) Cloning and expression analysis of the heavy-metal responsive gene PvSR2 from bean. Plant Sci 161:783–790

Chapter 10
Molecular Network of Nitrogen and Sulphur Signaling in Plants

Gurjeet Kaur, Asha Wadhwa, M.Z. Abdin, Maryam Sarwat, and Altaf Ahmad

Introduction

Mineral elements play an important role in growth and development of plants. There are 17 essential elements, of which six are required in large quantities. During the past half century, supply of mineral nutrients, especially N, P and K, through fertilizers has been considered as an important input in agriculture sector for crop production. Nitrogen occupies a unique position among these mineral elements since it forms an important component of many structural, genetic and metabolic compounds in plant cells. Nitrogen has a role in energy-transfer by being part of compounds, such as ATP (adenosine triphosphate) which allows cells to conserve and use the energy released in metabolism. It is a component of nucleic acids (DNA, RNA), proteins, vitamins and hormones. In addition to N, another important nutrient which has drawn considerable attention is sulphur (S). It is the tenth most abundant element in the universe. Despite its essentiality, it has been described as the neglected plant nutrient. The primary importance of S in crop nutrition arises from it being an essential component of amino acids, which act as building blocks in the synthesis of proteins. Further, S is a constituent of oil in oilseed crops. It has a role to play in increasing chlorophyll formation and aiding photosynthesis. S also plays a role in the activation of enzymes, nucleic acids and forms a part of Biotin and Thiamine (vitamins). Since, demand of N and S and their actual availability tend to vary in time, space and environmental conditions, the

G. Kaur • A. Wadhwa • A. Ahmad (✉)
Department of Botany, Faculty of Science, Jamia Hamdard, New Delhi 110062, India
e-mail: ahmadaltaf@rediffmail.com

M.Z. Abdin
Centre for Transgenic Plant Development, Department of Biotechnology, Faculty of Science, Jamia Hamdard, New Delhi, India

M. Sarwat
Pharmaceutical Biotechnology, Amity Institute of Pharmacy, Amity University, NOIDA, India

M. Sarwat et al. (eds.), *Stress Signaling in Plants: Genomics and Proteomics Perspective, Volume 1*, DOI 10.1007/978-1-4614-6372-6_10,

regulation of plant nitrogen and sulphur metabolism must be responsive to nutritional, metabolic and environmental cues. This article deals with the recent advances in our knowledge of the complex web of interactions in the regulation of nitrate and sulphate assimilation by internal and external signals and its coordination with the overall metabolism of the plant.

Nitrogen in Plants

Nitrogen is available in the form of molecular N_2, volatile ammonia/NOx, mineral (NO_3- and NH_4^+) and organic N (urea and amino acids) (vov Wiren et al. 1997). However, nitrate and ammonium are the preferred source of N by most plant species (Glass and Siddiqui 1995). In a typical aerobic agricultural soil, both nitrate and ammonium are present but nitrate is the major form. It is the most abundant source of N in many soils, especially those in cultivation with annual crops. In general, most crop plants prefer a mixture of ammonium and nitrate and usually take up a higher proportion of ammonium to nitrate than is present in the soil solution. In a study of 35 agricultural soil samples, the mean soil solution concentration of nitrate was found to be 6.0 mM compared to 0.77 mM for ammonium (Wolt 1994).

Nitrogen Uptake

First step of N acquisition and utilization in plants is the transport of nitrate from soil through plasmalemma of epidermal and cortical cells of the root (Daniel-Vedele et al. 1998). According to Scheible et al. (1997) nitrate appears to be a signal that can directly affect the expression of genes related to nitrate uptake, transport and assimilation. So, it acts as both a nutrient and a signal. Although most of the higher plants are capable of reducing nitrate in both roots and shoots, nitrate is reduced more efficiently in leaves than in roots because of the readily available reductants, energy and carbon skeletons produced by photosynthesis (Chen et al. 2004).

Biochemical Characterization

Nitrate uptake is an active process driven by the proton gradient or the proton motive force maintained by H^+-ATPase. Two or more protons are co-transported along with every nitrate ion (Crawford 1995; Santi et al. 1995). In many higher plants, a biphasic relationship between nitrate uptake rate and external nitrate concentration has been found. It has been suggested that there are at least two different types of transporter systems in higher plants namely High Affinity Transporter System (HATS) and Low Affinity Transporter System (LATS). HATS are further categorized into two groups, *Inducible High Affinity Transporter System* (*iHATS*) and *Constitutive High affinity transporters system* (*cHATS*). *Inducible High Affinity Transporter System* (*iHATS*) is substrate inducible and is responsible for uptake at low concentrations of NO_3^- (below ~1 mM). They are characterized by low K_m values (5–200 μM). *Constitutive High affinity transporters* (*cHATS*) is responsible for uptake at low concentrations of NO_3^- and provides a low capacity pathway in uninduced plants but operates simultaneously with *iHATS* in the induced state. Their activity becomes three-fold on exposure to nitrate (Crawford and Glass 1998). It is characterized by low values of both K_m (6–20 μM) and V_{max}. *Low affinity transporters* are constitutive transport system and is responsible for uptake at high external nitrate concentrations (>1 mM). Despite of showing linear kinetics, it appears to be an active H^+ dependent transport system (Kronzucker et al. 1995). They are characterized by high K_m values (>0.5 mM). It allows enough nitrate into the cell which is sufficient to induce the expression of transporter and assimilatory genes and presumably plays a physiological role in the nitrate uptake only above a certain threshold. Molecular biology studies of nitrate transporters in plant suggest that nitrate transporters belong to two different families, NNP and PTR (Forde 2000).

It has been suggested that both the high and low affinity transporters are functional during the early stages of growth but the high affinity systems are functional at later stages too, when the soil N concentration is low. Studies on oilseed rape showed that HATS accounted for about 89 % of the total nitrate uptake (18% and 79% for cHATS and iHATS, respectively) when no fertilizer was applied (Malagoli et al. 2004). They also found that LATS accounted for a minor proportion of the total nitrate uptake. It is proposed that NO_3^--inducible part of HATS functions chiefly as a sensor for root NO_3^- availability (Miller et al. 2007a, b).

Genomic Organization

Nrt1 and *Nrt2* gene families define two classes of membrane proteins, probably involved in low and high affinity nitrate transport, respectively (Forde 2000; Williams and Miller 2001). The first eukaryotic NO_3^- transporter gene was isolated over 16 years ago from the fungus, *Aspergillus nidulans* (Unkles et al. 1991). A few years later, a nitrate transporter gene (*AtNrt1:1*) was identified in *Arabidopsis thaliana* (Tsay et al. 1993). Later, this gene was used as a probe to isolate two more genes (*LeNrt1:1*) and (*LeNrt1:2*) in tomato (Lauter et al. 1996). Here, *Nrt1:2* is shown to be nitrate-inducible and its expression restricted to roots but not in the stem or leaves but *Nrt1:1* is not restricted to roots and is constitutively expressed. Nitrate transporter genes have been cloned from wide range of plants. *BnNrt1:2* is identified in *Brassica napus* also (Crawford and Glass 1998). *AtNrt1:4* has a very specific pattern of expression in leaf petiole where it plays a role during nitrate accumulation within these tissues (Chiu et al. 2004). *AtNrt1:3* expression was nitrate induced in the leaf, but in roots it was not found to be a significant contributor to LATS (Okamoto et al. 2003). According to Li et al. (2007), *AtNrt2:1* was found to be the major contributor to iHATS and cHATS.

Ammonium (NH_4^+) transport shows the normal homeostatic tendency but the range of the concentration at which absorption occurs is very limited due to the potential toxicity at elevated NH_4^+ concentrations. Like nitrate, ammonium is also transported by transporter protein located in the plasma membrane. AMT-type transporters handle NH_4^+ influx and they mediate the uniport of this ion. First NH_4^+

transporter *AtAMT1;1* was isolated from *A. thaliana* (Ninnemann et al. 1994) and then another five homologous sequences, *AtAMT1;1* to *AtAMT1;5* were discovered.

Regulation of Transporters

For the regulation of nitrate uptake, signals are derived from nitrate, which are involved in triggering widespread changes in gene expression; resulting in reprogramming of N metabolism to facilitate the uptake and assimilation of nitrate and its incorporation into amino acids. The nitrate assimilatory pathway is under tight regulation by the available nitrate and reduced N. In strawberry, increasing external nitrate concentration from 0 to 4 mM markedly increased the cumulative nitrate uptake (Taghavi and Babalar 2007). Several of the LATS- and HATS-related genes, apart from being root specific, are also inducible by nitrate and there is evidence that at least one HATS-related gene, *NpNrt2:1* is also repressible by reduced nitrogen (Quesada et al. 1997). In barley and white spruce, cHATS provides a high affinity, low capacity pathway for nitrate entry in uninduced plants. Nevertheless, cHATS activity is up regulated (approx three folds) by exposure to nitrate (Trueman et al. 1996). iHATS has been more extensively studied and is known to be induced by nitrate or nitrite. In barley, the fully induced iHATS flux was approximately 30 times higher than that resulting from the cHATS (Quesada et al. 1997). The increase in transcript is accompanied by increased rates of nitrate uptake (Ismande and Touraine 1994). The results on citrus seedlings suggest that LATS is under feedback control by the N status of plant (Cerezo et al. 2000). They also observed a decline in uptake rate by the addition of amino acids (Glu, Asp, Asn, Gln) to the external solution. The use of chemical inhibitors in physiological studies has suggested that protein synthesis is important for nitrate uptake (Aguera et al. 1990) and the transporters may turn over relatively slow. A degradation mechanism for transporter protein in *Arabidopsis* (*AtNrt2:1*) has been suggested (Cerezo et al. 2001). The presence of a number of conserved protein kinase C recognition motifs in the N and C domains of *HvNRT2:1* (Forde 2000) suggests that phosphorylation events are involved in regulating *AtNrt2:1* activity in response to environmental cues. Remans et al. (2006) found that under N-limited conditions, *AtNrt2:1* played a key role as a major NO_3^- uptake system and coordinated lateral root initiation and development with external NO_3^- availability.

Nitrate Assimilatory Enzymes

Nitrate Reductase (EC 1.6.6.1)

Biochemical Characterization and Localization

Nitrate reductase (NR) is a key enzyme involved in the first step of nitrate assimilation in plants (Crawford 1995). It catalyzes the reduction of nitrate to nitrite with pyrimidine nucleotide in higher plants. NR in plants is a soluble enzyme, located primarily in the cytosols of root epidermal and cortical cells and shoot mesophyll cells. It exists as homodimer metalloprotein of 110-kDa subunits. It catalyses reduction of nitrate to nitrite by transferring two electrons from NAD(P)H to nitrate via three redox centres composed of two prosthetic groups (flavin adenine dinucleotide[FAD] and heme) and a MoCo cofactor in a 1:1:1 stoichiometry per subunit. Each redox centre is associated with a functional domain of the enzyme that has activity independent of the other domains. NR is a substrate inducible enzyme and is thought to be the most limiting step in N assimilation. For this reason, NR activity is considered as a selection criterion for grain yield and N assimilation potential. However, because crop yield is the ultimate result of many factors operating at molecular and environmental levels, a correlation with any single factor is too much to expect (Abrol et al. 1984).

Genomic Organization

It is regulated in both shoots and roots. Most plants have two or more genes for NR. Clones of both genes have been isolated and mapped (Sivasankar and Oaks 1996). Molecular and genetic analyses have revealed that most plants have two or more structural genes for NR, the only known exception being *Nicotiana plumbaginifolia* which has a single NR gene that encodes an NADH-dependent NR (Caboche and Rouze 1990). In barley, the NADH specific NR is encoded by the *nar*I gene, while the NADPH bispecific NR is encoded by the *nar*7 gene. Clones of both gene have been isolated and mapped (Cheng et al. 1986; Miyazaki et al. 1991) and the induction properties have been compared (Sueyoshi et al. 1995). Although, the two proteins are distinct, the genes respond similarly to NO_3^-.

Regulation

Control of NR activity can be achieved either by altering the activity level of existing enzyme or by controlling the amount of enzyme by synthesizing new enzyme and degrading the old one. Many factors regulate this enzyme. Nitrate triggers transcription of inducible genes (NIA) encoding NR. *De novo* synthesis of new NR, stimulated by NO_3^- is one of the mechanisms for controlling enzyme level when combined with NR protein degradation (Stitt 1999). NO_3^- induced increase in the NR activity and NR protein is due to the enhanced steady state level of NR-mRNA (Cheng et al. 1986; Crawford et al. 1988; Miyazaki et al. 1991; Sueyoshi et al. 1995) and this induction is shown to be repressed by glutamine and asparagine (Vincentz et al. 1993). It appears that NR expression is regulated by light *via* phytochrome after it is triggered by NO_3^-. Role of light in induction is probably more related to the activity of the enzyme rather than to the activation of the NR gene. In another study, NR protein kinase was used to identify the apparent key Ser residue in spinach NR (Bachmann et al. 1996).

Many environmental (stress) factors trigger the modulation and post-translational regulation of NR. The existing amount of NR protein depends not only on the rate of synthesis, but also on the rate of degradation. Activity level can be controlled by mechanisms involving phosphorylation of the NR protein and binding of Mg^{2+} or another divalent cation and an inhibitor protein (Stitt 1999). Light has a very important effect though it is not a direct signal to activate NR as photosynthesis is required for NR activation. It has been shown that light and oxygen availability are the major external triggers for the rapid and reversible modulation of NR activity. Indeed, NR can be activated in the dark by feeding sugars to the leaves. Sugar and/or sugar phosphates are the internal signals regulating the protein kinase(s) and phosphatase. Roots usually do not change their reduction rate as rapidly as shoots. However, it was shown that during sudden anoxia the enzyme is rapidly modulated, being activated within minutes (Kaiser and Huber 2001).

Nitrite Reductase (EC 1.7.7.1)

Biochemical Characterization and Localization

The second enzyme in the sequence, nitrite reductase (NiR, ferredoxin nitrite oxidoreductase) catalyses the six-electron transfer reaction from reduced ferrodoxin to NO_2^- leading to the synthesis of NH_4^+. It is localized within chloroplasts in leaf and in plastids in root tissues (Sechley et al. 1992). It is a monomeric protein of about 63 kDa containing sirohaem and a 4Fe4S centre as prosthetic groups (Siegel and Wilkerson 1989). It obtains its reducing power from NADPH. Reduced ferredoxin serves as the electron donor in both leaves and roots. The ferredoxin in roots shares antigenic epitopes with the leaf protein, but is distinct protein (Oaks and Hirel 1985; Wada et al. 1989). But this enzyme can obtain its reducing power from NADPH also, generated by the oxidative pentose phosphate pathway located within the root plastid (Bowsher et al. 1989). Levels of both the ferredoxin and NADPH dependent ferredoxin:NADP-oxidoreductase in isolated pea root plastids increase in response to NO_3^- additions (Bowsher et al. 1993).

Genomic Organization

The gene for the NiR apoprotein has been cloned in at least six different plant species. There is one NiR apoprotein gene per haploid genome in barley and spinach, two in maize and four in *N. tabaccum* (Duncanson et al. 1993; Kronenberger et al. 1993). The four NiR apoprotein genes in tobacco are known to encode two distinct isoforms in shoots and a further two in roots, as indicated by the gene expression studies of Kronenberger et al. (1993). The promoter region of

NiR gene from has been fused to β-glucoronidase (GUS) and trans-gene has then been successfully introduced into tobacco (Ragtogi et al. 1993).

Regulation

This gene appears to be very responsive to NO_3^- additions and to the additions of sucrose, glutamine or asparagine. The experiments of Rastogi et al. (1993) provide evidence that induction of this gene by NO_3^- is a transcriptional event. However, addition of asparagine or glutamine results in a repression of induction, whereas sucrose enhances the induction (Sivasankar and Oaks 1995; Vincentz et al. 1993). Light is also an important environmental cue in the NiR induction (Wray 1993).

Glutamine Synthetase (EC 6.3.1.2)

Biochemical Characterization and Localization

Glutamine synthetase (GS) catalyses the ATP dependent conversion of inorganic N (NH_4^+) into an organic form (glutamine). This enzyme along with GOGAT represents the major pathway for incorporation of ammonia (toxic to plant function) into amino acids (Fei et al. 2003; Hirel and Lea 2001). There are two types of GS: type I-GS, which is dodecameric with subunits of about 52 kDa and type II-GS that is octameric and composed of about 40 kDa subunits. Type I is found mainly in bacteria and type II is best characterized in higher plants. It exists in plants as two major isoforms, a chloroplastic (GS2) and a cytosolic (GS1) enzyme (Scarpeci et al. 2007). During the plant development, N is moved into and out of proteins in different organs and transported through a limited number of transport compounds. Major portion of N is released as NH_3 and reassimilated via GS (Miflin and Habash 2002).

Genomic Organisation

Molecular analysis of GS genes reveal a multigene family whose individual members encode several distinct cytosolic GS (GS1) polypeptides and a single chloroplastic GS (GS2) polypeptide. Li et al. (1993) have identified five distinct cDNA clones of GS in maize. Six distinct genes encoding for GS in maize (Li et al. 1993) and five in sugarcane (Nogueira et al. 2005) were identified. It has been demonstrated that GS occurs in an organ specific manner; roots and nodules generally contain proportionally more cytosolic GS, while leaves contain more chloroplastic GS (Becker et al. 1993). Genetic study of GS has helped in explicating the role of each isoform. Chloroplastic GS is considered to be involved in the reassimilation of photorespiratory NH_4^+.

Regulation

Both GS2 and GS1 genes are regulated by external N application, but extent of this regulation depends on the plant species, N source and plant organ/tissue (Cren and Hirel 1999). Regulatory effects of N assimilation, NH_4^+ and/or NO_3^- on gene expression has been reported in many plants including rice, maize, tobacco, tomato, sunflower and mustard (Zozaya-Hinchliffe et al. 2005). It has been reported that N and C metabolites may also control the expression of GS in the leaf of Arabidopsis (Oliveira and Coruzzi 1999), and tobacco (Masclaux-Daubresse et al. 2005). According to Zozaya-Hinchliffe et al. (2005), light and metabolic factors associated with light (sucrose and carbon substrates) also regulate the expression of this enzyme. Detailed studies on *Pinus sylvestris* by Elmlinger et al. (1994) showed that light regulation of GS2 expression occurs crudely at transcriptional level but fine regulation occurs at post-transcriptional level. Similar observations were shown in tomato seedlings (Migge et al. 1998). Some mechanisms controlling the stability and activity of GS have been discovered. Finnermann and Schjoerring (2000) presented a tentative model for the reversible control of GS1 by phosphorylation and dephosphorylation incorporating the roles of ATP, Mg^{2+} and 14-3-3 binding. The model is based on central role that ATP/AMP ratio under light is an important factor. In that model, it is proposed that in dark, ATP/AMP levels are high, so GS1 is phosphorylated and binds 14-3-3 proteins, which protects it against degradation. Conversely, in the light, GS1 unphosphorylates and become susceptible to damage. Riedel et al. (2001) have also demonstrated that GS2 is phosphorylated in tobacco. Many workers have shown that important factors affecting GS activity are light, carbon status and N nutrition. Experiments with white, red, far-red or blue light by Becker et al. (1992) and Migge et al. (1998) have shown that the phytochrome and the blue light photoreceptors are involved in the positive response to light. C compounds important in stimulating GS1 and GS2 synthesis include sucrose and 2-oxoglutarate. Studies on dark-adapted *Arabidopsis* seedlings have shown that sucrose enhances the expression of GS2, thus mimicking the effect of light. Temperature is also an important environmental factor controlling the GS expression, as shown by the studies of Woodall et al. (1996) on pea and barley plants. Within 2 days of keeping the plants in 15°C instead of 25°C, they observed 50% reduction in GS2 activity while GS1 activity remained unaffected. There are indications that substrate availability (Ortega et al. 1999) or phosphorylation may be important factor controlling the enzyme turnover and activity respectively.

Glutamate Synthase (EC 1.4.1.13)

Biochemical Characterization and Localization

Glutamate synthase [glutamine (amide): 2-oxoglutarate aminotransferase, GOGAT] catalyses the reductive transfer of the amide group of glutamine (produced by GS) to 2-oxoglutarate (α-keto glutarate) to form two glutamate molecules (Ireland and Lea 1999). The discovery of NAD(P)H-dependent GOGAT in bacteria (Tempest et al. 1970), ferredoxin (Fd)-dependent GOGAT in pea chloroplast (Lea and Miflin 1974) and NAD(P)H-dependent GOGAT in carrot cell cultures (Dougall 1974) established a route; GS-GOGAT cycle, for the incorporation of NH_3 into organic compounds. The synthesized glutamate can be used either to replenish the glutamate pool for subsequent GS catalysis or to donate its amino group to form other N-containing compounds. One important fate of glutamate and glutamine is the synthesis of aspartate and asparagine. These amino acids are important N-transport compounds in many plants (Temple et al. 1998). In higher plants, GOGAT occurs as two distinct isoforms, NADH-GOGAT (EC 1.4.1.14) and Fd-GOGAT (EC 1.2.7.1) and these differ in molecular mass, subunit composition, enzyme kinetics and metabolic functions (Gregerson et al. 1993; Sakakibara et al. 1991). Fd-GOGAT, an iron-sulphur flavoprotein, generally functions as monomer with subunit molecular mass of 130–180 kDa. Maize roots contain a Fd-GOGAT isoform that is immunologically distinct from the enzyme found in leaves, suggesting that the two forms are encoded by distinct genes. The root isoform has been implicated in assimilation of NH_4^+ derived from soil NO_3^- (Redinbaugh and Campbell 1993). NADH-GOGAT is also an iron-sulphur flavoprotein and is found primarily in non-green tissues. In higher plants, it occurs as a monomer with a native subunit mass of 225–230 kDa and has a pH-optimum range from 7.5 to 8.5 (Lea et al. 1990).

Genetic Organisation

GOGAT is found in all type of organisms and its amino acid sequence is remarkably well conserved (Temple et al. 1998). The expression pattern of the genes encoding cytosolic GS and NADH-GOGAT appear to be coordinated in non-legumes, where the proteins function together in processes such as primary assimilation of NH_4^+ derived from soil NO_3^- and reassimilation of NH_4^+ released by amino acid catabolism (Lam et al. 1996). cDNA clones for Fd-GOGAT have been isolated from a number of species including barley (Avila et al. 1993); maize (Sakakibara et al. 1991) and *A. thaliana* (Coschigano et al. 1998). Full length cDNA and genomic clones of NADH-GOGAT have been isolated from alfalfa (Trepp et al. 1999) and rice (Goto et al. 1998).

Regulation

Light and a variety of metabolites exert major regulatory controls over metabolic pathways. Evidence by Suzuki and Rothstein (1997) indicates that light exerts a positive regulatory effect on the expression of Fd-GOGAT (GLU1). GLU2 expression is also induced by light but the induction of this gene by sucrose in dark indicates that light-induced expression may in part be caused by increased concentration of C metabolites (Oliveira et al. 1997). During the development and expansion of a new leaf, Fd-GOGAT activity has been shown to increase with onset of photosynthesis and photorespiration (Emes and Tobin 1993). In Barley, enzyme activity, protein and mRNA increased as the leaf emerged and expanded and decreased as the leaf aged (Pajuelo et al. 1997). Nitrate is also shown to act as a signal resulting in widespread changes in the expression of key genes in N metabolism pathway, including Fd-GOGAT (Scheible et al. 1997). Gene expression study in developing alfalfa nodules suggest that NADH-GOGAT is uniquely regulated as compared to other genes of N metabolism (Vance et al. 1995). Maximum expression of NADH-GOGAT occurred in effective nodules and in ineffective nodules and roots was only 12–20% of the maximum. These results show that active N fixation and NH_4^+ itself or a downstream product of its metabolism is required for maximum NADH-GOGAT gene expression. Fd-GOGAT proteins from *Arabidopsis* and maize contain a presequence with many of characteristics of plastid transit peptides. Similarly both rice and alfalfa (Gregerson et al. 1993) NADH-GOGATs contain presequences that are thought to be involved in plastid targeting. Interestingly, presequences are found in all characterized eukaryotic GOGAT proteins. As shown by the experiments of Yamaya et al. (1995) and Hirose et al. (1997) in rice seedlings, NH_4^+ ions and glutamine may act as a signal for the increase in transcription.

Glutamate Dehydrogenase (EC 1.4.1.2)

Biochemical Characterization and Localization

Glutamate dehydrogenase (GDH) is one of few enzymes capable of releasing amino nitrogen from amino acids to give keto-acid and NH_3 that can be separately recycled and used in respiration and amide formation, respectively. It is thought to be an alternative pathway for the formation of glutamate involving reductive amination of 2-oxoglutarate by NH_4^+. Its role in plant cells remains controversial (Miflin and Habash 2002). It is yet to be clearly demonstrated that the enzyme plays a significant role either in NH_3 assimilation or in carbon (C) recycling (Dubois et al. 2003). Studies show that it has a role in the deamination of glutamate to provide energy and return C from amino acids into the reactions of C metabolism during C or energy shortage (Miflin and Habash 2002). However, Dubois et al. (2003) have still argued that the physiological function of GDH in plants remains speculative. GDH is capable of synthesizing or de-aminating glutamate but the direction of activity depends on specific environmental cues (Pahlich 1996). One isoform of enzyme, localized in mitochondria in roots and leaves, uses NADH as the electron donor (Sechley et al. 1992). Another isoform, having specific requirement for NADPH is present in chloroplasts of photosynthetic tissues. The primary role of GDH could be replenishment of TCA cycle intermediates via its oxidation to 2-oxoglutarate. Glutamate is deaminated to

2-oxoglutarate in isolated mitochondria; however in the presence of aminooxyacetate, glutamate no longer contributes to mitochondrial respiration (Sechley et al. 1992). This observation indicates that GDH does not oxidize glutamate.

Nitrate as Signaling Molecule

The role of nitrate as a signal has been known for a long time, but the mechanism of nitrate sensing and the signaling events associated with it have not yet been fully understood. While the nitrate sensing protein proposed over a decade ago is yet to be identified, nitrate sensing by a cytokinin precursor followed by His-Asp phosphorelay has been proposed recently (Sugiyama and Sakakibara 2002). But it is not clear whether this constitutes nitrate signaling or a crosstalk with hormone signaling.

A few elements/events possibly associated with nitrate signaling have been characterized using pharmacological approaches. For example, Ca^{2+} and protein kinases/phosphatases have been implicated in mediating the nitrate signal for the expression of NR, NiR and GS2 mRNAs (Sakakibara et al. 1997; Sueyoshi et al. 1999). In addition to the kinases that post-translationally modulate NR, SPS or PEP carboxylase, Hartwell et al. (1999) described a Ca^{2+} independent PEPCase protein kinase, which is a novel member of the Ca^{2+} calmodulin-regulated group of protein kinases. Though a number of kinases/phosphatases involved in nitrate signaling have been described, their specific roles in mediating nitrate and other interacting signals have not been clearly delineated. A better understanding of the nitrate signaling cascade might emerge from the detailed characterization of mutants related to the signal transfer cascade from nitrate to the NR gene (Ogawa et al. 2000), revealing more intermediates and potential sites for manipulation of NUE.

Transcriptional regulation of several hundreds of nitrate responsive genes by nitrate as a signal requires cis-acting regulatory sequences or nitrate response elements (NRE) (Raghuram et al. 2006). One such sequence, originally reported to be comprised of an A[G/C]TCA core sequence motif preceded by a 7-bp AT rich region, based on promoter deletion analyses in nitrate and nitrite reductases from *A. thaliana* and birch (Warning and Hachel 2000). However, a genome-wide computational analysis of all the known nitrate responsive genes in *Arabidopsis* and rice indicated that these motifs were present almost randomly throughout these genomes and were neither specific nor common to nitrate responsive genes. These findings demand a fresh search for candidate sequences that qualify to be NREs in plants. The identification of putative *cis* elements that are responsive to carbon and nitrogen signaling interactions (Palenchar et al. 2004) also necessitate a search for different cis-regulatory elements that might work in concert. Identification of such regulatory elements provides an end point for nitrate signaling and provides new avenues for characterizing/manipulating the rest of the signaling pathway to enhance NUE.

Hormones and Nitrate Signaling

Several studies during the last decade point to the role of hormones in mimicking, mediating or modulating the nitrate response. For example, cytokinin metabolism and translocation could be modulated by the nitrogen nutrition status; in other words, cytokinin accumulation and translocation occurred after sensing a change in nitrogen availability (Samuelson and Larsson 1993; Takei et al. 2002). Application of cytokinin can mimic the nitrogen-dependent regulation of gene expression in photosynthesis, cell cycling and translational machinery (Takei et al. 2002). In maize and *A. thaliana* some response regulators of the His-Asp phosphorelay system have been found to be upregulated by both cytokinins and nitrate (Sakakibara et al. 1998, 1999; Imamura et al. 1999). These findings strongly suggest a role for cytokinins in communicating the availability of nitrogen from roots to leaves (Sugiyama and Sakakibara 2002).

The cross talk between various plant hormones also has implications for nitrogen sensing and response. For example, auxin synergistically affects cytokinin activity on cell division and organ development (Soni et al. 1995). On the other hand, ABA antagonises the cytokinin-mediated nitrogen signaling by means of negative regulation of cytokinin-inducible response regulator genes. Unlike cytokinins, which are positively regulated by nitrate as a signal, ABA biosynthesis is down regulated by nitrogen-sufficiency (Gawronska et al. 2003). These findings regarding the role of hormones in nitrogen signaling await further characterization of the complete signaling pathway. Gibberellins do not seem to play any role in the control of nitrate assimilation, at least in the vegetative stages of *Arabidopsis* (Bouton et al. 2002).

Nitrate Sensing and Light Signaling

While the role of nitrate as a signal and the range of responses it elicits have been well characterized, the mechanism of nitrate sensing and the exact signaling events that bring about signal-response coupling have not yet been understood. While the nitrate sensing protein proposed by Campbell (Redinbaugh and Campbell 1991) over a decade ago is yet to be identified, nitrate sensing by a cytokinin precursor followed by His-Asp phosphorelay has been proposed recently as discussed above (Sugiyama and Sakakibara 2002). However, some other elements of the signaling cascade have been suggested using pharmacological approaches. For example, Ca^{2+} and protein kinases/phosphatases have been implicated in mediating the nitrate signal for the expression of NR, NiR and GS2 mRNAs (Sakakibara et al. 1997; Sueyoshi et al. 1999). Other kinases that post-translationally modulate NR, SPS or PEP carboxylase have been purified and partially characterized. Hartwell et al. (1999) described a Ca^{2+} independent PEPCase protein kinase, which is a novel member of the Ca^{2+} calmodulin regulated group of protein kinases. Other kinases/ phosphatases have been described as well (reviewed by Krapp et al. 2002), but their specific roles in mediating nitrate and other interacting signals have not been clearly delineated. Mutants related to the signal transfer cascade from nitrate to the NR gene have been reported (Ogawa et al. 2000), and their detailed characterization may reveal more intermediates and potential sites for manipulation.

Light is an additional signal that regulates the expression of many nitrate responsive genes, though it has been studied in depth in only a few of them. The role of light in regulation of NR gene expression has often been reviewed (Raghuram and Sopory 1995; Chandok et al. 1997; Lillo and Appenroth 2001). At the transcriptional level, the expression of NR is regulated differently in green plants and etiolated seedlings and is mediated by different photoreceptors. Using pharmacological approaches, the phytochrome-mediated regulation of NR gene expression in maize was shown to be mediated through G-protein, PI cycle and protein kinase C. The effects of light in green plants are probably mediated more indirectly, through photosynthesis and sugars (Lillo and Appenroth 2001). At the post-translational level, light acts by modulating the phosphorylation status of the enzyme, in conjunction with 14-3-3 proteins.

14-3-3 Proteins and Metabolic Regulation

Efficient utilisation of available nitrate requires coordinated gene expression and/or post-translational regulation of the proteins/enzymes involved in nitrate transport and reduction, as well as those involved in carbon partitioning for amino acid synthesis. This is brought about in part by the criss-cross regulation of c-metabolising enzymes by nitrate (Stitt 1999) and N metabolising enzymes by sugars at the transcriptional level. At the post-translational level, it is becoming increasingly evident that regulatory binding proteins known as 14-3-3 proteins bring about this metabolic coordination. The plant cytosolic enzymes nitrate reductase, glutamine synthetase; sucrose-phosphate synthase, trehalose-phosphate synthase, glutamyl-tRNA synthetase, and an enzyme of folate metabolism have all been found to bind to 14-3-3s in a phosphorylation dependent manner. NR, for example, is inactivated by 14-3-3 following phosphorylation by protein kinases responding to light–dark transitions and changes in cellular energy status (Huber et al. 2002).

Recent experiments in transgenic potato plants indicate that repression of 14-3-3 proteins lead to significant increases in NR and SPS activities, and even higher levels of starch accumulation in the tuber. This indicates 14-3-3 regulation at the endpoint of signaling pathways, but 14-3-3 proteins are also implicated at earlier points in the same pathways. The 14-3-3 binding site in NR is known to be phosphorylated by at least two protein kinases; a calcium-dependent protein kinase (CDPK) and an SNF1-related kinase (SnRK1). It is striking that 14-3-3 proteins have been found to interact with both of these classes of kinase, including the *Arabidopsis* CPK1 isoform, and more significantly, the wheat SnRK1 homologue, WPK4 (Ikeda et al. 2000). Apart from interaction with protein kinases, 14-3-3 proteins also interact with other components of signaling pathways, for example with RGS3, a negative regulator of the G-alpha subunits of heterotrimeric G proteins. Thus, it would be interesting to examine whether 14-3-3 proteins form a link between G-protein signaling pathways and metabolic regulation in plants at the post translational level.

Sulphur in Plants

Sulphur, because of its many oxidation states, represents one of the most versatile elements in biology. These states range from +6 (SO_4^{2-}) to −2 (H_2S). The most oxidative and thus, the most stable of them is SO_4^{2-}. So, for the plants, SO_4^{2-} is the major source of S from soil. S metabolism in plants include uptake of ion from the environment, assimilation into organic compounds, and channeling into proteins and secondary substances. According to Droux (2004), assimilation of SO_4^{2-} could be summarized in four steps: (1) uptake of SO_4^{2-}; (2) activation of SO_4^{2-}; (3) reduction of SO_4^{2-} and (4) synthesis of cysteine.

Uptake of Sulphate

Sulphur uptake from soil is almost exclusively via roots and it is an energy independent process by proton/sulphate co-transporters (Leustek et al. 2000; Saito 2000). After transport of SO_4^{2-} across the plasma membrane, intracellular transport from roots to shoots occur by unload/download mechanism via xylem and finally transfer between tissues via the phloem.

Sulphate Transporters

Biochemical Characterization and Localization

Sulphate transport is driven by a proton gradient force and is a $3H+/SO_4^{2-}$ co-transport mechanism. Multiple transport steps through many membranes are involved. Many workers have worked for identification and functional characterization of SO_4^{2-} transporters. Plasma membrane SO_4^{2-} transporters are classified as proton/sulfate cotransporters. Thus, uptake mediated by this transporter is pH dependent, and the proton gradient is generated by plasma membrane proton ATPase. In yeast, Takahashi et al. (1997a, b) and Yoshimoto et al. (2002) have identified SO_4^{2-} transporters with high and low affinity for SO_4^{2-}. These observations were similar to those reported by Vidmar et al. (2000). The requirement for a pH gradient to drive uptake was also shown in higher plants using cultured tobacco cells (Jones and Smith 1981) and in an isolated vesicle system

(Hawkesford et al. 1993). A complex array of transporters differing with regard to localization, transport kinetics, and inducible expression contribute to facilitate effective transportation of SO_4^{2-}.

Genomic Organization

First plant sulphate transporter genes were isolated from the tropic legume *Stylosanthes hamata*, using a yeast complementation system (Smith et al. 1995). Since then, a large number of genes and cDNAs encoding sulphate transporters have been isolated from different plant species. About 12 SO_4^{2-} co-transporters like genes were identified and were divided into four groups in *A. thaliana* (Takahashi et al. 2000 and Yoshimoto et al. 2003) and *Brassica oleracea* (Buchner et al. 2004). In *Arabidopsis*, 14 genes encoding for SO_4^{2-} transporters have been reported (Yoshimoto et al. 2002; Hawkesford 2003). These are classified into five subfamilies (*AtSULTR 1–5*) based on their deduced amino acid sequences. These subgroups suggest specialized function and catalytic properties for the transport of SO_4^{2-} between compartments and tissues. Members in *SULTR1* are high-affinity transporters. *SULTR1;1* and *SULTR1;2* of *Arabidopsis* are localized to root epidermal cells and are inducible by SO_4^{2-} depletion, thus are responsible for initial SO_4^{2-} uptake from rhizosphere (Shibagaki et al. 2002). Low-affinity transporters of *SULTR2* and *SULTR3* are localized to vascular tissues and are thought to be involved in the uptake from plant apoplast into vascular cells. Transporters of *SULTR4* are responsible for efflux of SO_4^{2-} from vacuole to the cytoplasm (Kataoka et al. 2004). Phylogenetic analysis shows subdivision of transporters into four distinct groups. Group 1 is the best characterised and its analysis particularly in *A. thaliana* and *Hordeum vulgare*, suggests that these sulphate transporters are mainly responsible for sulphate uptake by the root (Smith et al. 1997; Takahashi et al. 2000; Vidmar et al. 2000; Yoshimoto et al. 2002). Sulphate transporters have been described in *S. hamata* and *A. thaliana*, which have a low affinity for sulphate, with Km values of 0.41 mM and more (Smith et al. 1995; Takahashi et al. 1997a, b, 2000). These sulphate transporters have been placed in group 2 because of phylogenetic analysis and are thought to be responsible for vascular sulphate transport. Group 3 contains five *Arabidopsis* SO_4^{2-} transporter genes but these are less well characterised (Takahashi et al. 1999). The sulphate transporters of the group 4 are characterised by a C-terminal plastidal transit peptide.

Regulation

Many studies have taken the impact of S-nutritional status on SO_4^{2-} influx capacity as a model for studying the regulation of S nutrition. Earlier studies indicated an increase in SO_4^{2-} uptake following a period of S limitation (Lee 1982; Clarkson et al. 1983). Indirect evidence using inhibitors showed a rapid turnover of SO_4^{2-} transporter proteins and thus, the importance of transcriptional regulation (Clarkson et al. 1992). Cytokinin-mediated regulation of gene expression is also indicated as cytokinins down-regulate the iHATS of SO_4^{2-} i.e. SULTR1;1 and SULTR1;2 in *Arabidopsis* (Maruyama-Nakashita et al. 2004). There is clear evidence that transcription of the genes encoding the transporters involved in initial uptake at the soil-root interface, cell-to-cell transfer, vascular transportation and vacuolar efflux transporter is controlled by plant S nutritional status (Buchner et al. 2004). The coordinated expression of this gene family helps in the optimum management of plant SO_4^{2-} under varying conditions of supply and demand. Smith et al. (1997) studied the impact of S status on the transcription of HATS of SO_4^{2-} in roots. Their observations were that following the removal of S, an increased abundance of mRNAs for high-affinity transporters was in parallel with decreasing tissue contents of SO_4^{2-}, cysteine and glutathione. Upon re-supply of S, a de-repression of activity and decrease in mRNA abundance occurred with hours. During the regulation of expression of SO_4^{2-} transporters, de-repression is mediated by feedback loops involving products of S-assimilation (Hawkesford et al. 2003). In addition to the quick responses to S-nutritional status in terms of transcriptional regulation and protein turnover, there are some additional levels of post-translational regulation acting on SO_4^{2-} transporters. The carboxy-terminal region contains a SO_4^{2-} transporter and anti-sigma antagonist (STAS) domain (Shibagaki and Grossman 2004 and Rouached et al. 2005). Mutations or deletions in this region affect function and plasma membrane targeting. This region contains a phosphorylation site and this region may be involved in protein:protein interactions, both of which contribute to the regulation.

Sulphate-Assimilatory Enzymes

Inorganic SO_4^{2-} is chemically very stable and therefore, has to be activated prior to reduction to sulfite. Reduction of SO_4^{2-} requires eight electrons and about twice as much energy as NO_3^- reduction. In plants, the high potential needed for SO_4^{2-} reduction is surmounted by the activation step i.e. the formation of adenosine 5′-phosphosulphate (APS) from SO_4^{2-}. APS is an energy rich mixed anhydride of phosphate and SO_4^{2-}.

ATP Sulphurylase (EC 2.7.7.4)

Biochemical Characterization and Localization

ATP sulphurylase (ATPS) catalyzes the first step in sulphate assimilation, the adenylation of sulphate to APS. This is the sole entry step for the metabolism of SO_4^{2-}. The formation of APS is an energetically unfavorable process, which is driven forward by the consumption of APS by subsequent reactions, reduction to sulphite by APS reductase or phosphorylation to PAPS by APS kinase (APK).

In this reaction, SO_4^{2-} is linked to a phosphate residue by an anhydride bond using ATP. The formation of APS is an energetically unfavorable reaction, which is driven by the consumption of pyrophosphate (Leustek et al. 2000). ATPS activity is found in chloroplasts and cytosol. In plants, ATPS activity was detected in chloroplasts and in the cytosol of spinach leaves (Lunn et al. 1990; Renosto et al. 1993). There are different functions of ATPS in the two compartments, sulphate reduction in the plastids and activation of sulphate for synthesis of sulphonated compounds in the cytosol (Rotte and Leustek 2000).

Genomic Organization

The initial step of SO_4^{2-} utilization has received considerable attention, since entry reactions are usually good candidates for the control of the pathway (Rotte and Leustek 2000). ATPS is encoded by small multigene families in all plant species. cDNAs encoding chloroplastic and cytosolic isoforms of ATPS have been isolated from potato (Klonus et al. 1994). On the other hand, four isoforms of ATPS were isolated from *Arabidopsis*, all of them containing a chloroplast transit peptide (Murillo and Leustek 1995; Logan et al. 1996; Hatzfeld et al. 2000). All four ATPS genes contain 5 exons and 4 introns and are localized on different chromosomes. In *Brassica juncea*, two isoforms of ATPS were cloned both containing an organelle-targeting peptide (Heiss et al. 1999a, b).

Regulation

Regulation of S assimilation by C metabolites has received very little attention. It has long been known that cysteine production from SO_4^{2-} is stimulated by light and the activities and mRNA levels of enzymes of SO_4^{2-} assimilation are higher in green leaves than in etiolated tissues (Hell et al. 1994). Furthermore, activity of ATPS was shown to be light induced in maize. As it is known that several products and intermediates of SO_4^{2-} assimilation are toxic, the pathway undergoes strict regulation by the S status of the plant (Brunold 1990 and Leustek et al. 2000). SO_4^{2-} assimilation seems to be controlled by demand. The pathway is repressed under normal levels of external SO_4^{2-} and derepressed by SO_4^{2-} limitation. During the investigations on the molecular mechanisms of feedback regulation of SO_4^{2-} assimilation by thiols, attention was first paid to ATPS. In *A. thaliana,* ATPS activity and mRNA levels of the APS1 isoform were decreased by GSH treatment. As external GSH supply increases the accumulation of cysteine, both GSH and cysteine might be responsible for the control of ATPS. Phloem sap analysis in *Brassica napus* and poplar indicated that GSH rather than cysteine was the acting signal. Research Studies in *Brassica* showed that blocking GSH synthesis relieved the repression of ATPS. By contrast, in maize, cysteine was able to regulate the levels of ATPS mRNA without the need for conversion to GSH.

APS Reductase (EC 1.8.99.2)

Biochemical Characterization and Localization

APS formed by the action of ATPS is further reduced to sulphite (SO_3^{2-}) by APS reductase (APR) by a thiol-dependent two-electron reduction (Bick and Leustek 1998; Bick et al. 1998). This enzyme was identified by chance while searching for cDNAs encoding plant PAPS reductase (Setya et al. 1996; Gutierrez-Marcos et al. 1996). Since the reaction equilibrium of ATPS favors the reverse direction, i.e. the formation of ATP and SO_4^{2-}, the products of the forward reaction, i.e. APS and Pi, must be further metabolized immediately by APR (Saito 2004). It is localized in chloroplasts. A key feature of this enzyme is that they are able to use a variety of reduced thiol compounds such as dithiothreitol and glutathione as a sole source of electrons. In vivo APR is present as a homodimer probably linked by a disulfide bond of conserved Cys residue. The APR enzymes are composed of three domains. At the amino terminus is a region that resembles a transit peptide that allows translocation of the mature protein to plastids and is cleaved from the protein once it is imported into the chloroplast. Adjacent to it is the amino-terminal domain of the mature protein that is homologous with PAPS reductase from a variety of organisms. At the carboxyl end is a domain that resembles thioredoxin. Thioredoxin is redox-active protein that functions with a number of different reductases. This fusion of reductase and cofactor into a single protein imply that thioredoxin-like domain may act as an exclusive electron donor for the reductase domain. Sulfite is generated by this enzyme through the addition of two electrons, through the reduced glutaredoxin/thioredoxin-like domain (Bick et al. 1998; Suter et al. 2000). APR is encoded by small multigene family of 2–3 isoforms in most plant species (Setya et al. 1996; Koprivova et al. 2001). Feeding experiments with $^{35}SO_4^{2-}$ for control flux analysis (Vauclare et al. 2002) to quantify the role of APR in control of SO_4^{2-} assimilation showed that flux control coefficient of APR was between 0.7 and 0.9 (equivalent to 70% and 90% of the total control). These observations indicated strong control of pathway by APR and thus it could be concluded that APR is indeed a key enzyme of SO_4^{2-} reduction pathway.

Genomic Organization

APS reductase cDNA was cloned from several plant species: *A. thaliana* (Gutierrez-Marcos et al. 1996; Setya et al. 1996), *Catharanthus roseus* (Prior et al. 1999), *Brassica juncea* (Heiss et al. 1999a, b), *Lemna minor* (Suter et al. 2000a, b), *Allium cepa*, *Zea mays*, *Populus tremula* × *P. alba*, and the green alga *Enteromorpha intestinalis* (Gao et al. 2000). APR is encoded by multigene families in most species. Three different cDNAs were cloned from *A. thaliana* (APR1, 2 and 3) that were able to complement the cysteine auxotrophy of an *E. coli* PAPS reductase mutant strain. The APR cDNAs encode individual members of a small, highly conserved gene family (Bick and Leustek 1998). At the level of nucleotide homology, the coding sequence of APR1 is more closely related to APR3 (78%) identity than to APR2 (68% identity). On the other hand, the genes of APR2 and APR3 have the same exon/intron organization with 3 introns, APR1 gene lacks the intron separating exon 2 and 3 in APR2 and APR3. APR was recently cloned from a cDNA library in soybean plant (Phartiyal et al. 2008) and they observed an abundant expression of the gene and activity of encoded protein in early developmental stages of soybean seed, which declined with seed maturity. Two cDNAs were obtained from *B. juncea* (Heiss et al. 1999a, b) and southern analysis revealed at least two APR genes in *L. minor* (Suter et al. 2000a, b), poplar, and several species of the genus *Flaveria* (Koprivova et al. 2001). In *A. thaliana* the corresponding genes were cloned and sequenced (Chen and Leustek 1998). In the APR1 gene sequence duplication is found at the 5′end of intron 2 and just before the thioredoxin active site in exon 3, which might possibly represent a remnant of an original intron separating the two domain-coding exons (Chen and Leustek 1998).

Regulation

Many experiments based on study of S assimilation have concentrated on APR as this enzyme was long known to be strongly regulated by various environmental factors, nutrient availability and stress (Brunold 1990). Kopriva et al. (1999), during the investigation of the control of SO_4^- assimilation by light, showed that APR activity undergo a diurnal rhythm with maximum activity 4 h after light onset and minimum activity at the beginning of night. Furthermore, the same study also revealed that sucrose was able to imitate the effect of light. It has been observed by many workers that carbohydrates induce APR mRNA accumulation and activity in the dark (Hesse et al. 2003; Kopriva et al. 1999, 2002). Nutritional stress also effect APR regulation at translational level. Withdrawal of N for 3 days in *A. thaliana* led to specific decrease of APR activity and these changes corresponded to changes in mRNA levels of all three isoforms and APR protein accumulation (Koprivova et al. 2000), showing that APR is primarily regulated at the level of transcription. S deficiency is one of the major problems connected with S metabolism and this aspect has been studied by many workers. Hirai et al. (2005) and Nikiforova et al. (2005) while working on *Arabidopsis* observed that more than 2,700 genes were affected by S starvation and as expected, genes induced by this nutritional stress included those coding for SO_4^- transporters and APR; other genes of SO_4^- assimilation were not significantly and/or consistently affected. Bick et al. (2001) revealed a post translational level of APR regulation by redox processes. Not much is known about the role of phytoharmones in control of S assimilation, but some experimental results indicate that this group of compounds is very important for regulation of S nutrition (Ohkama et al. 2002; Maruyama-Nakashita et al. 2004, 2005). Ohkama et al. (2002) using zeatin treatment on transgenic *Arabidopsis* plant showed that it resulted in increased APR mRNA accumulation. The hypothesis that SO_4^- assimilation is regulated by carbohydrates and not by light was confirmed by a finding in which *Lemna* plant cultivated in CO_2 free atmosphere showed rapid decrease in APR activity and mRNA level. This reduction in APR activity, but not in mRNA level, was attenuated by supplying sucrose to the nutrient solution.

Sulfite Reductase (EC 1.8.7.1)

Biochemical Characterization and Localization

Hydroden-sulfide:Fd oxidoreductase, commonly known as sulfite reductase (SiR) catalyzes the transfer of six electrons from ferrodoxin to SO_3^{2-} producing sulfide (S^{2-}) in the presence of light. This enzyme in plant cells consists of a homodimer of around 65 kDa and contains a siroheme and an iron-sulphur cluster [4Fe–4S] per subunit as co-factors. It is localized in plastids in both photosynthetic (electrons supplied from PSI) and non photosynthetic tissues (electrons supplied from NAD(P)H) (Saito 2004). SiR is represented by a single gene, and localized exclusively in the chloroplast (Hawkesford and De Kok 2006). It has an N-terminal cleavable extension peptide necessary for its plastid import. The proper combination of different isoforms of ferredoxin, ferredoxin-$NADP^+$ reductase and SiR is critical for efficient SO_3^{2-} reduction (Yonekura-Sakakibara et al. 2000). Fd and Fd-SiR form an electrostatically stabilized 1:1 protein-protein complex, with Fd supplying the negatively charged groups; this specific interaction is crucial for efficient electron transfer between the two proteins (Knaff 1996; Akashi et al. 1999). Fd seems to have both common and unique electrostatic interaction sites for SiR and FNR.

Genomic Organization

cDNAs have been cloned and complete amino acid sequences, deduced from the corresponding cDNAs, are available for Fd-SiRs from three higher plants, maize (Ideguchi et al. 1995), *Arabidopsis* (Brühl et al. 1996) and tobacco (Yonekura-Sakakibara 1998). The extent of amino acid sequence homologies among plant SiRs are very high (77–82 %). A cDNA and a single copy gene encoding SiR were isolated from *Arabidopsis* (Brühl et al. 1996; Bork et al. 1998). The gene consists of eight exons, with the first intron localized behind the first amino acid of the mature protein, and is localized on chromosome 5. Two SiR isoforms are present in leaves and non-photosynthetic organs of *Brassica rapa* and tobacco (Takahashi et al. 1997a, b; Yonekura-Sakakibara et al. 1998). One cDNA and corresponding gene encoding SiR were isolated from tobacco, the gene possessed the same exon/intron organization as the gene from *Arabidopsis* (Yonekura-Sakakibara et al. 1998a, b). Additional full-length cDNA clones for SiR were obtained from *Glycine max* and maize. SiR contains 19 % identical amino acids with nitrite reductase indicating that these genes may have the same evolutionary origin.

Serine Acetyltransferase (EC 2.3.1.30)

Biochemical Characterization and Localization

Biosynthesis of Cysteine (first thiol-containing amino-acid) is a two-step process. First step is condensation of L-serine and acetyl-coenzyme A catalysed by serine acetyl transferase (SAT) to form O-acetylserine (OAS). This OAS formed is used further in the second step. By contrast, to many amino acids whose synthesis takes place in plastids, this enzyme is associated with cytosol and mitochondria (Droux 2003). It has a molecular mass of 300–350 kDa. The association of SAT with OASTL is necessary for SAT stability and activity. They form a multienzyme complex in plants (Bogdanova and Hell 1997; Droux et al. 1998; Wirtz et al. 2001). Interaction with OAS-TL is a pre-requisite for SAT activity whereas OAS-TL is active as a free dimer (Droux et al. 1998; Wirtz et al. 2001).

Genomic Organization

SAT is encoded by multigene family. cDNA clones encoding Ser acetyltransferase have been isolated from watermelon (Saito et al. 1995), spinach (Noji et al. 2001), *A. thaliana* (Bogdanova et al. 1995; Hell and Bogdanova 1995; Howarth et al. 1997; Ruffet et al. 1995; Roberts and Wray 1996) and *Allium tuberosum* (Chinese chive) (Urano et al. 2000). In particular, cDNAs of three Ser acetyltransferase isoforms that exhibit different subcellular localization, SAT-c (cytosolic isoform), SAT-p (plastidic isoform) and SAT-m (mitochondrial isoform) have been cloned from *A. thaliana* (Noji et al. 1998). In *Arabidopsis,* SAT gene family consists of five members. The organization of the SAT family is still not very clear. SATc1 is a cytosolic isoform located on chromosome 1 (Gutierrez-Alcala et al. 2000; Ruffet et al. 1995). Another cytosolic isoform, SATc2, is associated with chromosome 5 and a mitochondrial SATm gene is located on chromosome 3 (Hell et al. 2002). Biochemical analyses of these SATs are still scarce as this enzyme is very instable (Droux et al. 1998).

Regulation

S-starvation induced the mRNA levels for the chloroplastic isoform of SAT (Barroso et al. 1995). Level of cysteine in plant compartments control SAT activity through feedback inhibition (Noji and Saito 2002). In soybean, regulation of cytosolic SAT in response to Cys levels, through phosphorylation has been studied involving a large family of calcium-regulated protein kinases (Yoo and Harmon 1997). Expression of SAT genes is modulated to some extent by nutritional conditions. In *A. thaliana*, short term S-deficiency increased OAS level but cysteine level remained unchanged so SAT activity did not change much but during long term deficiency, cysteine level was strongly affected, thus SAT activity was also derepressed (Nikiforova et al. 2003). H_2S has an effect on SAT, since it stabilizes the complex with OASTL and, thus increases the activity (Droux et al. 1998). Another potential signal in the regulation of S assimilation is OAS, which most probably acts as a transcriptional regulator because its addition increased the mRNA levels of cytosolic SAT (Koprivova et al. 2000). During the search for genes undergoing a circadian control of expression (Harmer et al. 2000), it was observed that mRNA levels of SAT started rising at the beginning of light period.

O-Acetyl Serine (Thiol) Lyase (EC 4.2.99.8)

Biochemical Characterization and Localization

This is the final step of cysteine biosynthesis where sulfide is incorporated into amino acid skeleton of OAS to form cysteine and acetate by the action of OAS (thiol) lyase. It is a β-replacement reaction as OASTL belongs to a large family of enzymes catalyzing the reaction of β-substitution of amino acids (Saito 2004). This step i.e. the formation of cysteine marks the almost exclusive entry of reduced S into organic compounds in plants (Hell 1997). It exists as a homodimer of 60–70 kDa and is a pyridoxal-dependent enzyme. The cellular activity of OASTL is 100–300 fold in excess over SAT (Droux 2003; Hell et al. 2002; Saito 2000). OASTL seems to be required in all cellular compartments that carry out protein biosynthesis. Like SAT, this enzyme is found associated with not only plastids but also cytosol and mitochondria of plant cells (Hell et al. 2002; Droux 2003). The OASTL bound to SAT is inactive in the synthesis of Cys, but triggers SAT stabilization. Free OASTL, as an auxillary enzyme, consumes OAS in the presence of S^{2-} to achieve full capacity for cys synthesis (Droux et al. 1998).

Genomic Organization

The binding site of PLP cofactor was determined by site-directed mutagenesis of conserved lysine residues (Saito et al. 1993a, b) and by identification of the cofactor-binding partial peptide fragment (Rolland et al. 1996). It was first cloned by Römer et al. (1992) and Saito et al. (1992a, b), but meanwhile cDNAs encoding OASTL have been isolated from spinach (Rolland et al. 1992); watermelon (Noji et al. 1994); *A. thaliana* (Barroso et al. 1995; Hesse and Altmann 1995).

Regulation

S-starvation induced the mRNA levels for the cytosolic isoform of OASTL (Barroso et al. 1995). Contrary to the above study, Warrilow and Hawkesford

(1998), observed reduction and Takahashi and Saito (1996) observed no changes in the activity of OASTL by S deficiency. The activity of this enzyme is strongly affected by the nutritional status of N. An upregulation of mitochondrial OASTL in spinach has been reported in response to N-deprivation (Takahashi and Saito 1996).

References

Abrol YP, Kumar PA, Nair TVR (1984) Nitrate assimilation and grain nitrogen accumulation. Adv Cereal Sci Technol 6:1–48

Aguera E, Haba PD, Fontes JM, Maldonando JM (1990) Nitrate and nitrite uptake and reduction by intact sunflower plants. Planta 182:149–154

Akashi T, Matsumura T, Ideguchi T, Iwakiri K, Kawakatsu T, Taniguchi I, Hase T (1999) Comparison of the electrostatic binding sites on the surface of ferredoxin for two ferredoxin dependent enzymes, ferredoxin-NADP1 reductase and sulphite reductase. J Biol Chem 274:29399–29405

Avila C, Marquez AJ, Pajeulo P, Cannell ME, Wallsgrove RM, Forde BG (1993) Cloning and sequence analysis of a cDNA for barley ferredoxin-dependent glutamate synthase and molecular analysis of photorespiratory mutants deficient in the enzyme. Planta 189:475–483

Bachmann M, Huber JL, Liao PC, Gage DA, Huber SC (1996) The inhibitor protein of phosphorylated nitrate reductase from spinach leaves is a 14-3-3 protein. FEBS Lett 387:127–131

Barroso C, Vega JM, Gotor C (1995) A new member of the cytosolic O-acetylserine (thiol)lyase gene family in Arabidopsis thaliana. FEBS Lett 363:1–5

Becker TW, Caboche M, Carrayol E, Hirel B (1992) Nucleotide sequence of a tobacco cDNA encoding plastidic glutamine synthetase and light inducibility, organ specificity and diumal rhythmicity in the expression of the corresponding genes of tobacco and tomato. Plant Mol Biol 19:367–379

Becker TW, Nef-Campa C, Zehnacker C, Hirel B (1993) Implication of the phytochrome in the light regulation of the tomato gene(s) encoding ferrodoxin-dependent glutamate synthase. Plant Physiol Biochem 31:725–729

Bick JK, Leustek T (1998) Plant sulphur metabolism—the reduction of sulphate to sulphite. Curr Opin Plant Biol 1:240–244

Bick JA, Aslund F, Chen Y, Leustek T (1998) Glutaredoxin function for the carboxyl terminal domain of the plant-type 5'-adenylylsulfate (APS) reductase. Proc Natl Acad Sci USA 95: 8404–8409

Bick JA, Setterdahl AT, Knaff DB, Chen Y, Pitcher LH, Zilinskas BA, Leustek T (2001) Regulation of the plant-type 5′-adenylyl sulfate reductase by oxidative stress. Biochemistry 40:9040–9048

Bogdanova N, Hell R (1997) Cysteine synthesis in plants: protein-protein interactions of serine acetyltransferase from Arabidopsis thaliana. Plant J 11:251–262

Bogdanova N, Bork C, Hell R (1995) Cysteine biosynthesis in plants: isolation and functional identification of a cDNA encoding a serine acetyltransferase from Arabidopsis thaliana. FEBS Lett 358:43–47

Bork C, Schwenn JD, Hell R (1998) Isolation and characterization of a gene for assimilatory sulfite reductase from Arabidopsis thaliana. Gene 212:147–153

Bouton S, Leydecker MT, Meyer C, Truong HN (2002) Role of gibberellins and of the RGA and GAI genes in controlling nitrate assimilation in Arabidopsis thaliana. Plant Physiol Biochem 40:939–947

Bowsher CG, Hucklesby DP, Emes MJ (1989) Nitrite reduction and carbohydrate metabolism in plastids purified from roots of Pisum sativum L. Planta 177:359–366

Bowsher CG, Hucklesby DP, Emes MJ (1993) Induction of ferrodoxin –NADP+ oxidoreductase and ferrodoxin synthesis in pea root plastids during nitrate assimilation. Plant J 3:463–467

Brühl A, Haverkamp T, Gisselmann G, Schwenn JD (1996) A cDNA clone from Arabidopsis thaliana encoding plastidic ferredoxin: sulfite reductase. Biochim Biophys Acta 1295:119–124

Brunold C (1990) Reduction of sulfate to sulfide. In: Rennenberg H, Brunold C, De Kok LJ, Stulen I (eds) Sulphur nutrition and sulphur assimilation in higher plants. SPB Academic Publishing, The Hague, p 13.31

Buchner P, Stuiver CE, Westerman S, Wirtz M, Hell R, Hawkesford MJ, De Kok LJ (2004) Regulation of sulphate uptake and expression of sulfate transporter genes in Brassica oleracea as affected by atmospheric H2S and pedospheric sulphate nutrition. Plant Physiol 136:3396–3408

Caboche M, Rouze P (1990) Nitrate reductase: a target for molecular and cellular studies in higher plants. Trends Genet 6:187–192

Cerezo M, Flors V, Legaz F, Garcia-Agustin P (2000) Characterization of the low affinity transport system for NO_3-uptake by Citrus roots. Plant Sci 160:95–104

Cerezo M, Tillard P, Filleur S, Munos S, Daniel-Vedele F, Gojon A (2001) Major alterations of the regulation of root NO_3-uptake are associated with the mutation of Nrt2.1 and Nrt2.2 genes in Arabidopsis. Plant Physiol 127:262–271

Chandok MR, Raghuram N, Sapory SK (1997) Deciphering the molecular events downstream of phytochrome photoactivation in nitrate reductase regulation in maize. In: Tewari KK, Singhal GS (eds) Plant molecular biology and biotechnology. Narosa Publishing House, London, pp 77–85

Chen Y, Leustek T (1998) Three genomic clones from Arabidopsis thaliana encoding 5′-adenylyl sulfate reductase (Acc. Nos. AF016282, AF016283, and AF016284). Plant Physiol 116:869

Chen BM, Wang ZH, Li SX, Wang GX, Song HX, Wang XN (2004) Effects of nitrate supply on plant growth, nitrate accumulation, metabolic nitrate concentration and nitrate reductase activity in three leafy vegetables. Plant Sci 167:635–643

Cheng CL, Dewdney J, Kleinhofs A, Goodman HM (1986) Cloning and nitrate induction of nitrate reductase mRNA. Proc Natl Acad Sci USA 83:6825–6828

Chiu CC, Lin CS, Hsia AP, Su RC, Lin HL, Tsay YF (2004) Mutation of a nitrate transporter, AtNRT1:4, results in a reduced petiole nitrate content and altered leaf development. Plant Cell Physiol 45:1139–1148

Clarkson DT, Smith FW, Vandenberg PJ (1983) Regulation of sulfate transport in a tropical legume, Macroptilium atropurpureum cv Sirato. J Exp Bot 34:1463–1483

Clarkson DT, Hawkesford MJ, Davidian JC, Grignon C (1992) Contrasting responses of sulfate and phosphate-transport in barley (Hordeum vulgare L) roots to protein-modifying reagents and inhibition of protein-synthesis. Planta 187:306–314

Coschigano KT, Oliveira RM, Lim J, Coruzzi GM (1998) *Arabidopsis* gls mutants and distinct Fd-GOGAT genes: implications for photorespiration and primary nitrogen assimilation. Plant Cell 10:741–752

Crawford NM (1995) Nitrate: nutrizent and signal for plant growth. Plant Cell 7:859–868

Crawford NM, Smith M, Bellissimo D, Davis RW (1988) Sequence and nitrate regulation of the Arabidopsis thaliana mRNA encoding nitrate reductase, a metalloflavoprotein with three functional domains. Proc Natl Acad Sci USA 85:5006–5010

Cren M, Hirel B (1999) Glutamine synthetase in higher plants: regulation of gene and protein expression from the organ to the cell. Plant Cell Physiol 40:1187–1193

Daniel-Vedele F, Filleur S, Caboche M (1998) Nitrate transport: a key step in nitrate assimilation. Curr Opin Plant Biol 1:235–239

Dougall DK (1974) Evidence for the presence of glutamate synthase in extracts of carrot cell cultures. Biochem Biophys Res Commun 58:639–646

Droux M (2003) Plant serine acetyltransferase: new insights for regulation of sulphur metabolism in plant cells. Plant Physiol Biochem 41:619–627

Droux M (2004) Sulphur assimilation and the role of sulphur in plant metabolism: a survey. Photosynthesis Res 79:331–348
Droux M, Ruffet ML, Douce R, Job D (1998) Interactions between serine acetyltransferase and O-acetylserine (thiol) lyase in higher plants - structural and kinetic properties of the free and bound enzymes. Eur J Biochem 255:235–245
Dubois F, Terce-Laforgue T, Gonzalez-Moro MB, Estavillo JM, Sangwan R, Gallais A, Hirel B (2003) Glutamate dehydrogenase in plants: is there a new story for an old enzyme? Plant Physiol Biochem 41:565–576
Duncanson E, Gilkes AF, Kirk DW, Sherman A, Wary JJ (1993) nir1, a conditional lethal mutation in barley causing a defect in nitrite reduction. Mol Gen Gene 236:275–282
Elmlinger MW, Bolle C, Batschauer A, Qelmuller R, Mohr H (1994) Co-action of blue light and light absorbed by phytochrome in control of glutamine synthatase gene expression in scots pine (Pinus sylvestris L.) seedlings. Planta 192:189–194
Emes MJ, Tobin A (1993) Control of metabolism and development in higher plant plastids. Int Rev Cytolol 145:149–216
Fei H, Chaillou S, Mahon JD, Vessey JK (2003) Overexpression of a soyabean cytosolic glutamine synthetase gene linked to organ-specific promoters in pea plants grown in different concentrations of nitrate. Planta 216:467–474
Finnermann J, Schjoerring JK (2000) Post translational regulation of cytosolic glutamine synthetase by reversible phosphorylation and 14-3-3 protein interaction. Plant J 24:171–181
Forde BG (2000) Nitrate transporters in plants: structure, function and regulation. Biochim Biophys Acta 1465:219–235
Gao Y, Schofield OM, Leustek T (2000) Characterization of sulfate assimilation in marine algae focusing on the enzyme 5′-adenylyl sulfate reductase. Plant Physiol 123:1087–1096
Gawronska H, Deji A, Sakakibara H, Sugiyama T (2003) Hormone-mediated nitrogen signaling in plants: implication of participation of abscissic acid in negative regulation of cytokinin-inducible expression of maize response regulator. Plant Physiol Biochem 41:605–610
Glass ADW, Siddiqui MY (1995) Nitrogen absorption by plant roots. In: Srivastava HS, Singh RP (eds) Nitrogen nutrition in higher plants. Associate Publishers, New Delhi, pp 21–56
Goto S, Akagawa T, Kojima S, Hayakawa T, Yamaya T (1998) Organization and structure of NADH-dependent glutamate synthase gene from rice plants. Biochim Biophys Acta 1387:298–308
Gregerson RG, Miller SS, Twary SN, Gantt JS, Vance CP (1993) Molecular characterization of NADH-dependent glutamate synthase from alfalfa nodules. Plant Cell 5:215–226
Gutierrez-Alcala G, Gotor C, Meyer AJ, Fricker M, Vega JM, Romero LC (2000) Glutathione biosynthesis in Arabidopsis trichome cells. Proc Natl Acad Sci USA 97:11108–11113
Gutierrez-Marcos JF, Roberts MA, Campbell EI, Wray JL (1996) Three members of a novel small gene-family from Arabidopsis thaliana able to complement functionally an Escherichia coli mutant defective in PAPS reductase activity encode proteins with a thioredoxin-like domain and .APS reductase activity. Proc Natl Acad Sci USA 93:13377–13382
Harmer SL, Hogenesch JB, Straume M, Chang HS, Han B, Zhu T, Wang X, Kreps JA, Kay SA (2000) Orchestrated transcription of key pathways in Arabidopsis by the circadian clock. Science 290:2110–2113
Hartwell J, Gill A, Wilkins MB, Jenkins GI, Nimmo HG (1999) Phosphoenol-pyruvate carboxylase kinase is a novel protein kinase regulated at the level of expression. Plant J 20:333–342
Hatzfeld Y, Lee S, Lee M, Leustek T, Saito K (2000) Functional characterization of a gene encoding a fourth ATP sulfurylase isoform from Arabidopsis thaliana. Gene 248:51–58
Hawkesford MJ (2003) Transporter gene families in plants: the sulphate transporter gene family—redundancy or specialization? Physiol Plantarum 117:155–163
Hawkesford MJ, Buchner P, Hopkins L, Howarth JR (2003) The plant sulfate transporter family: specialised function and integration with whole plant nutrition. In J-C Davidian, D Grill, LJ De Kok, I Stulen, MJ Hawkesford, E Schnug, H Rennenberg, eds, 5th Workshop on Sulfur

Transport and Assimilation: Regulation, Interaction, Signalling. Backhuys Publishers, Leiden, The Netherlands, pp 1–10

Hawkesford MJ, De Kok LJ (2006) Managing sulphur metabolism in plants. Plant Cell Environ 29:382–395

Hawkesford MJ, Davidian JC, Grignon C (1993) Sulfate proton cotransport in plasma membrane vesicles isolated from roots of Brassica napus L. Increased transport in membranes isolated from sulfur-starved plants. Planta 190:297–304

Heiss S, Schäfer HJ, Haag-Kerwer A, Rausch T (1999a) Cloning sulfur assimilation genes of Brassica juncea L.: cadmium differentially affects the expression of a putative low-affinity sulfate transporter and isoforms of ATP sulfurylase and APS reductase. Plant Mol Biol 39:847–857

Heiss S, Schäfer HJ, Haag-Kerwer A, Rausch T (1999b) Cloning sulfur assimilation genes of Brassica juncea L.: cadmium differentially affects the expression of a putative low-affinity sulfate transporter and isoforms of ATP sulfurylase and APS reductase. Plant Mol Biol 39:847–857

Hell R (1997) Molecular physiology of plant sulfur metabolism. Planta 202:138–148

Hell R, Bork C, Bogdanova N, Frolov I, Hauschild R (1994) Isolation and characterization of two cDNAs encoding for compartment specific isoforms of O-acetylserine(thiol)lyase from Arabidopsis thaliana. FEBS Lett 351:257–262

Hell R, Jost R, Berkowitz O, Wirtz M (2002) Molecular and biochemical analysis of the enzymes of cysteine biosynthesis in the plant Arabidopsis thaliana. Amino Acids 22:245–257

Hesse H, Altmann T (1995) Molecular cloning of a cysteine synthase cDNA from Arabidopsis thaliana. Plant Physiol 108:851–852

Hesse H, Trachsel N, Suter M, Kopriva S, von Ballmoos P, Rennenberg H et al (2003) Effect of glucose on assimilatory sulphate reduction in roots of roots of Arabidopsis thaliana. J Exp Bot 54:1701–1709

Hirai MY, Klein M, Fujikawa Y, Yano M, Goodenowe DB, Yamazaki Y et al (2005) Elucidation of gene-to-gene and metabolite- to-gene networks in Arabidopsis by integration of metabolomics and transcriptomics. J Biol Chem 280:25590–25595

Hirel B, Lea PJ (2001) Ammonia assimilation. In: Lea PJ, Morat-Gaudry JF (eds) Plant nitrogen. Springer, Berlin, pp 79–99

Hirose N, Hayakawa T, Yamaya T (1997) Inducible accumulation of mRNA for NADH-dependent glutamate synthase in rice roots in response to ammonium ions. Plant Cell Physiol 38:1295–1297

Howarth JR, Roberts MA, Wray JL (1997) Cysteine biosynthesis in higher plants: a new member of the Arabidopsis thaliana serine acetyltransferase small gene family obtained by functional complementation of an Escherichia coli cysteine auxotroph. Biochim Biophys Acta 1350:123–127

Huber SC, MacKintosh C, Kaiser WM (2002) Metabolic enzymes as targets for 14-3-3 proteins. Plant Mol Biol 50:1053–1063

Ideguchi T, Akashi T, Onda Y, Hase T (1995) cDNA cloning and functional expression of ferredoxin-dependent sulfite reductase from maize in E. coli cells. In: Mathis P (ed) Photosynthesis: from light to biosphere, vol II. Kluwer Academic Publishers, Dordrecht, pp 713–716

Ikeda Y et al (2000) Specific binding of a 14-3-3 protein to autophosphorylated WPK4, an SNF1-related wheat protein kinase, and to WPK4-phosphorylated nitrate reductase. J Biol Chem 275: 31695–31700

Imamura A, Hanaki N, Nakamura A, Suzuki T, Taniguchi M, Kiba T, Ueguchi C, Sugiyama T, Mizuno T (1999) Compilation and characterization of Arabidopsis thaliana response regulators implicated in His–Asp phosphorelay signal transduction. Plant Cell Physiol 40:733–742

Ireland RJ, Lea PJ (1999) The enzymes of glutamine, glutamate, asparagines and aspartate metabolism. In: Singh BK (ed) Plant amino acids: biochemistry and biotechnology. Marcel Dekker, New York, pp 49–109

Ismande J, Touraine B (1994) N demand and the regulation of nitrate uptake. Plant Physiol 105:3–7

Jones SL, Smith IK (1981) Sulfate transport in cultured tobacco cells. Effects of calcium and sulfate concentration. Plant Physiol 67:445–448

Kaiser WM, Huber SC (2001) Post-translational regulation of nitrate reductase: mechanism, physiological relevance and environmental triggers. J Exp Bot 52:1981–1989

Kataoka T, Watanabe-Takahashi A, Hayashi N, Ohnishi M, Mimura T, Buchner P, Hawkesford MJ, Yamaya T, Takahash H (2004) Vacuolar sulfate transporters are essential determinants controlling internal distribution of sulfate in Arabidopsis. Plant Cell 16:2693–2704

Klonus D, Hofgen R, Willmitzer L, Riesmeier JW (1994) Isolation and characterization of two cDNA clones encoding ATP-sulfurylases from potato by complementation of a yeast mutant. Plant J 6:105–112

Knaff DB (1996) Ferrodoxin and ferrodoxin-dependent enzymes. In: Ort DR, Yocum CF (eds) Advances in photosynthesis, vol. 4. Oxygenic photosynthesis: the light reactions. Kluwer Academic, Dordretch, pp 331–361

Kopriva S, Muheim R, Koprivova A, Trachsler N, Catalano C, Suter M, Brunold C (1999) Light regulation of assimilatory sulfate reduction in Arabidopsis thaliana. Plant J 20:37–44

Kopriva S, Büchert T, Fritz G, Suter M, Benda R, Schünemann V, Koprivova A, Schürmann P, Trautwein AX, Kroneck PM, Brunold C (2002) The presence of an iron-sulfur cluster in adenosine 5′-phosphosulfate reductase separates organisms utilizing adenosine 5′-phosphosulfate and phosphoadenosine 5′-phosphosulfate for sulfate assimilation. J Biol Chem 277:21786–21791

Koprivova A, Suter M, Op den Camp R, Brunold C, Kopriva S (2000) Regulation of sulfate assimilation by nitrogen in Arabidopsis. Plant Physiol 122:737–746

Koprivova A, Melzer M, von Ballmoos P, Mandel T, Brunold C, Kopriva S (2001) Assimilatory sulfate reduction in C3, C3-C4, and C4 species of Flaveria. Plant Physiol 127:543–550

Kronenberger J, Lepingle A, Caboche M, Vaucheret H (1993) Cloning and expression of distinct nitrite reductases in tobacco leaves and roots. Mol Gene Genet 236:203–208

Kronzucker HJ, Siddiqi MY, Glass ADM (1995) Kinetics of nitrate influx in spruce. Plant Physiol 99:1124–1133

Lam HM, Coshigano K, Oliveira I, Melo-Oliveira R, Coruzzi G (1996) The molecular genetics of nitrogen assimilation into amino acids in higher plants. Annu Rev Plant Physiol Plant Mol Biol 47:569–593

Lauter FR, Ninnemann O, Bucher M, Riesmeier JW, Frommer WB (1996) Preferential expression of an ammonium transport and of two putative nitrate transporters in root hairs of tomato. Proc Natl Acad Sci USA 93:8139–8144

Lea PJ, Miflin BJ (1974) An alternative route for nitrogen assimilation in higher plants. Nature 251:614–616

Lea PJ, Robinson SA, Stewart GR (1990) The enzymology and metabolism of glutamine, glutamate and asparagine. In: Miflin BJ, Lea PJ (eds) The biochemistry of plants. Academic Press, New York, pp 121–159

Lee RB (1982) Selectivity and kinetics of ion uptake by barley plants following nutrient deficiency. Ann Bot 50:429–449

Leustek T, Martin MN, Bick JA, Davies JP (2000) Pathways and regulation of sulfur metabolism revealed through molecular and genetic studies. Annu Rev Plant Physiol Plant Mol Biol 51:141–165

Li M, Villemur PJ, Hussey CD, Silflow JS, Gantt DP, Snustad DP (1993) Differential expression of six glutamine synthetase genes in Zea mays. Plant Mol Biol 23:401–407

Li W, Wang Y, Okamoto M, Crawford NM, Siddiqi MY, Glass ADM (2007) Dissection of the AtNRT2.1:AtNRT2.2 inducible high affinity nitrate transporter gene cluster. Plant Physiol 143:425–433

Lillo C, Appenroth KJ (2001) Light regulation of nitrate reductase in higher plants: which photoreceptors are involved? Plant Biol 3:455–465

Logan HM, Cathala N, Grignon C, Davidian JC (1996) Cloning of a cDNA encoded by a member of the Arabidopsis thaliana ATP sulfurylase multigene family. Expression studies in yeast and in relation to plant sulfur nutrition. J Biol Chem 271:12227–12233

Lunn J, Droux M, Martin J, Douce R (1990) Localization of ATP sulfurylase and O-acetylserine (thiol) lyase in spinach leaves. Plant Physiol 94:1345–1352

Malagoli P, Laine P, Le Deunff E, Rossato L, Ney B, Ourry A (2004) Modeling nitrogen uptake in oilseed rape cv capitol during a growth cycle using influx kinetics of root nitrate transport systems and field experimental data. Plant Physiol 134:388–400

Maruyama-Nakashita A, Nakamura Y, Yamaya T, Takahashi H (2004) A novel regulatory pathway of sulfate uptake in Arabidopsis roots: implication of CRE1/WOL/AHK4-mediated cytokinin-dependent regulation. Plant J 38:779–789

Maruyama-Nakashita A, Nakamura Y, Watanabe-Takahashi A, Inoue E, Yamaya T, Takahashi H (2005) Identification of a novel cis-acting element conferring sulfur deficiency response in Arabidopsis roots. Plant J 42:305–314

Masclaux-Daubresse C, Carrayol E, Valadier MH (2005) The two nitrogen mobilisation- and senescence-associated GS1 and GDH genes are controlled by C and N metabolites. Planta 221:580–588

Miflin BJ, Habash DZ (2002) The role of glutamine synthetase and glutamate dehydrogenase in nitrogen assimilation and possibilities for improvement in nitrogen utilization of crops. J Exp Bot 53:979–987

Migge A, Carrayol E, Hirel B, Lohmann M, Meya G, Becker TW (1998) Regulation of the subunit composition of plastidic glutamine synthetase of the wild type and of the phytochrome deficient aurea mutant of by blue/UV-A or UV-B light. Plant Mol Biol 37:689–700

Miller AJ, Fan X, Orsel M, Smith SJ, Wells DM (2007a) Nitrate transport and signaling. J Exp Bot 58:2297–2306

Miller AJ, Fan X, Shen Q, Smith SJ (2007b) Amino acids and nitrate as signals for the regulation of nitrogen acquisition. J Exp Bot 58:2297–2306

Miyazaki J, Juricek M, Angelis K, Schnorr KM, Kleinhofs A, Warner RL (1991) Characterization and sequence of a novel nitrate reductase from barley. Mol Gen Genet 228:329–334

Murillo M, Leustek T (1995) Adenosine-5′-triphosphate-sulfurylase from Arabidopsis thaliana and Escherichia coli are functionally equivalent but structurally and kinetically divergent: nucleotide sequence of two adenosine-5′-triphosphate-sulfurylase cDNAs from Arabidopsis thaliana and analysis of a recombinant enzyme. Arch Biochem Biophys 323:195–204

Nikiforova V, Freitag J, Kempa S, Adamik M, Hesse H, Hoefgen R (2003) Transcriptome analysis of sulphur depletion in Arabidopsis thaliana:interlacing of biosynthetic pathways provides response specificity. Plant J 33:633–650

Nikiforova VJ, Kopka J, Tolstikov V, Fiehn O, Hopkins L, Hawkesford MJ et al (2005) Systems rebalancing of metabolism in response to sulfur deprivation, as revealed by metabolome analysis of Arabidopsis plants. Plant Physiol 138:304–318

Ninnemann O, Jauniaux JC, Frommer WB (1994) Identification of a high affinity ammonium transporter from plants. EMBO J 13:3464–3471

Nogueira EDM, Olivares FL, Japiassu JC, Vilar CV, Vinagre F, Baldini JI, Hemerly AS (2005) Characterisation of glutamine synthetase genes in sugarcane genotypes with different rates of biological nitrogen fixation. Plant Sci 169:819–832

Noji M, Saito K (2002) Molecular and biochemical analysis of serine acetyltransferase and cysteine synthase towards sulfur metabolic engineering in plants. Amino Acids 22:231–243

Noji M, Murakoshi I, Saito K (1994) Molecular cloning of a cysteine synthase cDNA from Citrullus vulgaris (watermelon) by genetic complementation in an Escherichia coli Cys-auxotroph. Mol Gen Genet 244:57–66

Noji M, Inoue K, Kimura N, Gouda A, Saito K (1998) Isoform-dependent differences in feedback regulation and subcellular localization of serine acetyltransferase involved in cysteine biosynthesis from Arabidopsis thaliana. J Biol Chem 273:32739–32745

Noji M, Saito M, Nakamura M, Aono M, Saji H, Saito K (2001) Cysteine synthase overexpression in tobacco confers tolerance to sulfur containing environmental pollutants. Plant Physiol 126:973–980

Oaks A, Hirel B (1985) Nitrogen metabolism in roots. Annu Rev Plant Physiol 36:345–365

Ogawa K, Soutome R, Hiroyama K, Hagio T, Ida S, Nakagawa H, Komamne A (2000) Co-regulation of nitrate reductase and nitrite reductase in cultured spinach cells. J Plant Physiol 157:299–306

Ohkama N, Takei K, Sakakibara H, Hayashi H, Yoneyama T, Fujiwara T (2002) Regulation of sulfur-responsive gene expression by exogenously applied cytokinins in Arabidopsis thaliana. Plant Cell Physiol 43:1493–1501

Okamoto M, Vidmar JJ, Glass ADM (2003) Regulation of NRT1 and NRT2 gene families of Arabidopsis thaliana: responses to nitrate provision. Plant Cell Physiol 44:304–317

Oliveira IC, Coruzzi GM (1999) Carbon and amino acids reciprocally modulate the expression of glutamine synthetase in Arabidopsis. Plant Physiol 121:301–310

Oliveira IC, Coschigano K, Lam HM, Melo-Oliveira R, Coruzzi G (1997) Molecular-genetic dissection and metabolic engineering of nitrogen assimilation in plants. Plant Physiol Biochem 35:185–198

Ortega JL, Roche D, Sengupta-Gopalan C (1999) Oxidative turnover of soybean root glutamine synthetase: in vitro and in vivo studies. Plant Physiol 119:1483–1495

Pahlich E (1996) Remarks concerning the dispute related to the function of plant glutamate dehydrogenase: commentary. Can J Bot 74:512–515

Pajuelo P, Pajuelo E, Forde BG, Marquéz AJ (1997) Regulation of the expression of ferredoxin glutamate synthase in barley. Planta 203:517–525

Palenchar PM, Kouranov A, Lejay LV, Coruzzi GM (2004) Genome-wide patterns of carbon and nitrogen regulation of gene expression validate the combined carbon and nitrogen (CN)-signaling hypothesis in plants. Genome Biol 5:1–15

Phartiyal P, Kim WS, Cahoon RE, Jez JM, Krishnan HB (2008) The role of 5′-adenylylsulfate reductase in the sulphur assimilation pathway of soybean: molecular cloning, kinetic characterization, and gene expression. Phytochemistry 69:356–364

Prior A, Uhrig JF, Heins L, Wiesmann A, Lillig CH, Stoltze C, Soll J, Schwenn JD (1999) Structural and kinetic properties of adenylyl sulfate reductase from Catharanthus roseus cell cultures. Biochim Biophys Acta 1430:25–38

Quesada A, Krapp A, Trueman L, Daneil-Vedele F, Fernendez E, Forde B, Caboche M (1997) PCR-identification of a Nicotiana plumbaginifolia cDNA homologous to the high affinity nitrate transporters of the crnA family. Plant Mol Biol 34:265–274

Raghuram N, Sopory SK (1995) Light regulation of NR gene expression: mechanism and signal-response coupling. Physiol Mol Biol Plants 1:103–114

Raghuram N, Pathak RR, Sharma P (2006) Signaling and the molecular aspects of N-use efficiency in higher plants. In: Singh RP, Jaiwal PK (eds) Biotechnological approaches to improve nitrogen use efficiency in plants

Rastogi R, Back E, Schneiderbauer A, Bowsher C, Moffatt B, Rothstein SJ (1993) A 330 bp region of the spinach nitrite reductase gene promoter directs nitrate-inducible tissue specific expression in transgenic tobacco. Plant J 4:317–326

Redinbaugh MG, Campbell WH (1991) Higher plant responses to environmental nitrate. Physiol Plant 82:640–650

Redinbaugh MG, Campbell WH (1993) Glutamine synthetase and ferredoxin-dependent glutamate synthaseexpression in the maize (Zea mays) root primary response to nitrate. Plant Physiol 101:1249–11255

Remans T, Nacry P, Pervent M, Girin T, Tillard P, Lepetit M, Gojon A (2006) A central role for the nitrate transporter NRT2.1 in the integrated morphological and physiological responses of the root system to nitrogen limitations in Arabidopsis. Plant Physiol 140:909–921

Renosto F, Patel HC, Martin RL, Thomassian C, Zimmerman G, Segel IH (1993) ATP sulfurylase from higher plants: kinetic and structural characterization of the chloroplast and cytosol enzymes from spinach leaf. Arch Biochem Biophys 307:272–285

Riedel J, Tischner R, Mack G (2001) The chloroplastic glutamine synthetase (GS2) of tobacco is phosphorylated and associated with 14-3-3 proteins inside the chloroplast. Planta 213:396–401

Roberts MA, Wray JL (1996) Cloning and characterization of an Arabidopsis cDNA clone encoding an organellar isoform of serine acetyltransferase. Plant Mol Biol 30:1041–1049

Rolland N, Droux M, Douce R (1992) Subcellular distribution of o-acetylserine(thiol) lyase in cauliflower (Brassica oleracea) inflorescence. Plant Physiol 98:927–935

Rolland N, Ruffet ML, Job D, Douce R, Droux M (1996) Spinach chloroplast O-acetylserine(thiol) lyase exhibits two catalytically nonequivalent pyridoxal-5-phosphate-containing active sites. Eur J Biochem 236:272–282

Römer S, d.Harlingue A, Camara B, Schantz R and Kuntz M (1992). Cysteine synthase from Capsicum annuum chromoplasts. J Biol Chem 267:17966–17970

Rotte C, Leustek T (2000) Differential subcellular localization and expression of ATP sulfurylase and 5′-adenylylsulfate reductase during ontogenesis of Arabidopsis leaves indicates that cytosolic and plastid forms of ATP sulfurylase may have specialized functions. Plant Physiol 124:715–724

Rouached H, Berthomieu P, El Kassis E, Cathala N, Catherinot V, Labesse G, Davidian JC, Fourcroy P (2005) Structural and functional analysis of the C-terminal STAS (sulfate transporter and anti-sigma antagonist) domain of the Arabidopsis thaliana sulfate transporter SULTR1.2. J Biol Chem 280:15976–15983

Ruffet ML, Lebrun M, Droux M, Douce R (1995) Subcellular distribution of serine acetyltransferase from Pisum sativum and characterization of an Arabidopsis thaliana putative cytosolic isoform. Euro J Biochem 227:500–509

Saito K (2000) Regulation of sulfate transport and synthesis of sulphurcontaining amino acids. Curr Opin Plant Biol 3:188–195

Saito K (2004) Sulfur assimilatory metabolism. The long and smelling road. Plant Physiol 136:2443–2450

Saito K, Miura N, Yamazaki M, Hirano H, Murakoshi I (1992a) Molecular cloning and bacterial expression of cDNA encoding a plant cysteine synthase. Proc Natl Acad Sci USA 89:8078–8082

Saito K, Miura N, Yamazaki M, Hirano H, Murakoshi I (1992b) Molecular cloning and bacterial expression of cDNA encoding a plant cysteine synthase. Proc Natl Acad Sci USA 89:8078–8082

Saito K, Kurosawa M, Murakoshi I (1993a) Determination of a functional lysine residue of a plant cysteine synthase by site-directed mutagenesis, and molecular evolutionary implication. FEBS Lett 328:111–114

Saito K, Tatsuguchi K, Murakoshi I, Hirano H (1993b) cDNA cloning and expression of cysteine synthase B localized in chloroplasts of Spinacia oleracea. FEBS Lett 324:247–252

Saito K, Yokoyama H, Noji M, Murakoshi I (1995) Molecular cloning and characterization of a plant serine acetyltransferase playing a regulatory role in cysteine biosynthesis from watermelon. J Biol Chem 270:16321–16326

Sakakibara H, Matanabe M, Hase T, Sugiyama T (1991) Molecular cloning and characterization of complementary DNA encoding for ferredoxin dependent glutamate synthase in maize leaf. J Biol Chem 266:2028–2035

Sakakibara H, Kobayashi K, Deji A, Sugiyama T (1997) Partial characterization of signaling pathway of nitrate-dependent expression of the genes for nitrogen assimilatory enzymes using detached maize leaves. Plant Cell Physiol 38:837–843

Sakakibara H, Suzuki M, Takei K, Deji A, Taniguchi M, Sugiyama T (1998) A response-regulator homologue possibly involved in nitrogen signal transduction mediated by cytokinin in maize. Plant J 14:337–344

Sakakibara H, Hayakawa A, Deji A, Gawronski S, Sugiyama T (1999) His–Asp phosphotransfer possibly involved in a nitrogen signal transduction mediated by cytokinin in maize: molecular cloning of cDNAs for two-component regulatory factors and demonstration of phosphotransfer activity in vitro. Plant Mol Biol 41:563–573

Samuelson ME, Larsson CM (1993) Nitrate regulation of zeatin riboside levels in barley roots: effects of inhibitors of N assimilation and comparison with ammonium. Plant Sci 93:77–84

Santi S, Locci G, Pinton R, Cesco S, Varanini Z (1995) Plasma membrane H+−ATPase in maize roots induced for NO_3-uptake. Plant Physiol 109:1277–1283

Scarpeci TE, Marro ML, Bortolotti S, Boggio SB, Valle EM (2007) Plant nutritional status modulates glutamine synthetase levels in ripe tomatoes (Solanum lycopersicum cv. Micro-Tom). J Plant Physiol 164:137–145

Scheible WR, Lauerer M, Schulze ED, Caboche M, Stitt M (1997) Accumulation of nitrate in the shoot acts as a signal to regulate shoot-root allocation in tobacco. Plant J 11:671–691

Sechley KA, Yamaya T, Oaks A (1992) Compartmentation of nitrogen assimilation in higher plants. Int Rev Cytol 134:85–163

Setya A, Murillo M, Leustek T (1996) Sulfate reduction in higher plants—molecular evidence for a novel 50-adenylylsulfate reductase. Proc Nat Acad Sci USA 93:13383–13388

Shibagaki N, Grossman AR (2004) Probing the function of STAS domains of the Arabidopsis sulfate transporters. J Biol Chem 279:30791–30799

Shibagaki N, Rose A, McDermott JP, Fujiwara T, Hayashi H, Yomeyama T, Davis JP (2002) Selenateresistant mutants of Arabidopsis thaliana identify Sultr1;2, a sulfate transporter required for efficient transport of sulfate into roots. Plant J 29:475–486

Siegel LM, Wilkerson JO (1989) Structure and function of spinach ferrodoxin-nitrite reductase. In: Wray JL, Kinghorn JR (eds) Molecular and genetic aspects of nitrate assimilation. Oxford Science Publ, Oxford, pp 263–283

Sivasankar S, Oaks A (1995) Regulation of nitrate reductase during early seedling growth: a role for asparagine and glutamine. Plant Physiol 107:1225–1231

Sivasankar S, Oaks A (1996) Nitrate assimilation in higher plants: the effect of metabolite and light. Plant Physiol Biochem 34:609–620

Smith FW, Ealing PM, Hawkesford MJ, Clarkson DT (1995) Plant members of a family of sulfate transporters reveal functional subtypes. Proc Nat Acad Sci USA 92:9373–9377

Smith FW, Hawkesford MJ, Ealing PM, Clarkson DT, van den Berg PJ, Belcher AR, Warrilow AG (1997) Regulation of expression of a cDNA from barley roots encoding a high affinity sulphate transporter. Plant J 12:875–884

Soni R, Carmichael JP, Shah ZH, Murray JAH (1995) A family of cyclin D homologs from plants differentially controlled by growth regulators and containing the conserved retinoblastoma protein interaction motif. Plant Cell 7:85–103

Stitt M (1999) Nitrate regulation of metabolism and growth. Curr Opin Plant Biol 2:178–186

Sueyoshi K, Kleinhofs A, Warner RL (1995) Expression of NADH-specific and NAD(P)H-bispecific nitrate reductase genes in response to nitrate in barley. Plant Physiol 107:1303–1311

Sueyoshi K, Mitsuyama T, Sugimoto T, Kleinhofs A, Warner RL, Oji Y (1999) Effects of inhibitors of signaling components on the expression of genes for nitrate reductase and nitrite reductase in excised barley leaves. Soil Sci Plant Nutr 45:1015–1019

Sugiyama T, Sakakibara H (2002). Regulation of carbon and nitrogen assimilation through gene expression. In: Foyer CH, Noctor G (eds) Photosynthetic nitrogen assimilation and associated carbon and respiratory metabolism. Kluwer Academic, Dordrecht, pp 227–238

Suter M, von Ballmoos P, Kopriva S, Op den Camp R, Schaller J, Kuhlemeier C, Schürmann P, Brunold C (2000) Adenosine 5′-phosphosulfate sulfotransferase and adenosine 5′-phosphosulfate reductase are identical enzymes. J Biol Chem 275:930–936

Suzuki A, Rothstein S (1997) Structure and regulation of ferredoxin dependent glutamate synthase from Arabidopsis thaliana: cloning of cDNA, expression in different tissues of wild-type and gltS mutant strains, and light induction. Eur J Biochem 243:708–718

Taghavi TS, Babalar M (2007) The effect of nitrate and plant size on nitrate uptake and in vitro nitrate reductase activity in strawberry (Fragaria X ananassa cv. Selva). Scientia Horticult 112:393–398

Takahashi H, Saito K (1996) Subcellular localization of spinach cysteine synthase isoforms and regulation of their gene expression by nitrogen and sulfur. Plant Physiol 112:273–280

Takahashi H, Yamazaki M, Sasakura N, Watanabe A, Leustek T, de Almeida Engler J, Engler G, Van Montagu M, Saito K (1997a) Regulation of sulfur assimilation in higher plants: a sulfate transporter induced in sulfate-starved roots plays a central role in Arabidopsis thaliana. Proc Natl Acad Sci USA 94:11102–11107

Takahashi S, Yip WC, Tamura G (1997b) Purification and characterization of ferredoxin-sulfite reductase from turnip (Brassica rapa) leaves and comparison of properties with ferredoxin-sulfite reductase from turnip roots. Biosci Biotechnol Biochem 61:1486–1490

Takahashi H, Sasakura N, Kimura A, Watanabe A, Saito K (1999) Identification of two leaf-specific sulfate transporters in Arabidopsis thaliana (Accession, AB012048 and AB004060) (Pgr.99-154). Plant Physiol 121:686

Takahashi H, Watanabe-Takahashi A, Smith FW, Blake-Kalf M, Hawkesford MJ, Saito K (2000) The roles of three functional sulphate transporters involved in uptake and translocation of sulphate in Arabidopsis thaliana. Plant J 23:171–182

Takei K, Takashaki T, Sugiyama T, Yamaya T, Sakakibara H (2002) Multiple routes communicating nitrogen availability from roots to shoots: a signal transduction pathway mediated by cytokinin. J Exp Bot 53:971–977

Tempest DW, Meers JL, Brown CM (1970) Synthesis of glutamate in Aerobacter aerogenes byan hitherto unknown route. Biochem J 117:405–407

Temple SJ, Bagga S, Sengupta-Gopalan C (1998) Down regulation of specific members of the glutamine synthetase gene family in alfalfa by antisense RNA technology. Plant Mol Biol 37:535–547

Trepp GB, Plank DW, Gantt JS, Vance CP (1999) NADH-glutamate synthase in alfalfa root nodules. Immunocytochemical localization. Plant Physiol 119:829–837

Trueman LJ, Onyeocha I, Forde B (1996) Recent advances in molecular biology ofa afamily of eukaryotic high affinity nitrate transporters. Plant Physiol Biochem 34:621–627

Tsay YF, Schroeder JI, Feldmann KA, Crawford NM (1993) The herbicide sensitivity gene CHLI of Arabidopsis encodes a nitrate-inducible nitrate transporter. Cell 72:705–713

Unkles S, Hawker K, Grieve C, Campbell E, Montague P, Kinghorn J (1991) crnA encodes a nitrate transporter in Aspergillus nidulans. Proc Natl Acad Sci USA 88:204–208

Urano Y, Manabe T, Noji M, Saito K (2000) Molecular cloning and functional characterization of cDNAs encoding cysteine synthase and serine acetyltransferase that may be responsive for high cellular cysteine content in Allium tuberosum. Gene 257:269–277

Vance CP, Miller SS, Gregerson RG, Samac DA, Robinson DL, Gantt JS (1995) Alfalfa NADH-dependent glutamate synthase: structure of the gene and importance in symbiotic N2 fixation. Plant J 8:345–358

Vauclare P, Kopriva S, Fell D, Suter M, Sticher L, von Ballmoos P, Krähenbühl U, Op den Camp R, Brunold C (2002) Flux control of sulphate assimilation in Arabidopsis thaliana: adenosine 5′-phosphosulphate reductase is more susceptible to negative control by thiols than ATP sulphurylase. Plant J 31:729–740.

Vidmar JJ, Zhuo D, Siddiqui MY, Schjoerring JK, Touraine B, Glass ADM (2000) Plant Physiol 123:307–318

Vincentz M, Moureaux T, Leydecker MT, Vaucheret H, Caboche M (1993) Regulation of nitrate and nitrite reductase expression in Nicotiana plumbaginifolia leaves by nitrogen and carbon metabolites. Plant J 3:315–324

Vov Wiren N, Gazzarrini S, Frommer WB (1997) Regulation of mineral nitrogen uptake in plants. Plant Soil 196:191–199

Wada K, Onda M, Matsubara H (1989) Amino acid sequences of ferrodoxin isoproteins from radish roots. J Biochem 105:619–625

Warning HO, Hachel W (2000) Functional analysis of a nitrite reductase promoter from birch in transgenic tobacco. Plant Sci 155:141–151
Warrilow AGS, Hawkesford MJ (1998) Separation, subcellular location and influence of sulphur nutrition on isoforms of cysteine synthase in spinach. J Exp Bot 49:1625–1636
Williams LE, Miller AJ (2001) Transporters responsible for the uptake and partitioning of nitrogenous solutes. Ann Rev Plant Physiol Mol Biol 52:659–688
Wirtz M, Berkowitz O, Droux M, Hell R (2001) The cysteine synthase complex from plants. Mitochondrial serine acetyltransferase from Arabidopsis thaliana carries a bifunctional domain for catalysis and protein-protein interaction. Eur J Biochem 268:686–693
Wolt JD (1994) Soil solutions chemistry. Wiley, New York
Woodall J, Havill DC, Pearson J (1996) Developmental changes in glutamine synthetase isoforms in Sambucus nigra and Trientalis europaea. Plant Physiol Biochem 34:697–706
Wray JL (1993) Molecular biology, genetics and regulation of nitrite reductase in higher plants. Plant Physiol 89:607–612
Yamaya T, Tanno H, Hirose N, Watanabe S, Hayakawa T (1995) A supply of nitrogen causes an increase in the level of NADH-dependent glutamate synthase protein and the activity of the enzyme, in the roots of rice. Plant Cell Physiol 36:197–1204
Yonekura-Sakakibara K, Ashikari T, Tanaka Y, Kusumi T, Hase T (1998) Molecular characterization of tobacco sulfite reductase: enzyme purification, gene cloning, and gene expression analysis. J Biochem 124:615–621
Yonekura-Sakakibara K, Onda Y, Ashikari T, Tanaka Y, Kusumi T, Hase T (2000) Analysis of reductant supply systems for ferredoxin-dependent sulfite reductase in photosynthetic and nonphotosynthetic organs of maize. Plant Physiol 122:887–894
Yoo BC, Harmon AC (1997) Regulation of recombinant soybean serine acetyltransferase by CDPK. Plant Physiol (Suppl) 114:267
Yoshimoto N, Takahashi H, Smith FW, Yamaya T, Saito K (2002) Two distinct high-affinity sulfate transporters with different inducibilities mediate uptake of sulfate in Arabidopsis roots. Plant J 29:465–473
Yoshimoto N, Inoue E, Saito K, Yamaya T, Takahashi H (2003) Phloem-localizing sulfate transporter, Sultr1;3, mediates re-distribution of sulfur from source to sink organs in Arabidopsis. Plant Physiol 131:1511–1517
Zozaya-Hinchliffe M, Potenza C, Ortega JL, Sengupta-Gopalan C (2005) Nitrogen and metabolic regulation of the expression of plastidic glutamine synthetase in alfalfa (Medicago sativa). Plant Sci 168:1041–1052

Index

M. Sarwat et al. (eds.), *Stress Signaling in Plants: Genomics and Proteomics Perspective, Volume 1*, DOI 10.1007/978-1-4614-6372-6,

Q

R

MIX
Papier aus verantwortungsvollen Quellen
Paper from responsible sources
FSC® C105338

If you have any concerns about our products,
you can contact us on
ProductSafety@springernature.com

In case Publisher is established outside the EU,
the EU authorized representative is:
Springer Nature Customer Service Center GmbH
Europaplatz 3, 69115 Heidelberg, Germany

Printed by Libri Plureos GmbH
in Hamburg, Germany